Innovations in Science Education and Technology

Volume 24

Series editor
Karen C. Cohen
Weston, MA, USA

As technology rapidly matures and impacts on our ability to understand science as well as on the process of science education, this series focuses on in-depth treatment of topics related to our common goal: global improvement in science education. Each research-based book is written by and for researchers, faculty, teachers, students, and educational technologists. Diverse in content and scope, they reflect the increasingly interdisciplinary and multidisciplinary approaches required to effect change and improvement in teaching, policy, and practice and provide an understanding of the use and role of the technologies in bringing benefit globally to all.

More information about this series at http://www.springer.com/series/6150

Yehudit Judy Dori • Zemira R. Mevarech
Dale R. Baker
Editors

Cognition, Metacognition, and Culture in STEM Education

Learning, Teaching and Assessment

Springer

Editors
Yehudit Judy Dori
Faculty of Education in Science and Technology
Technion
Haifa, Israel

The Samuel Neaman Institute for National Policy Research
Haifa, Israel

Zemira R. Mevarech
David Yellin Academic College of Education
Bar-Ilan University
Jerusalem, Israel

Dale R. Baker
Mary Lou Fulton Teachers College
Educational Leadership and Innovation
Arizona State University
Tempe, AZ, USA

ISSN 1873-1058 ISSN 2213-2236 (electronic)
Innovations in Science Education and Technology
ISBN 978-3-319-66657-0 ISBN 978-3-319-66659-4 (eBook)
https://doi.org/10.1007/978-3-319-66659-4

Library of Congress Control Number: 2017956580

Printed on acid-free paper

This Springer imprint is published by Springer Nature
The registered company is Springer International Publishing AG
The registered company address is: Gewerbestrasse 11, 6330 Cham, Switzerland

Foreword

One of the most prevalent and dangerous misconceptions about education in the twenty-first century is that because so much information is easily accessible "at the tip of our fingers," knowledge construction should no longer be viewed as a major goal of schooling. Supporters of this view often maintain that current schools should focus on teaching generic skills rather than content-specific knowledge. This book, edited by Dori, Mevarech, and Baker (2017), contributes to refuting this misconception. It is therefore highly significant that the three general, "big" constructs discussed throughout the book, namely, cognition, metacognition, and culture, are addressed in a variety of content-rich frameworks and across all four components of STEM education: science, technology, engineering, and mathematics. Although in other educational contexts technology and engineering education had sometimes been pushed aside in favor of science and mathematics education, in this book, technology and engineering education take center stage, as they should, in current STEM education.

More specifically, there are debates in the literature about whether and to what extent cognitive and metacognitive skills are general, domain transcending, or context bound (e.g., Perkins and Salomon 1989; Veenman 2011). According to a recent review on metacognition in science education (Zohar and Barzilai 2013), the development of learners' conceptual understanding has long been a central focus of science education research. However, previous reviews of metacognition in science education have claimed the role of metacognition in developing learners' understanding of specific science concepts has not been sufficiently studied in such contexts. Therefore, the role of metacognition in developing conceptual understanding deserves increased attention from researchers and educators. In contrast, the data of the more recent 2013 review showed that, in recent years, the potential contribution of metacognition to the understanding of science content has become a central area of research. Many of the reviewed studies addressed metacognition in rich scientific subject-specific contexts. This trend firmly places metacognitive research in a central junction of science education interest, namely, advancing deep understanding of scientific concepts and reasoning skills. This trend also opens the door for new research- and practice-oriented questions.

One such important question pertains to the generality versus content specificity of metacognition in STEM education. When designing STEM curriculum in an integrated way, an issue that becomes central is determining the aspects of metacognition that can and should be taught in general, across all four components of STEM education, and the aspects that need to be tailored to each area separately. How can we best use metacognition to deepen the understanding of specific concepts throughout the STEM curricula? Finally, how can we use evidence from previous research to support the adaptation to specific content areas?

The first step in answering these questions is to create an inventory of studies that examine metacognitive learning and instruction in all STEM areas. At a later stage, an analysis of many studies might form the foundation for devising integrated STEM evidence-based guidelines for how to best apply metacognition across and within areas of STEM education. While at this point there are many studies about metacognition in mathematics and science education, similar research in technology[1] education and engineering education is less developed. Although there are an increasing number of studies about the application of metacognition in learning technologies (e.g., Azevedo and Aleven 2013), the area of metacognition in engineering education is in its infancy. The idea promoted in the book by Dori, Mevarech, and Baker (2017) proposes to look at all four STEM subjects together with respect to metacognition. The book begins to uncover the similarities among them in terms of metacognitive instruction and is therefore a step in the right direction. The addition of culture is also significant.

Finally, the review of studies of metacognition in science education (Zohar and Barzilai 2013) revealed that there are an insufficient number of studies of teachers' (both preservice and in-service) knowledge and professional development regarding metacognition. This research gap is a serious limitation in the ability to carry out large-scale implementation of metacognitive instruction in authentic classrooms. The book edited by Dori, Mevarech, and Baker attends to that research gap as seven of the chapters address teachers or instructors, thereby contributing to the literature in this important field.

Overall, the book constitutes an important landmark in discussing cognition, metacognition, and culture in STEM education, adding significant value to the body of literature on these fundamental subjects.

Anat Zohar
Besen Family Chair for Integrated Studies in Education
The Seymour Fox School of Education
The Hebrew University of Jerusalem
Jerusalem, Israel

[1]As used throughout this book, technology, the T in STEM, is discussed here in the context of educational technology.

References

Azevedo, R., & Aleven, V. A. (2013). *International handbook of metacognition and learning technologies* (Vol. 26). New York: Springer.

Perkins, D. N., & Salomon, G. (1989). Are cognitive skills context-bound? *Educational Researcher, 18*, 16–25.

Veenman, M. V. J. (2011). Learning to self-monitor and self-regulate. In R. E. Mayer & P. A. Alexander (Eds.), *Handbook of research on learning and instruction* (pp. 197–218). New York: Routledge.

Zohar, A., & Barzilai, S. (2013). A review of research on metacognition in science education: Current and future directions. *Studies in Science Education, 49*(2), 121–169.

Preface

To my late parents, Moshe (Polack) and Risia (Dudman) Haisraeli, whose history and passion for education made me who I am.

Education is the pinnacle of civilization activities. It is not just how we accept values, knowledge, and insights about the world we live in; education is by far the most effective means to instill peace and harmony among people and peoples. Science satisfies humans' insatiable curiosity and drive to know more and understand better, providing the underpinning for engineering and technology. If used correctly, following universal values of social justice and the right to pursue happiness for all, science and technology are the foundations for humans' welfare, health, and well-being. Unfortunately, people have not always applied these values as torches of progress. Throughout history, and especially in the twentieth century, there were terrible deviations from using science and technology to benefit humans, leading to a murderous regime that killed millions of innocents in the name of some false racist theory. In the current century, religion often leads to similar arguments and outcomes, though not yet at the same scale.

Education is the only single means through which it is possible to eradicate these movements and encourage people and peoples to respect each other's right to live their lives the way they wish, exercising their cultures, so long as it does not infringe on others' lives. Moreover, only science, technology, engineering, and mathematics (STEM) can stop the deterioration of our planet's environment while increasing the economic pie, so nobody has to fight over shrinking resources.

STEM education must evolve through research to gain better understanding of how our brains process and assimilate new information and turn it to knowledge and comprehension. Metacognition is a relatively new frontier in education in general and in STEM education in particular. This book attempts to fuse cognition, metacognition, and culture to enhance STEM education. As this book involves authors from diverse backgrounds, different countries and continents, representing a plethora of cultures and approaches, it is my sincere hope that the book will make its modest contribution to encourage dialogue among STEM educators and teachers.

I wish to thank my co-editors and the chapter authors for their contributions to this book and their ongoing involvement. Thanks to Rea Lavi, my PhD student, whose editorial help was indispensable. Last but not least, thanks to my dear family, and especially to my husband, Dov Dori, for being there for me.

Faculty of Education in Science and Technology
Technion
Haifa, Israel
The Samuel Neaman Institute for National
Policy Research
Haifa, Israel

Yehudit Judy Dori

Contents

1 **Introduction** 1
Yehudit Judy Dori, Zemira R. Mevarech, and Dale R. Baker

2 **Teacher Cognition of Engaging Children in Scientific Practices** 9
Barbara A. Crawford and Daniel K. Capps

3 **Students' Metacognition and Metacognitive Strategies in Science Education** 33
Shirly Avargil, Rea Lavi, and Yehudit Judy Dori

4 **Reconsidering Different Visions of Scientific Literacy and Science Education Based on the Concept of *Bildung*** 65
Jesper Sjöström and Ingo Eilks

5 **Designing for Collaborative Problem Solving in STEM Cyberlearning** 89
Kent J. Crippen and Pavlo D. Antonenko

6 **Technology, Culture, and Young Science Teachers: A Promise Unfulfilled and Proposals for Change** 117
Randy Yerrick, Michael Radosta, and Kelsey Greene

7 **Technology, Culture, and Values: Implications for Enactment of Technological Tools in Precollege Science Classrooms** 139
Noemi Waight and Fouad Abd-El-Khalick

8 **Engineering Cognition: A Process of Knowledge Acquisition and Application** 167
Senay Purzer, Tamara J. Moore, and Emily Dringenberg

9 **Metacognition and Meta-assessment in Engineering Education** 191
Niva Wengrowicz, Yehudit Judy Dori, and Dov Dori

10 The Impact of Culture on Engineering and Engineering Education 217
Adam R. Carberry and Dale R. Baker

11 Engineering Education in Higher Education in Europe 241
M. Sencer Corlu, Kjeld Svidt, Dorina Gnaur, Rea Lavi, Oğuz Borat, and M. Ali Çorlu

12 Cognition, Metacognition, and Mathematics Literacy 261
Zemira R. Mevarech and Lianghuo Fan

13 Promoting Mathematics Teachers' Pedagogical Metacognition: A Theoretical-Practical Model and Case Study 279
Zehavit Kohen and Bracha Kramarski

14 Mathematical Modeling and Culturally Relevant Pedagogy 307
Cynthia O. Anhalt, Susan Staats, Ricardo Cortez, and Marta Civil

15 Discussion 331
Yehudit Judy Dori, Zemira R. Mevarech, and Dale R. Baker

Correction to: Reconsidering Different Visions of Scientific Literacy and Science Education Based on the Concept of *Bildung* C1

Editor Biographies 353

Author Biographies 357

Author Index 369

Subject Index 373

Chapter 1
Introduction

Yehudit Judy Dori, Zemira R. Mevarech, and Dale R. Baker

The fields of research on cognition, metacognition, and culture in learning and teaching Science, Technology, Engineering, and Mathematics—STEM—have been growing rapidly in recent years, attracting considerable interest among scholars and educators. This book addresses the point where these three fields intersect. The main purpose of this book is to identify, map, and analyze the research in cognition, metacognition, and culture. We have aimed to identify and characterize commonalities and differences in research pursuits and findings across each of the STEM areas, pointing out and discussing discipline-dependent nuances. To this end, we have solicited chapters from leading researchers in these areas and asked them to elaborate on aspects pertaining to cognition, metacognition, and culture in their respective domains of STEM education expertise.

While there has been research on metacognition in science and mathematics education, the studies related to metacognition and culture in engineering and technology education is almost nonexistent. This book is thus likely the first one to tackle the interaction between the domains of engineering and technology education on the one hand and the metacognitive and cultural aspects on the other hand.

We are still at early stage of research on the intersection and interaction between the four STEM domains, specifically their T and E components, and the three themes: cognition, metacognition, and culture. This state of affairs raises questions

Y.J. Dori (✉)
Faculty of Education in Science and Technology, Technion, Haifa, Israel

The Samuel Neaman Institute for National Policy Research, Haifa, Israel
e-mail: yjdori@technion.ac.il

Z.R. Mevarech
David Yellin Academic College of Education, Bar-Ilan University, Jerusalem, Israel

D.R. Baker
MaryLou Fulton Teachers College, Educational Leadership and Innovation,
Arizona State University, Tempe, AZ, USA

Y.J. Dori et al. (eds.), *Cognition, Metacognition, and Culture in STEM Education*, Innovations in Science Education and Technology 24,
https://doi.org/10.1007/978-3-319-66659-4_1

related to these themes across the fields of STEM education research, schools of thoughts, and cultural boundaries. Hence, in this book, we review theoretical background and cutting-edge research on how various forms of cognitive and metacognitive instruction in different cultures may enhance learning and thinking in STEM classrooms. We expect that this approach will lay down the foundations needed for a more formal attempt at defining and testing clear-cut definitions of the concepts we investigate in this book.

Most of the authors in this book investigate high school and university students—or both—in different countries, while a few investigate younger students. The focus in this book is mostly on students' learning, with an emphasis on cognition, metacognition, and culture in STEM. Nonetheless, there are chapters which view these topics through a lens of teaching, making pedagogical content knowledge another topic discussed in the book. Some authors review the existing literature, while others described their own framework or models, case studies, or empirical studies, using qualitative or quantitative methods. The variety of author nationalities and the cultures they investigated are testimonial of the multifaceted nature and robustness of this book.

Table 1.1 specifies each chapter's research population region, academic class, educational focus, research type and tools, and domain of study. The book includes four parts, one for each of the STEM domains: science education, technology education, engineering education, and mathematics education. As the field of STEM is fraught with ambiguous definitions for concepts used by various researchers, chapter authors made an effort to provide a clear definition of the cognitive, metacognitive, and cultural components addressed by them.

Collectively, the book is not just an account of the state of the art in research about cognition, metacognition, and culture in the four STEM domains; it also provides a wide, integrating perspective on what researchers are investigating with respect to these topics. To this end, Chap. 15 synthesizes the findings and views expressed in the invited chapters, makes general observations and insights stemming from analyzing the chapters, and suggests concrete future research directions for each topic(s) presented in each of the 13 chapters of the book.

1.1 Overview of Chapters in the Four Book Parts

1.1.1 Part I: Science Education

In Chap. 2, *Teacher Cognition of Engaging Children in Scientific Practices*, Crawford and Capps are concerned with defining the required teacher knowledge for engaging children in scientific practices. The authors base their definition on existing educational frameworks and on their own study. The study involved 30 elementary science teachers who took part in a professional development course. They present qualitative findings concerning two of the teachers who took part in this study. The authors provide the definition of *pedagogical content knowledge of*

Table 1.1 Chapter characteristics

Chapter	Authors	Population			Type and tools	STEM education domain
		Age group	Focus	Region		
2	Crawford and Capps	Upper elementary school	Teachers	USA	Review and two case studies	Science education
3	Avargil, Lavi, and Dori	Elementary, middle, high school, and postsecondary school	Students	Western world	Review	Science education
4	Sjöström and Eilks	High school	Teachers	Europe	Model/framework	Science education
5	Crippen and Antonenko	Undergraduate	Instructors	USA	Framework	Technology education
6	Yerrick, Radosta, and Greene	University	Teachers	USA	Empirical qualitative study	Technology and science education
7	Waight and Abd-El-Khalick	Precollege	Students and teachers	USA	Framework and review	Technology and science education
8	Purzer, Moore, and Dringenberg	Precollege	Teachers	USA	Model/framework	Engineering education
9	Wengrowicz, Dori, and Dori	Undergraduate and graduate	Students	Israel and USA	Empirical quantitative and qualitative study	Engineering education
10	Carberry and Baker	University	Programs and their graduates	Western world	Review	Engineering education
11	Corlu, Svidt, Gnaur, Lavi, Borat, and Çorlu	University	Programs and their graduates	Europe	Comparison of three national programs	Engineering education
12	Mevarech and Fan	High school	Students	Israel and Singapore	Description of two national approaches; quantitative empirical study	Mathematics education
13	Kohen and Kramarski	University	Preservice teachers	Israel	Review; two case studies	Mathematics education
14	Anhalt, Staats, Cortez, and Civil	University	Preservice high school teachers	USA	Review; task examples	Mathematics education

scientific practices (PCK-SP), which is affected by *subject matter knowledge* (SMK), scientific practices, *nature of science* (NoS), and pedagogical knowledge. They describe and analyze each teacher's PCK-SP prior to and following the professional development course the teachers took part in.

In Chap. 3, *Students' Metacognition and Metacognitive Strategies in Science Education*, Avargil, Lavi, and Dori present literature review findings on empirical research on metacognition with focus on science education. They divided the papers they had reviewed into three categories: (a) metacognition-based pedagogical intervention, (b) assessment tools for metacognition, and (c) metacognitive learning processes. The authors have identified development of assessment tools and their validation as the largest gap in research. Presenting the survey outcomes of the assessment tools described in these papers, they derive key criteria for developing and comparing such tools. Finally, the authors provide suggestions for educators and researchers concerning assessment and nurture of metacognition in science education.

In Chap. 4, *Recinsidering Different Visions of Scientific Literacy and Science Education based on the Concept of Bildung*, Sjöström and Eilks present three humanistic approaches to *scientific literacy* (SL), termed *visions*. Each vision pertains to a different set of SL aspects and represents different perspectives on science and society. The simplest vision, Vision I, pertains to SL as science knowledge or "pipeline science"—what is required to know in order to do science. Vision II pertains to SL with respect to economic and everyday requirements or "science for all"—what is required to know in order to live with science, whether one is a scientist or a nonscientist. Finally, Vision III, the most complex one, argued by the authors as the one that the educational systems should teach. It pertains to SL as critical, reflexive, and transformative or "science for transformation"—the kind of science required in order to change society through scientific literacy. This vision highlights the concept of *Bildung*, which, in this context, refers to the self-development of the individual into a socio-critical being, who takes responsibility for the world around him or her and transforms it through scientific knowledge and skills.

1.1.2 Part II: Technology Education

In Chap. 5, *Designing for Collaborative Problem Solving in STEM Cyberlearning*, Crippen and Antonenko present their self-developed cyberlearning software for scaffolding students' collaborative process of solving authentic problems and for developing their twenty-first century skills, dubbed ECLIPSE: Environment for Collaborative Learning Integrating Problem Solving Experiences. The authors applied a framework within ECLIPSE for scaffolding the problem-solving process, named Define, Explore, Explain, Present, Evaluate, and Reflect (DEEPER), which focuses not only on the planning and argumentation skills for problem solving but also on stakeholder analysis, information discrimination, and solution communication. The authors review problem solving and critical learning, authenticity, and collaboration as parts of STEM education and propose cyberlearning as an aid to problem solving. They argue that cyberlearning can transform scaffolding of students, as it facilitates problem solving in STEM education by making it authentic and situated in the real world.

In Chap. 6, *Technology, Culture, and Young Science Teachers – a Promise Unfulfilled and Proposals for Chnage*, Yerrick, Radosta, and Greene describe teachers who are digital natives, i.e., born into the digital age, as tending to use technology for teaching in the same traditional ways as their non-digital native counterparts. The authors argue that programs for supporting elementary school science teachers in integrating digital learning into the classroom must address teachers' knowledge regarding (a) students' and teachers' understanding of scientific concepts, (b) pedagogical strategies for teaching science, and (c) the teachers' own past as science learners and the effect this had on their pedagogical approach. Based on these principles, the authors developed and administered a graduate program for a group of digital native elementary school science teachers. The program focused on inquiry and reflection and used new technologies and technology-based teaching methods. Using a sociocultural lens, they investigated teachers' participation in and responses to this program. The authors report that digital video was the most effective technology for achieving the learning outcomes. Teachers considered it to be the most important, while social networking and flipped classroom were the least effective.

In Chap. 7, *Technology, Culture, and Values: Implications for Enactment of Technological Tools in Precollege Science Classrooms*, Waight and Abd-El-Khalick discuss the impact of technology on the culture and values of precollege students in technology-supported scientific inquiry environments. The authors present the role of Nature of Technology (NoT) and associated culture and values of science teaching and learning. Dimensions of NoT include notions of technological progression, technology as part of systems, technological diffusion, technology as a "fix" for social problems, and technology as expertise. They highlight how various factors at different stages of technology adoption and implementation influence culture and values and examine empirical investigations of the enactment of technology-supported inquiry environments. Finally, the authors suggest investigating the interaction of technology with values, beliefs, knowledge, and skills of both teachers and students.

1.1.3 Part III: Engineering Education

In Chap. 8, *Technology, Culture, and Values: Implications for Enactment of Technological Tools in Precollege Science Classrooms*, Purzer, Moore, and Dringenberg present their concept of *engineering design cognition*. The authors discuss the following different definitions of design: (a) design as ill-defined problem solving, (b) design as a set of strategies, and (c) design as abductive reasoning, i.e., searching for a logical inference which is based on the simplest explanation. They then continue to compare different processes of engineering design and provide a diagram that summarizes the common elements of these three design processes. The authors also present a conceptual diagram of engineering design cognition as an iterative interaction of knowledge acquisition with knowledge application, which produces knowledge through trade-off decisions within a specific design context. Finally, the authors provide explanations for and examples of how to teach, assess, and carry out research within the engineering design cognition framework.

In Chap. 9, *Metacognition and Meta-assessment in Engineering Education*, Wengrowicz, Dori, and Dori describe the implementation of peer assessment and meta-assessment in two project-based learning engineering courses to foster students' higher-order thinking and metacognitive skills. The courses, taught at leading technological institutes, were a large, undergraduate course in Israel and a small graduate course in the USA. Benefits reported include diminishing the instructors' assessment burden, releasing time to better mentor teams as they engage in conceptual modeling of systems, and fostering students' higher-order thinking and metacognitive skills by assessing the work of their peers. The authors suggested implementing their pedagogical approach, method, and the set of assessment tools for conducting further research on cognition and metacognition in large-scale PBL engineering courses.

In Chap. 10, *The Impact of Culture on Engineering and Engineering Education*, Carberry and Baker explore how culture shapes engineering and engineering education. They review a variety of research studies worldwide that document the powerful influence of culture. They conclude that engineering has its own cultural norms, but it is influenced by the wider Western culture that views engineering as a masculine endeavor. They provide evidence for how culture influences experiences that shape engineers' understanding, identity, interest, and solutions to engineering problems. In particular, they address the impact of culture on engineering education and the need to present engineering as a field that supports society and improves the lives of people all over the world. They encourage engineering educators to rethink how they teach engineering and to prioritize cultural considerations when preparing engineering students for real-world activities and educating engineers to solve global problems.

In Chap. 11, *Engineering Education in Higher Education in Europe*, Corlu, Svidt, Gnaur, Lavi, Borat, and Çorlu discuss the historical development of engineering education in Europe, from the first industrial revolution to the present day, linking engineering education to innovation in Europe. The authors outline the historical development of the Anglo-American and Continental European engineering education traditions. They proceed to describe in more depth the historical development of engineering education in two European countries with Continental tradition: Denmark and Turkey. Finally, the authors compare the engineering education systems in Denmark, Turkey, and the UK—an Anglo-American tradition country—across several dimensions, including innovation scores of the European Commission, prevalent teaching methods, and advantages and challenges of each system. The authors explain the differences in innovation scores between the three countries and suggest how policy makers, educators, and researchers can improve innovation in engineering education.

1.1.4 Part IV: Mathematics Education

In Chap. 12, *Cognition, Metacognition, and Mathematics Literacy*, Mevarech and Fan define mathematics literacy based on the PISA and Chinese approaches. The authors pose and respond to the question of what kinds of mathematical knowledge are important for citizens in the modern world? Their mathematics literacy definition revolves around the use of mathematics in various contexts on top of using algorithms and practicing the manipulation of numbers and symbols. The authors provide examples of Complex, Unfamiliar and Non-routine (CUN) tasks that have the potential to enhance mathematics learning. Exposing students to complex, unfamiliar, and nonroutine (CUN) tasks is necessary but not sufficient. Thus, the chapter describes two metacognititve approaches that have the potential to promote mathematics literacy: the IMPROVE method, which has been implemented and investigated in Israel, and the Pentagon framework, which has been implemented in Sigapore. IMPROVE includes introducing the new materials to the whole class by modeling, metacognitive questioning in small groups, practicing the metacognitive questioning and reviewing, obtaining mastery on lower and higher cognitive processes, verification, and enrichment. The Pentagon framework fosters conceptual understanding, skills proficiency, mathematical thinking and metacognitive processes, and students' attitudes.

In Chap. 13, *Promoting Mathematics Teachers' Pedagogical Metacognition – a Theoretical-practical Model and Case Study*, Kohen and Kramarski present a theoretical and practical model of pedagogical metacognition in teacher education, called Cognition/Metacognition and Teaching instruction (Cog/Meta_T). They then provide a case study of two preservice teachers, in which they apply this model in a microteaching course embedded with Web-based learning and supported by reflective discussions using the Cog/Meta-T model. The case documents the impact of the model in the microteaching course through videotaped 5-min teaching episodes of the two mathematics teachers. Kohen and Kramarski's line-by-line videotape analysis of the teachers' cognitive/metacognitive instructional strategies supports the claim that the model can be an effective pedagogical tool in the preparation of mathematics teachers, as it stimulates students' engagement in learning.

In Chap. 14, *Mathematical Modeling and Culturally Relevant Pedagogy*, Anhalt, Staats, Cortez, and Civil provide a method to contextualize the teaching of mathematics. They propose a way to combine a knowledge-based approach with the strategies of mathematical modeling. In order to make mathematics relevant to students' lives, these strategies include problem solving and sense making, developing problem-solving tools, interpretation and validation of results, and model improvement. They provide a charming example of how neighborhood fences and gates can serve as a modeling activity to describe mathematical functions, connecting to culture and community. The authors demonstrate the impact of their approach with preservice teachers by presenting and discussing their questions and concerns about teaching secondary mathematics concepts and including culturally relevant aspects. Finally, they document the preservice teachers' difficulties in making critical conscious connections between mathematics, relevance to culture, and students' daily lives.

In Chap. 15, the discussion chapter, we summarize these definitions, but we do not attempt to fixate a specific one or provide an all-encompassing set of definitions of the subjects this book focuses on. Our modest objective has been to frame the subjects in context and expose readers to the interactions among the orthogonal notions of STEM domains on the one hand and the aspects of cognition, metacognition, and culture on the other hand.

The chapters in this book lay out much of the groundwork needed to analyze and map the research on cognition, metacognition, and culture in STEM education. While the book discusses important aspects of these topics in four parts, some chapters discuss how learners learn and others how teachers teach. This leaves room for further research topics, which we present at the end of the discussion chapter. Considering the width and depth of this book's scope, and the wealth of newly reported research in unchartered territory, it is our sincere hope that the book will be instrumental in grounding and advancing the emerging field of research and practice in cognition, metacognition, and culture in science, technology, engineering, and mathematics education.

Chapter 2
Teacher Cognition of Engaging Children in Scientific Practices

Barbara A. Crawford and Daniel K. Capps

2.1 Proposing a Framework of Teacher Cognition of Engaging Children in Scientific Practices

In this chapter, we propose a framework of teacher cognition of engaging children in scientific practices, formerly referred to as inquiry teaching consistent with recent education reforms in the United States (US). We address the knowledge teachers need to acquire in order to engage children of all ages in inquiry and scientific practices in the classroom. The historical basis of shifting from inquiry to that of scientific practices in the US is described below. In developing this framework, we draw on theoretical and empirical literature related to teachers' knowledge bases and use of scientific practices, and our own work with teachers. Instruction related to students carrying out and thinking about scientific work, as translated into the classroom, is complex and sophisticated (Crawford 2000). We need to know more about what teachers understand, and what many do not understand, about scientific practices, inquiry-oriented pedagogy, and how to translate these ideas into the classroom (Abell 2007; Crawford 2014; van Driel et al. 2014).

Classrooms that support students in learning through engaging in scientific practices promote critical thinking skills that empower these students to become lifelong learners of science (Llewellyn 2005; NRC 2000, 2012). Understanding aspects of teacher cognition related to scientific practices is important in light of recent education reforms, as in the *Next Generation Science Standards* (NGSS Lead States 2013) in the US, as well as reforms in countries around the world (e.g., Barnea et al. 2010; Department for Education and Skills/Qualification and Curriculum Authority 2004; Ministry of Education of Singapore 2007). New standards re-emphasize the goal of

B.A. Crawford (✉) • D.K. Capps
Department of Mathematics and Science Education, College of Education, University of Georgia, Athens, GA, USA
e-mail: barbarac@uga.edu

Y.J. Dori et al. (eds.), *Cognition, Metacognition, and Culture in STEM Education*, Innovations in Science Education and Technology 24,
https://doi.org/10.1007/978-3-319-66659-4_2

achieving scientific literacy in young people. In the US the *NGSS* advocate for teachers integrating three dimensions, Scientific Practices (SP), Disciplinary Core Ideas (DCI) and Crosscutting Concepts (CC) using a three-Dimensions (3-D) teaching approach. Although these dimensions, by themselves, were part of earlier reform movements (e.g., NRC 1996), the explicit call for the integration of the three is novel. Achieving scientific literacy includes gaining critical, higher order thinking skills (Zohar 2004) versus acquiring superficial knowledge of discrete science concepts and terminology, and limited lower order thinking skills. We claim that teaching SP in the classroom enhances higher-order thinking and contributes towards developing scientific literacy in young people. Specifically, we are interested in what teachers *need to know* in order to enact the kind of instruction we refer to as, SP pedagogy; that engages young people in sophisticated scientific practices and in developing higher order thinking. We draw on the empirical literature related to teacher knowledge of reform-based pedagogy, and our own work with teachers, to better understand what might constitute the cognition needed to engage students in scientific practices in the classroom.

2.2 Teacher Cognition

Teacher cognition has been referred to as, "somewhat ambiguous, because researchers invoke the term to refer to different products, including teachers' interactive thoughts during instruction; thoughts during lesson planning; implicit beliefs about students, classrooms, and learning; reflections about their own teaching performance; automatized routines and activities that form their instructional repertoire; and self-awareness of procedures they use to solve classroom problems" (Kagan 1992, p. 420). Teacher cognition has been defined simply as the knowledge structures for classroom teaching (Finley et al. 1992). Calderhead's definition of cognition includes not only the knowledge, but the beliefs, and thinking of teachers (Calderhead 1996). We view teacher cognition as encompassing more than simple knowledge structures. Our view it is that teacher cognition related to scientific practices includes the explicit thought processes of teachers as they take into account their own knowledge and experiences both teaching science and engaging in scientific practices, combined with their experiences with and their views of science, and their beliefs about how children learn and their self-awareness of analyzing their teaching.

Although it is commonly agreed that knowledge and beliefs are central to teacher cognition, Pajares (1992) pointed out that there is confusion between where these two constructs intersect. A variety of words have been used to define knowledge and beliefs including, but not limited to perspectives, teachers' conceptions, personal knowledge, practical knowledge, and personal practical knowledge. Due to the multiplicity of terms employed and the blending of beliefs and knowledge, it is difficult to pinpoint where knowledge ends and beliefs begin, at times making the boundaries quite arbitrary. Although we recognize the important role beliefs play in teacher cognition of scientific practices (e.g., see Crawford 2007 for a more thorough discussion on the matter), in this chapter we focus on the nature of the knowledge teachers need to acquire in order to engage children in scientific practices. We emphasize subject matter knowledge, knowledge of scientific practices, knowledge of nature of science as particularly important aspects of teacher cognition related to teaching science in this way. In the next section we review a variety of conceptions, frameworks, and models of teacher knowledge discussed in the literature.

2.3 Theoretical Frameworks for a Knowledge Base of Teaching

During the past several decades the knowledge base of teaching has been defined and represented using various frameworks and models. Researchers have described categories of teacher knowledge, including subject matter or content knowledge (Grossman 1990; Shulman 1987), theoretical knowledge (Alexander 1984); craft knowledge (Calderhead 1996; Fenstermacher 1994), personal practical knowledge (Connelly et al. 1997; Connelly and Clandinin 1990; Tamir 1991), case knowledge (Shulman 1986), and pedagogical content knowledge (Shulman 1986, 1987). Some of these categories of knowledge, such as subject matter or content knowledge and theoretical knowledge, can be seen as more formal knowledge of one's discipline including factual propositions, structural knowledge of one's field, and knowledge of how problems are solved in the discipline. Others of these are less formal and more personal to the teacher and may begin to blur the line between knowledge and beliefs. Although formal knowledge alone, does not constitute teacher cognition of how to teach science by engaging children in scientific practices, our interest here is to consider some of the more formal types of knowledge teachers need to acquire to engage their students in scientific practices.

Lee Shulman's seminal knowledge base for teaching model appeared in 1986 and 1987. The model involves various kinds of knowledge, including subject matter knowledge (SMK), pedagogical knowledge (PK), and knowledge of one's teaching context. In his model Shulman theorized a form of specialized professional knowledge he coined, pedagogical content knowledge (PCK). Using the construct of PCK, the content knowledge of science concepts and principles that might be known by a professional scientist does not necessarily include the integrated knowledge necessary for teaching science. Pamela Grossman's model of teacher knowledge (1990) categorized knowledge as: (a) general pedagogical knowledge, (b) subject matter knowledge, (c) pedagogical content knowledge, and (d) knowledge of context. Later Cochran-Smith and Lytle (1999) described several types of teacher knowledge related to teaching. Included in these were the more formal *knowledge for practice* and the more practical *knowledge in practice*. Both the Grossman and Cochran-Smith & Lytle models offer different insights into necessary and related forms of teacher knowledge.

In the recent *Handbook of Research on Science Education Vol II,* van Driel et al. (2014) reviewed definitions and models of teacher knowledge and those published in the last few years, with a focus on disciplinary-based teacher knowledge, such as *science* teacher knowledge. For example, van Driel et al. highlighted the mathematical knowledge for teaching (MKT) model. In this model Deborah Ball and colleagues address specialized mathematical content knowledge (Ball et al. 2008). van Driel and colleagues noted limited attention in the literature, as of their writing, to *specialized science* knowledge for teaching. Consideration of the MKT model and how it might apply to teaching science, may be useful in hypothesizing what knowledge is needed to engage students in scientific practices, given this knowledge is unique to the work of scientists.

In summary, decades ago Shulman addressed the question, what are the characteristics of a professional knowledge base of teaching, with respect to reform-based teaching (Shulman 1987)? We raise a similar question in this chapter. However, we extend this question to what knowledge base is needed to teach science in the twenty-first century? Specifically, what foundational knowledge is needed by teachers to teach students about how to engage in *scientific practices.*

2.4 Science Inquiry/Practices as Related to Teaching Science

Classrooms that support students in learning through engaging in scientific practices promote critical thinking skills that empower these students to become lifelong learners of science. As a brief historical background to the teaching of scientific practices, the concept of inquiry has been advocated as an instructional approach of science teaching for over hundred years (e.g. Dewey 1910). Yet, there exists much confusion about what inquiry means in classrooms. Different definitions of teaching science as inquiry appear in the literature (e.g., National Research Council 2007). Further, confusion over the meaning of teaching science as inquiry appears to exist in the minds of many teachers and researchers (Anderson 2007; Capps and Crawford 2013a; Crawford 2014; Wheeler 2000). For example, the first author vividly recalls a personal experience at a science education conference, when Sandra Abell, a well-known researcher in the US, called out the question, "Just what is it? Inquiry? What are we all talking about?" In working towards a definition of inquiry in science education, Anderson (2002) proposed three variations of inquiry: (1) *scientific inquiry* (the various ways in which scientists study the natural world): (2) *inquiry learning* (a process by which children acquire knowledge of science concepts and learn about nature of science); and (3) *inquiry teaching* (broadly defined as the pedagogy by which teachers engage students in inquiry. Unfortunately, some researchers conflate these three meanings.

One of the most common misconceptions held by teachers is that inquiry equates to using "hands-on" activities. For example, teachers may view students building a DNA model from beads and pipe cleaners as inquiry. However, this is limited view of inquiry. "Hands-on experience is important but does not guarantee meaningfulness"

(AAAS 1993, p. 319), nor does it guarantee inquiry is taking place. Another misconception is that inquiry can be devoid of science content. Just using what might be called science process skills, with no testable question or connection with scientific concepts and principles, does not count as teaching children inquiry. There is also the issue of the amount of teacher direction versus student initiation. "The dual issues of a) the amount of teacher direction, and b) the quality of the cognitive activities, are important to consider when assessing the nature of an inquiry experience in a given classroom" (Crawford 2014).

In the US the *Framework for K-12 Science Education* (NRC 2012) and *NGSS* (Lead States 2013) have replaced the use of the term inquiry, as articulated in previous reform documents, with eight, "Scientific and Engineering Practices". In the NGSS there is a renewed focus on the thinking processes of inquiry, the intellectual work, associated with scientific work. We refer to teaching science as inquiry, inquiry teaching, and engaging children in scientific practices to mean the following, "Teaching science as inquiry involves engaging students in using critical thinking skills, that includes asking questions, designing and carrying out investigations, interpreting data as evidence, creating arguments, building models, and communicating findings, in the pursuit of deepening their understanding by using logic and evidence about the natural world," (Crawford 2014, p. 515). From this point forward in the chapter, we will use the phrase *engaging children in Scientific Practices*, in place of inquiry, with the added emphasis on the pursuit of deepening understanding and use of logic and evidence about the natural world.

We believe teachers need to have a *deep and integrated knowledge* of foundational science concepts and principles, scientific practices, nature of science (NOS), and pedagogy, as well as take a metacognitive stance towards their teaching, in order to expertly *engage their students in scientific practices*, which leads to use of logic and critical thinking. In other words, we argue for an expanded view of teacher cognition, including knowledge acquisition of core science concepts, of the nature of scientific inquiry, of what science is, and what science is not, and of how one translates these ideas into the teaching of children in K-12 classrooms. Although explicit thought processes are a component of our view of cognition (see discussion above), we believe it is important to emphasize teachers need to take a metacognitive stance about their teaching and about science.

2.5 Teachers' Subject Matter Knowledge and Knowledge of Engaging Students in Scientific Practices

It is an old adage that 'you cannot teach what you do not know'. At the very least teachers need a certain depth of discipline-specific subject matter knowledge (SMK), in order to engage children of any grade level in scientific practices. A common assumption is that teachers who know more, make better teachers. Over the past several decades, studies have tested the assumption that teachers with more

SMK are better teachers. Though it is not always the case, in general, studies suggest a positive relationship between SMK and accomplished teaching practice (Abell 2007). A handful of studies included findings related to teachers' use of reform-based teaching practices in their classrooms. Pioneering work in this area used coursework as a proxy for SMK, comparing the number of science courses teachers took in college to their reform-based teaching practice. These studies found that teachers with more science coursework were more likely to engage their students in science processes and student-conducted experiments (Dobey and Schafer 1984; Smith and Cooper 1967). More recently, studies have used other methods to measure teacher SMK including concept maps (Hashweh 1987), card sorting tasks (Carlsen 1993), observation and interviews (Capps and Crawford 2013a; Henze et al. 2009), and SMK assessments (Usak et al. 2011). Findings from studies like these suggest that teachers with greater SMK are less likely to stick to the script of a lesson and are more responsive to student ideas during class discussion; whereas teachers with less SMK tend to lean more heavily on instructional resources and spend more time lecturing and less time on student ideas (Abell and Roth 1992; Carlsen 1993; Lee 1995; Newton and Newton 2001). Thus, these studies provide further evidence for the important role teacher SMK plays in engaging students in more sophisticated instructional approaches like inquiry.

In addition to SMK teachers need knowledge of how scientists carry out their work, including how to ask testable questions, design and carry out scientific investigations, analyze data, build models and develop arguments, maintain skepticism, and communicate and justify conclusions. We will refer to this as knowledge of Scientific Practices (SPK). We posit that teachers will also need to have knowledge of Nature of Science (NOSK), and that NOSK is an important component of what students need to know in order to deeply understand the nature of scientific knowledge. Thus, teachers likely need more and perhaps different kinds of knowledge than is traditionally learned in a standard set of science college courses in physics, chemistry, biology, or earth sciences. It is this expanded content knowledge, and the *integration* of knowledge of pedagogy and of classroom and school context that we intend to explore. We will delve into the question, what kind of knowledge do teachers need in order to engage their students in scientific practices and the use of logic and evidence to understand science? Further, we will address implications for teacher education and professional development (PD).

2.6 Model for Teacher Cognition for Engaging Students in Scientific Practices

Our model for teacher cognition for engaging students in *scientific practices* is built on over 5 years of work with teachers of fifth to ninth grade children (ages 10–15) in the US involved in an authentic science professional development program. To illustrate the necessary knowledge base that may serve to foster teaching science as inquiry we compare the knowledge of a research scientist to that of a seventh grade science teacher. First, consider the knowledge of a research scientist holding a Ph.D. in ecology who studies the impact of invasive plants on freshwater ecosystems in Tanzania. This scientist, no doubt, has deep SMK. SMK is an important aspect of teacher cognition related to engaging students in scientific practices. Although necessary, SMK alone is not sufficient. This scientist also, presumably, has deep SPK in that she understands what inquiry is and is able to competently design and carry out research investigations. She may or may not have deep NOSK (see Schwartz et al. 2004). What kind of PK does this research scientist have related to teaching life science to grade 7 children (ages 12–13)? Of course, it is difficult to say, for sure. However, if this research scientist has never stepped foot into a seventh grade classroom as a teacher, it is safe to say, she would have limited knowledge of: the developmental capabilities of these seventh graders, effective pedagogy, the curriculum, cultural backgrounds, community aspects, or the performance indicators for engaging children in Scientific Practices. Thus, it is likely this research scientist may have limited ability to effectively engage these students in scientific practices in the classroom.

Now, let's consider the knowledge of a typical seventh grade teacher. In order to teach science concepts and principles, naturally the teacher needs to have a certain depth of SMK. Traditional teacher preparation programs will likely afford science teachers with minimum subject matter competencies as demonstrated on standardized teacher certification tests, and through taking required introductory science courses in college. To engage children in scientific practices it is likely this teacher needs to know about how scientists do their work [SPK]. There is some evidence that many science teachers may not possess this kind of knowledge. For example, a study by Capps et al. (2016) demonstrated that even highly-motivated science teachers lacked an understanding of the practices of science, which would make it difficult to engage children in scientific practices in their classrooms. Further, many teachers do not have knowledge of the way scientists work and think; what it means to be a critical thinker, or of how scientists think about the world and develop their explanations. This is not surprising given that very few teachers actually have experience engaging in the practice of science, either on their own or as part of a team. Thus, SPK may be a limiting factor. There is also evidence that many teachers do not possess adequate NOSK (Abd-El Khalick and BouJaoude 1997; Capps and Crawford 2013a; Carey and Stauss 1970) without additional PD. However, unlike the research scientist, this seventh grade teacher will likely be an expert in many general strategies of teaching (PK), including knowing the developmental capabilities of seventh students, effective pedagogy, knowledge of curriculum, cultural backgrounds, and influence of school community. Yet, given her limitations in SPK and perhaps, in SMK, it is likely this seventh grade teacher will have limited success in effectively engaging students in scientific practices.

In summary, there are many types of knowledge teachers need to acquire to successfully engage their students in scientific practices in the classroom. Teaching is complex. Among these, SMK is just one type. We acknowledge that SMK is an important aspect of teacher cognition to engage students in scientific practices (van

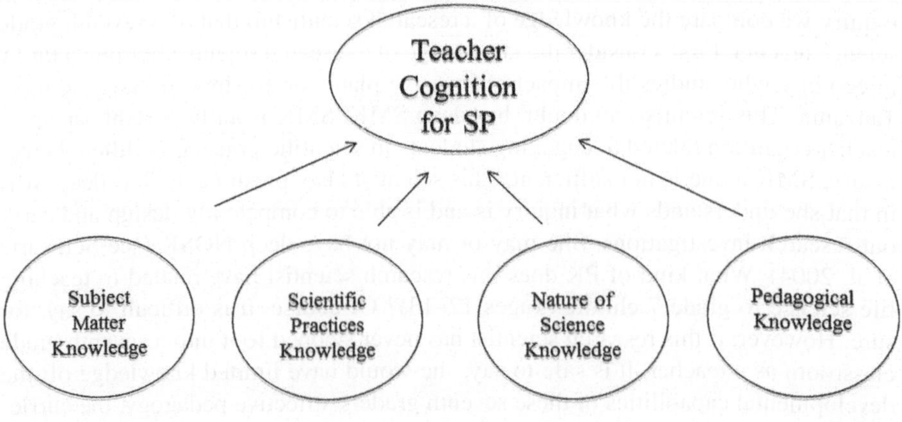

Fig. 2.1 Model representing aspects of knowledge that comprise teacher cognition for engaging students in scientific practices

Driel et al. 2014). Also important, however, are teachers' conceptions or understandings of scientific practices and of the way scientists work and think (SPK), knowledge of how scientific knowledge is constructed or nature of science (NOSK), and PK. These types of knowledge must be developed together in order for teachers to have some level of confidence in how to engage their students in a scientific investigation, even when they do not know the answers themselves. All these kinds of knowledge relate to teacher cognition for scientific practices (see Fig. 2.1). The knowledge gained *in* the instructional practice of teaching about scientific practices is a necessary component of a teacher's ability to shape their SMK, SPK, NOSK, based on the context of the classroom to the specific needs of the students and contribute to a teacher's beliefs and disposition. A mechanism for doing this is a teacher's reflective or metacognitive stance towards teaching.

2.7 Reflection and Metacognition Related to Engaging Students in Scientific Practices

Reflection and metacognition are both thinking processes. Yet, reflection and metacognition differ in some ways. The act of reflecting includes making sense of an event or phenomenon and in the case of teaching, trying to grasp the meaning of a particular science teaching episode. Reflection on one's teaching involves thinking about a classroom event, an entire lesson or unit, or interactions and behaviors of students. Reflection can span a range of thinking processes from the simple recall of events of the lesson to systematic inquiry into the extent to which students learned during the lesson and contemplating how to revise that lesson (Schon 1987). Standard teacher education practice involves asking new teachers to reflect on their teaching. For example, teacher candidates might respond to these questions: How did the lesson unfold? What happened during the lesson? What did the students do? How many students participated in a discussion? What kinds of questions did you as the teacher ask? What is your evidence your students learned? How might the lesson be revised? Generally, novice teachers have fewer experiences and a lower level of cognition of reform-based, sophisticated teaching methods then experienced, highly qualified teachers. Therefore, reflection by novice teachers may begin by dealing with superficial aspects of the lesson, such as how did the teacher perceive herself in front of her students?

Metacognition is often defined as "thinking about one's thoughts". We base our view of metacognition on that of Flavell's 1979 model and interpretations and applications to education theory and practice (Flavell 1979; Hacker et al. 1998). Metacognition is a deliberate, planful, and goal-oriented mental process, using higher level thinking skills applied to one's thoughts and experiences (Hacker et al. 1998). The thinking process is tied to a person's own internal mental representations of that reality. Metacognition is associated with how a teacher relates to her environment. Thus, taking a metacognitive stance involves thinking about one's teaching, at the same time as having an awareness of one's self as an actor in his or her environment. We are referring here to the work of teaching science- in the sense of having educational objectives, a philosophy that drives one's teaching, and enactment of a teaching plan. We suggest that a teacher who thinks about and reflects on her teaching and takes a metacognitive stance by being aware of the complexities of the classroom and her role as a teacher, is likely to achieve a higher ability to translate inquiry/scientific practices into the classroom than a teacher who does not (see Capps and Crawford 2013b).

2.8 Subject Matter Knowledge (SMK) Related to Teaching Scientific Practices

In our own work, we recruited a group of 30, fifth to ninth grade teachers from different parts of the US to participate in an inquiry-based professional development (PD) program. Prior to teachers' participation in our PD, we obtained baseline data of the nature of their teaching practices by observing their classroom lessons or by viewing videotaped recordings of what teachers identified as their best inquiry-based lessons. In analyzing these baseline lessons, we determined that the nature of these lessons consisted mostly of teacher-led discussions, hands-on activities, or confirmatory laboratories, even though many teachers believed they were using inquiry-based approaches (Capps and Crawford 2013a). Only a handful of these lessons were truly investigative in nature, allowing students to actively engage in scientific practices. Although there was no single factor uniting the four teachers who fostered more investigative environments in their classrooms, one commonality was that each of the teachers either scored at the ceiling, or reached the ceiling of our SMK assessment by the end of the program. This was the case, even if earth science was not the disciplinary area in which they were trained. For instance, one teacher had an undergraduate degree in geology and attained a master's degree in geology while he was teaching. In addition, this teacher actively sought out science professional development and research opportunities throughout his career. Another teacher had an engineering degree and regularly engaged in science-like activities in his free time. Two teachers with less formal training in science (i.e., the equivalent of a college minor) actively sought out professional development and research opportunities throughout their careers (Capps and Crawford 2013a). Thus, findings from our work concur with other studies that suggest that teacher SMK is important in enacting sophisticated teaching approaches, like engaging students in scientific practices that require a high amount of teacher involvement (i.e. Crawford 2000).

However, not all the teachers in our PD with strong SMK engaged their children in scientific practices. This point is important. In other words, strong SMK may be necessary, but not sufficient, to enable teachers to engage students in scientific practices. For example, one teacher, Kendra had a wealth of background experiences, including majoring in science, undergraduate research experience, and work in a lab prior to obtaining her teaching certification. Yet, we determined that Kendra did not teach science in an investigative way (note: we will return to Kendra later on). This suggests that although SMK is important, and even essential, it is likely not sufficient by itself, to promote a teacher using an investigative approach.

2.9 Teachers' Knowledge of Scientific Practices Related to Engaging Children in the Classroom

Another type of knowledge, related to SMK, is a teacher's knowledge of scientific practices. Teachers' knowledge of how scientists work and think, is rarely discussed in the literature. Most studies that have done this have looked at how teachers' knowledge align with inquiry/scientific practices as conceived in reform-based documents (e.g., Brown et al. 2006; Capps et al. 2016; Lotter et al. 2006). Similar to SMK, it seems likely that teachers who have a better understanding of scientific practices will likely have more success in engaging their students in scientific practices, than teachers who do not. At first, this statement may seem simplistic. In our work we have explored the relationship between a teacher's conception of scientific practices and his or her use of scientific practices in a classroom (Capps and Crawford 2013a). To gain an understanding of teachers' conceptions of scientific practices (SPK), we assessed teachers' knowledge prior to experiencing our PD. We asked teachers to describe scientific inquiry and inquiry-based instruction (related to Scientific Practices), through writing and in interviews. We also conducted observations of some of their best, self-identified lessons to see if there was correspondence between their knowledge and their teaching practice. We found that most of these teachers held limited SPK when compared to *NSES & INSES*. Conceptions included thinking of engaging children in scientific practices in a limited way, as "hands-on and minds-on teaching". A few teachers described science teaching in terms of scientific practices, such as using data as evidence to investigate a question. We found teachers who did explain this kind of teaching in terms of scientific practices were more likely to engage their students in this type of instruction than those who did not (Capps and Crawford 2013a). Thus, it appears that the ability to operationalize teaching Scientific Practices in a succinct form that still bears resemblance to scientific practice, which we have referred to as one's summary conception of scientific practices, may be important to teaching science in this way (Capps and Crawford 2013a; Capps et al. 2016). This is not the only kind of knowledge of scientific practice of which we should be concerned, but we do think it is an important part of teachers' overall conception of this kind of sophisticated teaching.

A second aspect of teacher knowledge related to engaging children in scientific practices is an understanding of how to carry out scientific investigations. There are a variety of sources where teachers might learn about carrying out scientific inquiries including Research Experiences for Teachers (RETs), Research Experiences for Undergraduates (REUs), a college science or methods course, a website, or through an informal connection with a scientist, just to name a few. These experiences have the potential to enhance participants' knowledge both of designing and carrying out investigations, and possibly deepen their conceptual knowledge of scientific practices (Blanchard et al. 2009; Crawford 2000; Hunter et al. 2006). It is likely that the majority of teachers have not had the opportunity to participate in these kinds of science research experiences, thus many teachers may not bring this type of knowledge into the classroom.

2.10 Teachers' Knowledge of NOS Related to Engaging Children in Scientific Practices

An additional component of knowledge related to teaching scientific practices that appears important is knowledge of NOS (NOSK). NOS refers to an understanding of science as a way of knowing (Abd-El-Khalick et al. 1998). Empirical studies and literature reviews suggest many teachers do not hold adequate NOSK (Abd-El Khalick and BouJaoude 1997; Carey and Stauss 1970; Capps and Crawford 2013a; Lederman 1992). In the literature there is speculation that NOSK arises from engaging in inquiry and that NOSK is necessary for teachers to engage their students in inquiry. The former is well described in the research literature in articles by Lederman (1992, 1999), Schwartz et al. (2004), and others, while the latter is less well described. Considering the latter, Rutherford (1964) argued, "Science teachers must come to understand just how inquiry is in fact conducted in the sciences. Until science teachers have acquired a rather thorough grounding in the history and philosophy of the sciences they teach, this kind of understanding will elude them, in which event not much progress toward the teaching of science as inquiry can be expected" (p. 84). Here, Rutherford argued that inadequate views of NOS held by teachers may prevent them from engaging students in inquiry (NRC 1996). We explored the relationship between teachers' views of NOS and their use of scientific practices in the classroom. In doing so, we found that teachers generally held limited conceptions of NOS, but teachers with more robust NOSK were more likely to engage their students in scientific practices than teachers with less informed conceptions of NOS (Capps and Crawford 2013a). This concurs with Rutherford's speculation that there is a relationship between teachers' NOSK and their ability to engage their students in scientific practice.

2.11 Pedagogical Knowledge Related to Engaging Children in Scientific Practices

We will briefly address general pedagogical knowledge (PK) necessary for teaching children about scientific practices and how to carry them out. Knowledge of how to manage a classroom, from the organizing and preparing materials to the general management of the classroom itself is essential and cannot be taken for granted. Walking into a poorly managed classroom and seeing educational materials in disarray, students wandering around in the classroom or just messing around, and being unproductive; or worse, seeing students carrying out unsafe practices, underscores the fact that teachers absolutely need adequate management skills to successfully engage children in scientific practices. A lack of classroom management skills can lead to disaster and discouragement in new teachers trying out new lessons. Different kinds of pedagogical knowledge including how to design science lessons that flow well, how to elicit students' prior knowledge, and various teaching strategies, such as designing group work, are all equally important. Suffice it to say, we assume teachers have a general grasp of pedagogical knowledge as a prerequisite to engaging children in scientific practices and use of logic and evidence.

2.12 Research Related to Developing Our Model of Teacher Cognition for Engaging Students in SP

In this section we present images from the field of our model of cognition related to engagement of students in SP. Over the last several years we have been involved in a research and development project aimed at supporting teachers in engaging students in scientific practice and learning foundational evolution and geological concepts. Through this project, we are invested in learning about how we can best support teachers in learning how to carry out reform-based teaching. Embedded in our work are theoretical underpinnings of how teachers acquire knowledge. These theoretical constructs include the situated nature of knowledge and a community of practice (CoP) [Wenger 1998]. We ask, how do teachers gain knowledge of scientific inquiry and the use of logic and evidence? Further, how do teachers translate this knowledge into their classrooms? An assumption is that the knowledge required to create this kind of investigative classroom is sophisticated, complex, and multifaceted. This explains, in part, why this kind of teaching is rare in classrooms.

We worked with a group of talented and motivated fifth to ninth grade teachers in the US in the context of an authentic scientific investigation appropriate for classrooms. The Fossil Finders investigation centers on understanding how organisms in a shallow Devonian sea might have changed in response to environmental changes. As part of this investigation we designed curricular materials enriched with scientific practices and we designed a series of PD summer institutes. During the PD teachers worked through the curricular materials and participated in the actual paleontological investigation as learners. Teachers later translated the curricular materials and investigation into their classrooms. Teachers had the support of scientists and education researchers both during the PD and later as they enacted Fossil Finders in their classrooms. Throughout the 2-year PD we prompted each cohort of teachers to formally reflect in writing or in conversation on their newly acquired knowledge of science concepts, principles, and practices of paleontologists, in a CoP environment (Lave and Wenger 1991; Wenger 1998).

In addition to designing curricular materials and carrying out PD experiences, we conducted research on teacher knowledge and pedagogical practices related to carrying out Fossil Finders in their classrooms. We observed that most, if not all, teachers were excited to share their experiences with their students and enact new lessons in their classrooms. Although they expressed positive feelings about trying out the new instructional approach, more than a few teachers struggled. After the summer institutes, teachers brought back samples to their classrooms of typical Devonian fossils they had found in the field. Following the background lessons, scientists shipped actual scientific samples to the classrooms, so that students could carry out the scientific investigation. All the teachers conscientiously had their students scrutinize the rock samples for fossils. Moreover, teachers went to great lengths to have their students collect data, which included identifying the fossil taxa found in the samples, and measuring and recording the sizes, fragmentation, and color of rock. Further, the students shared classroom data with scientists using a database on the projects website. Yet, only a handful of the teachers went deeper into the investigation by having students analyze and make sense of the data they collected (the Scientific Practices of Analyzing and Interpreting data). At first, we were disappointed, as we perceived that many students lost out on what we considered the key learning outcome—students engaged in the kinds of thinking scientists carry out, through carrying out Scientific Practices related to answering a scientific question. In other words, students fell short of experiencing the full extent of what science truly is. Upon reflection, we asked the question, "Why did some teachers and not others involve their children in the more sophisticated aspects of Scientific Practices?"

Over the course of the Fossil Finders project we gathered a great deal of data on all the 30 teachers. Empirical data included pre and post-tests of knowledge of geology and NOSK as well as SPK, and how to engage children in this kind of learning. Further, we captured many videos of classroom lessons, both before teachers participated in Fossil Finders and as they translated Fossil Finders into their classrooms following the PD. We also conducted interviews with most teachers, asked them to write reflections about their experiences enacting the curriculum, and collected teaching and learning artifacts. We purposively selected two of these teachers and developed contrasting cases. Both teachers increased their knowledge of inquiry and NOSK. Yet, these two teachers demonstrated different levels of enactment of inquiry in their science classrooms.

One of these teachers, we will call Brit, represented a teacher who entered the program having limited SMK, SPK, and NOSK. However, Brit demonstrated solid PK. The other teacher, we will call Kendra, represented a teacher with a strong SMK background, fairly robust knowledge of SP, and NOSK, and strong general PK (see Capps and Crawford 2013b). To say these teachers were worlds apart in their knowledge base for teaching science would be a misrepresentation. Both teachers were effective teachers by general standards. However, following PD, only Kendra was able to successfully engage her students in Scientific Practices, going beyond the basic lesson script of the Fossil Finders program. Below we expand on these cases and present an argument for the reasons why one teacher enacted this kind of teaching and the other did not.

2.12.1 Brit: Before Participating in Fossil Finders

Brit taught sixth grade science in a public school in a suburban area in the Midwest of the US. She had 5 years of teaching experience when we met her and was a well-respected teacher in her school district. Brit's curriculum coordinator noted that she was "One of our best." Brit was your typical upper elementary teacher in that she was a generalist. She had taken only three college courses in science and little additional experience beyond the occasional district workshop.

When she began the Fossil Finders program Brit's SMK, as measured on our assessment, was lower than the group average. Yet, Brit was in the middle of the pack for the upper elementary teachers (Capps and Crawford 2013b). Brit conceptualized inquiry/engaging children in scientific practices as, "students inquiring or asking their own questions and then going about answering those questions." In describing what it might look like in her classroom Brit stated, "Students talking to other students, working in groups, using computers, hands-on activities, and/or asking me questions." Prior to her participation in the PD we visited her classroom. We observed only limited aspects of engaging children in scientific practices in her lessons. We concluded Brit's classroom practices aligned with her developing but limited conception of how to teach children about scientific practices, that is, using hands-on activities and group work (Capps and Crawford 2013a).

2.12.2 *Kendra: Before Participating in Fossil Finders*

Kendra had 6 years of experience teaching seventh grade in a public school. Kendra taught science in a town adjacent to Brit's school. Using the barometer of "No complaining parents", Kendra's principal recommended Kendra as a very good science teacher. Different from Brit, Kendra had a strong background in science, including a Bachelor's of Science degree and an Undergraduate Research Experience (URE). Further, Kendra had spent time working in a lab after graduating from college. Even though her disciplinary background was in biology, Kendra had a fairly strong background in earth science. Her measured score on the pre-assessment was among the highest in the group. She scored near the ceiling. Kendra's knowledge of SP and NOS were also quite high compared with the rest of the Fossil Finders teachers. Although Kendra affirmed she understood what it meant to teach SP and that she used this kind of teaching in her classroom, in actuality, she described it as "hands-on" activities following the 5E structure. This view of teaching SP was limited, in that she made little mention of the importance of data driven lessons, involving students in explanation construction and justification, aligned with NGSS standards of engaging students in scientific practices. We visited Kendra's classroom several times prior to her involvement in the Fossil Finders PD. We confirmed Kendra's principal's assessment that she was a very good science teacher. We affirmed that Kendra had good PK related to generic teaching methods. However, our observations did not confirm that she engaged her students in scientific practices aligned with critical thinking and use of data as evidence. Although she had expressed that she believed she was teaching science in this way, in actuality she was not (Capps and Crawford 2013a).

Modeling Brit and Kendra's Cognition for Engaging Students in SP before Fossil Finders. Prior to the Fossil Finders experience, Brit and Kendra's professional backgrounds were quite different from one another. Kendra clearly had a richer background in traditional science experiences. If we were to model the components of knowledge that contributed to these two teachers' respective cognitions for engaging students in SP it might look something like Fig. 2.2. Kendra had greater resources in terms of SMK, SP, and NOS upon which she could draw. Based on this, one might presume Kendra would demonstrate greater levels of cognition than Brit. Although their knowledge bases were different, both Brit and Kendra's cognition for how to teach their students about SP were what we considered to be inert. There might be a variety of explanations for this situation. We posit that both teachers did not have sufficient knowledge in one of more areas to activate teaching students about scientific practices.

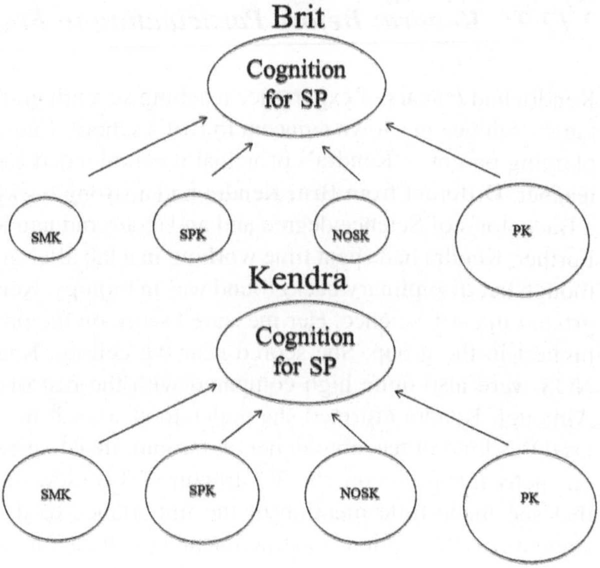

Fig. 2.2 Model representing aspects of knowledge that comprised Brit's and Kendra's cognition for engaging students in scientific practice prior to participating in the Fossil Finders professional development

2.12.3 *Brit: After Participating in Fossil Finders*

Brit's experience during the Fossil Finders summer institute supported her growth in SMK. Impressively, by the end of the institute, her SMK score more than doubled on the assessment. In fact, her score increased to be on par with many of the middle school teachers who had backgrounds in science. We observed gains in Brit's knowledge of inquiry and NOS, as well. In observing Brit teach the Fossil Finders lessons in her classroom, we were struck by the emphasis she placed on teaching NOS. The idea of NOS and the need to explicitly teach aspects of NOS to her students was something new for her. In an interview she shared, "I guess you know when we did it I thought well, yeah, but how much do I emphasize it [NOS] with my students? Probably, not enough. You know we think that, okay, here is a scientist and they said that, and case closed, let's move on. So that's really a key point to I think emphasize with my students" (Capps and Crawford 2013b, p. 1966). We did not however, see changes with respect to her use of scientific practices. Brit, like many of the teachers dutifully followed the Fossil Finders lesson plans. Once she reached the actual investigation, she and her students carefully counted and measured the fossils in the sample shipped to their classroom. Following procedures Brit had students enter their data into their data sheets. Brit transferred the information herself to the project's database. After this, Brit and her students were done. Brit wrapped up the investigation by telling them how well they had done and they moved onto another unit.

2.12.4 Kendra: After Participating in Fossil Finders

Though Kendra began the program with an initial higher level of knowledge, Kendra's experiences during the summer institute promoted added growth in her SMK and SPK and NOSK. Similar to Brit, her new understandings of NOS prompted reflection on her teaching. Kendra realized that prior to the PD, she was not teaching her students about NOS, and she began to do so explicitly. When we visited her classroom we observed that Kendra had hung up posters about NOS and regularly referred to these in her teaching. A notable difference between Kendra and Brit was that Kendra's experience in Fossil Finders was directly translated into her teaching practice. Kendra was able to contrast the Fossil Finders experience with her former classroom teaching, and recognized that the two conceptions of engaging studenrs in scientific practices were not congruent. This had a major impact on her thinking, and we have evidence from her reflections on teaching. Reflecting on her teaching before Fossil Finders she said, "I think I was doing a lot of hands-on science teaching before, but didn't necessarily have aspects of inquiry (Scientific Practices)". This recognition was something we heard her express both during the summer institute and in subsequent interviews. Based on these data we believed Kendra was poised to take on the challenging role of a teacher supporting children in engaging in scientific practices. We anticipated we might see this enactment of teaching scientific practices when we later visited her classroom.

Visiting Kendra's classroom, it became clear that her new understanding of scientific practices had definitely impacted her instruction. For example, she showed us some labs she had already "tweaked" to make them more enhanced (i.e., she made sure she began the lesson with an investigable question, had students grapple with data, and provide some sort of conclusion). Although using strategies to engage her students in scientific practices was new for her, she embraced this way of teaching. Unlike many of her colleagues who ended the investigation when data were entered into the database, Kendra took things a step further. As her students wrapped up data collection and data entry, Kendra challenged them to use the data they collected to answer the overarching question, "How do organisms respond to changes in the environment?" (Classroom observation, 11-18-09). Kendra informed her students that they would be sharing their interpretations with scientists who would be visiting the class virtually on Skype. Elaborating on this Kendra said, "Not only are you going to look at the numbers, but you will need to bring information from sixth grade, looking at basic needs of organisms, ecology, like predator-prey relationships…to take a guess at what the circumstances might have been…it could be food it could be any number of things. To help you back up your answers, you will need to choose two graphs and explain what they mean and explain a possible way the environment might have changed" (Classroom observation, 11-18-09). She set up the parameters of the investigation, but Kendra let students decide what they wanted to do. Some worked with graphs made within the database, constraining the questions they could ask. Others opted to download the data and transform it, so they could explore other questions. As students put their ideas together, Kendra provided logistical support and challenged students' interpretations, much like the scientists would do when the visited virtually.

Modeling Brit and Kendra's Cognition of Engaging Students in SP after Fossil Finders. Figure 2.3 compares the relative growth of Brit and Kendra's cognition for engaging students in SP prior to, and following their participation in Fossil Finders. Dashed lines in the figure represent knowledge after the experience. As described above, both teachers' knowledge bases increased, even though Kendra's knowledge base was quite high to begin with. Another interesting comparison is that both Brit and Kendra became more articulate about NOS and made explicit connection to their teaching practices related to NOS. Although their measured levels of knowledge of NOS were different, both teachers began to explicitly teach about NOS following the PD. One difference was their use of scientific practices in the Fossil Finders investigation (and at least for Kendra in other places in her teaching). For Kendra something clicked. It was as if there was an "activation energy" or threshold involved, and once that was reached, Kendra began to engage her students in the practices of science. However, this was not the case for Brit. It is unclear if that

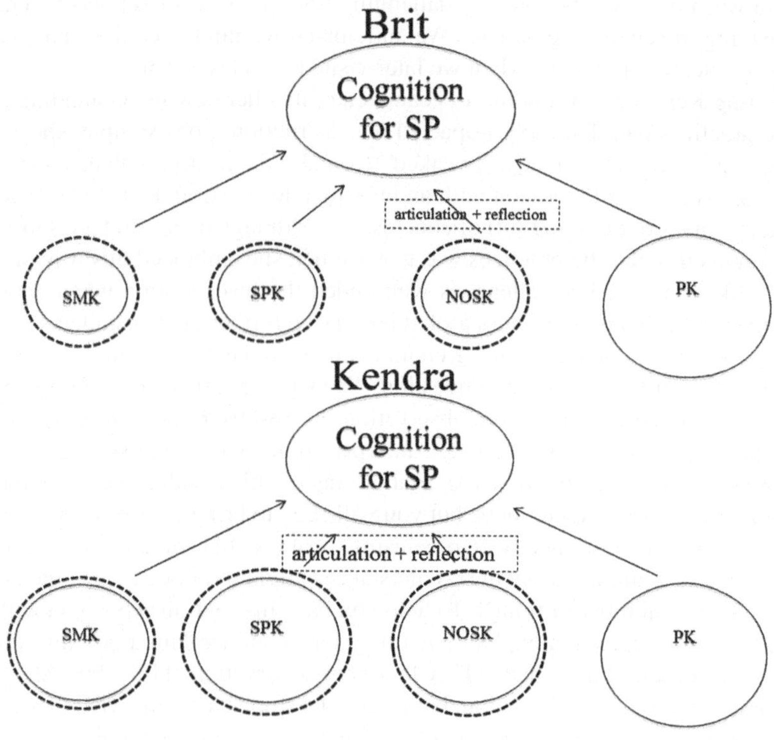

Fig. 2.3 Model representing aspects of knowledge that comprised Brit's and Kendra's cognition for engaging students in scientific practice after participating in the Fossil Finders professional development. *Dashed lines* represent growth in knowledge from pre-Fossil Finders levels

threshold was the "amount" of knowledge or the ability to access the knowledge through reflection. Our thinking is that both are likely important.

2.13 Summary and Conclusions

We have grappled with the question: What constitutes what teachers need to know in order to engage children of all ages in scientific practices in the classroom? We have drawn from previous scholars' frameworks of teacher knowledge and presented cases from our own research and work with teachers. It is important to make it clear that we have the greatest respect for all teachers who aim to change their teaching, and endeavor to engage their students in higher order thinking through carrying out scientific practices in the classroom, yet might struggle at first with doing so. Teaching inquiry/scientific practices in today's science classrooms is very challenging and entails a sophisticated pedagogical approach (Crawford 2000, 2007). Teaching science in this way depends on the context, and requires a teacher to orchestrate a dance between school-wide curricular standards, expectations of parents and community, pressures from high stakes testing and teacher accountability, often based on simple metrics, and perhaps, most importantly, students who may resist taking ownership of their learning. It is also clear from previous research that many misconceptions exist about what it means to engage children in scientific practices as they carry out inquiries in the classroom. While teaching science, important questions teachers should ask are: first – is there higher order thinking going on? Second – who is doing the thinking – is it the teacher, the students, or – ideally – both teachers and students? From our work with teachers we can make several assertions about teachers' cognition for how to engage children in scientific practices.

First, it is quite obvious that acquiring robust knowledge of how to engage students in SP is difficult. Political documents in science education (e.g., the *Framework* and *NGSS*) emphasize performance expectations that students engage in eight practices of science, integrated with disciplinary core ideas and crosscutting concepts using a 3D approach. As science teacher educators, we have witnessed our newest teachers look through all the US reform documents, charts, and tables with wide eyes and skepticism. Our prospective teachers' first line of defense is, "my mentor teacher says, we don't have to teach in this way! Our school does not advocate it." As far as we know, there are relatively few science teachers who exemplify this high standard of teaching.

One conclusion from the state of affairs is that engaging children in scientific practices is not an easy way to teach science. Development of cognition for engaging children in scientific practices is complex. Second, it is apparent that multiple kinds of knowledge are needed (science concepts and principles, context, culture, nature of science, scientific practices, pedagogy, assessment) and that these kinds of knowledge need to be *integrated*, or to come together, synergistically. Third, in order to develop cognition for engaging children in SP a teacher needs to deeply reflect on what science is, and what science is not, and on student learning. Further, teachers need to take a metacognitive stance on their teaching. In our view taking a *metacognitive stance* on one's science teaching involves thinking in a more sophisticated way about one's teaching practice as compared with simple reflection, while taking into consideration the role of the teacher and the particular context of their teaching. The ability to take a metacognitive stance is dependent on one's cognition including knowledge and beliefs. Not only would a metacognitive teacher think about how a lesson or unit unfolded, but this teacher would draw upon the knowledge she acquired of children's developmental abilities, current reforms, the context of the school, combined with her array of teaching experiences accumulated during the years with her mental awareness of pedagogical strategies to engage children in scientific practices. Taking a metacognitive stance involves teachers asking questions such as: What should I be spending time on in my classroom (e.g., content OR teaching students how to be scientists)? How does my current classroom practice support students in learning how to be scientists? For teachers who do not have a strong background in these areas (e.g., subject matter, scientific practice, NOS, etc.) it is reasonable to expect difficulties in responding to children's questions. Instead of engaging their students in science teachers may end up directing students to measure and identify a bunch of fossils algorithmically, and never take the time to engage children in making sense of the data. Kendra is an example of a teacher who was poised to gain the necessary knowledge base, in regards to her SMK and knowledge of scientific practices. However, there were initially gaps in her various kinds of knowledge. She did not have a way to articulate what teaching through engaging children in scientific practices actually means, related to her own pedagogical practice. Initially, Kendra did not explicitly reflect on her former teaching. To support deep reflection, teachers need authentic science experiences and a way to think about what engaging students in scientific practices looks like in real classrooms. Rich images from the field are needed. For example, we have a plethora of research articles and practitioner journal articles offering lesson plans, templates, steps, examples, and even written scenarios designed to guide teachers in teaching in this way. Yet, what does this kind of teaching really look like?

A fourth conclusion is that acquiring robust and enduring cognition of SP in the classroom will require sustained and long-term support for teachers; certainly beyond a one-hour workshop, or even a year-long PD. This need for long term and sustained support is especially true given that many science educators must teach outside of their area of expertise, often teaching multiple science disciplines. Acquisition of knowledge of how to engage children in scientific inquiry is not a general form of knowledge that can easily be transferred from one discipline to the next. Instead, it is likely more nuanced and will take time and effort to develop. The support should include development of a CoP involving many players contributing towards learning how to teach in this way. Kendra acquired the threshold level of knowledge of SMK, PK, SPK, and NOSK, through an intensive, resource-rich 2 year PD experience. Yet, as we discovered later, Kendra was teetering on the edge of fully embracing the myriad roles of expert science teaching that engages students in scientific practices. When we visited Kendra's classroom later in the year, she confessed her curriculum coordinator had mandated a new approach to teaching called "Mastery Learning". This method involved direct teaching of concepts and vocabulary, drill and practice. Her newly acquired knowledge was shaky and not evident in her lessons later in the year. Newly acquired knowledge is fragile, and before knowledge can be truly integrated as a part of one's regular classroom pedagogical repertoire, the knowledge needs to be worked with to become part of a teacher's practice. Otherwise, a teacher might shift away from it, in favor of easier and more familiar teaching strategies.

Our conclusions align with those of Crippen and Antonenko in Chap. 5 (this book, 2018), in that the problem solving process in STEM education requires authentic practices and development of collaborative skills at the cognitive and metacognitive levels. Similarly, our conclusions about the need to situate teachers in the kinds of learning in which they will engage their students aligns with conclusions of Yerrick, Radosta, and Greene (Chap. 6 in this book, 2018). In Chap. 6 Yerrick and colleagues claim teachers need to engage in rich and meaningful learning experiences, and the importance of reflection as a regular practice for teachers.

In summary, we believe teachers need to acquire a deep and *integrated* knowledge of foundational science concepts and principles, scientific practices, nature of science, and pedagogy, as well as take a metacognitive stance towards their teaching, in order to expertly engage their students in *scientific practices*, including teaching the use of logic and evidence and development of critical thinking.

2.14 Recommendations

How can science educators help teachers develop expertise in robustly supporting children in engaging in scientific practices, in higher-order thinking and the use of logic and evidence, much like a scientist? We have several recommendations. (1) We need to provide teachers (both prospective and practicing teachers) with rich, integrated and authentic science experiences; in which to engage as learners. In addition, we need to scaffold teachers in how to reflect on these experiences, through which teachers can more fully develop their cognition for engaging students in SP. (2). We need further research on identifying if there might be some threshold of knowledge that needs to be acquired for teachers to begin to engage students in scientific practices. Is there some minimal required level of understanding upon which to build? Determining if such a threshold exists has implications for differentiating the kind of PD experiences we might provide for various teachers, based on their initial knowledge base. (3) We also need a valid means of assessing a teacher's cognition for how to engage students in SP, in order to track changes and progress. (4). Finally, more research is needed on developing and testing a viable theoretical model of the knowledge base of teaching SP. All of these recommendations, if successfully carried out could contribute to better-designed and more effective teacher education programs for prospective teachers and professional development experiences for practicing teachers.

References

Abd-El Khalick, F., & BouJaoude, S. (1997). An exploratory study of the knowledge base for science teaching. *Journal of Research in Science Teaching, 34*, 673–699.

Abd-El-Khalick, F., Bell, R. L., & Lederman, N. G. (1998). The nature of science and instructional practice: Making the unnatural natural. *Science Education, 82*, 417–436.

Abell, S. K. (2007). In S. K. Abell & N. G. Lederman (Eds.), *Science teacher knowledge*. Mahwah: Lawrence Erlbaum Associates, Inc.

Abell, S. K., & Roth, M. (1992). Constraints to teaching elementary science: A case study of a science enthusiast student teacher. *Science Education, 76*(6), 581–595.

Alexander, R. J. (1984). *Innovation and continuity in the initial teacher education curriculum*. London: Holt, Rinehart & Winston.

American Association for the Advancement of Science. (1993). *Benchmarks for science literacy*. New York: Oxford University Press.

Anderson, R. D. (2002). Reforming science teaching? What research says about inquiry. *Journal of Science Teacher Education, 13*(1), 1–12.

Anderson, R. D. (2007). Inquiry as an organizing them for science curricula. In S. K. Abell & N. G. Lederman (Eds.), *Handbook of research on science education* (pp. 808–830). New York: Routledge.

Ball, D. L., Thames, M. H., & Phelps, G. (2008). Content knowledge for teaching: What makes it special? *Journal of Teacher Education, 59*(5), 389–407.

Barnea, N., Dori, Y. J., & Hofstein, A. (2010). Development and implementation of inquiry-based and computerized-based laboratories: Reforming high school chemistry in Israel. *Chemical Education Research and Practice, 11*(3), 218–228.

Blanchard, M. R., Southerland, S. A., & Granger, E. M. (2009). No silver bullet for inquiry: Making senseof teacher change following an inquiry-based research experience for teachers. *Journal of Research in Science Teaching, 93*(2), 322–360.

Brown, P. L., Abell, S. K., Demir, A., & Schmidt, F. J. (2006). College science teachers' views of classroom inquiry. *Science Education, 90*(5), 784–802.

Calderhead, J. (1996). Teachers: Beleifs and knowledge. In D. C. Berliner & R. C. Calfee (Eds.), *Handbook of educational psychology*. New York: Macmillan.

Capps, D. K., & Crawford, B. A. (2013a). Inquiry-based instruction in science classrooms: Is it happening? *Journal of Science Teacher Education, 24*(3), 497–526.

Capps, D. K., & Crawford, B. A. (2013b). Inquiry-based professional development: What does it take to support teachers in learning about inquiry and nature of science? *International Journal of Science Education, 35*(12), 1947–1978.

Capps, D. K., Shemwell, J. T., & Young, A. M. (2016). Over reported and misunderstood? A study of teachers' reported enactment and knowledge of inquiry-based science teaching. *International Journal of Science Education, 38*(6), 934–959.

Carey, R. L., & Stauss, N. G. (1970). An analysis of experienced science teachers' understanding of the nature of science. *School Science and Mathematics, 70*(5), 366–376.

Carlsen, W. S. (1993). Teacher knowledge and discourse control: Quantitative evidence from novice biology teachers' classrooms. *Journal of Research in Science Teaching, 30*, 471–481.

Cochran-Smith, M., & Lytle, S. L. (1999). Relationships of knowledge and practice: Teacher learning in communities. *Review of Research in Education, 24*, 249–305.

Connelly, F. M., & Clandinin, D. J. (1990). Stories of experience and narrative inquiry. *Educational Researcher, 19*(5), 2–14.

Connelly, F. M., Clandinin, D. J., & Hee, M. F. (1997). Teachers' personal practical knowledge on the professional knowledge landscape. *Teaching and Teacher Education, 13*(7), 665–674.

Crawford, B. A. (2000). Embracing the essence of inquiry: New roles for science teachers. *Journal of Research in Science Teaching, 37*(9), 916–937.

Crawford, B. A. (2007). Learning to teach science as inquiry in the rough and tumble of practice. *Journal of Research in Science Teaching, 44*(4), 613–642.

Crawford, B. A. (2014). From inquiry to scientific practices in the science classroom. In N. Lederman & S. Abell (Eds.), *Handbook of research on science education* (Vol. II). New York: Routledge.

Crippen, K., & Antonenko, P. D. (2018). Designing for collaborative problem solving in STEM. In Y. J. Dori, Z. Mevarech, & D. Baker (Eds.), *Cognition, metacognition, and culture in stem education* (pp. 89–116). Springer.

Department for Education and Skills/Qualification and Curriculum Authority. (2004). *Science – the national curriculum for England*. London: HMSO.

Dewey, J. (1910). Science as subject matter and as method. *Science, 31*, 121–127.

Dobey, D. C., & Schafer, L. E. (1984). The effects of knowledge on elementary science inquiry teaching. *Science Education, 68*, 39–51.

Fenstermacher, G. D. (1994). The knower and the known: The nature of knowledge in research on teaching. *Review of Research in Education, 20*, 3–56.

Finley, F., Lawrenz, F., & Heller, P. (1992). A summary of research in science education- 1990. *Science Education, 76*, 239–254.

Flavell, J. H. (1979). Metacognition and cognitive monitoring. *American Psychologist, 34*, 906–911.

Grossman, P. L. (1990). *A tale of two hamlets. In the making of a teacher: Teacher knowledge and teacher education*. New York: Teachers College Press.

Hacker, D. J., Dunlosky, J., & Graesser, A. C. (1998). *Metacognition in educational theory and practice*. Mahwah: Erlbaum.

Hashweh, M. Z. (1987). Effects of subject matter knowledge in the teaching of biology and physics. *Teaching and Teacher Education, 3*, 109–120.

Henze, I., van Driel, J. H., & Verloop, N. (2009). The change of science teachers' personal knowledge about teaching models and modelling in the context of science education reform. *International Journal of Science Education, 29*, 1819–1846.

Hunter, A. B., Laursen, S. L., & Seymour, E. (2006). Becoming a scientist: The role of undergraduate research in students' cognitive, personal, and professional development. *Science Education, 91*, 36–74.

Kagan, D. M. (1992). Implication of research on teacher belief. *Educational Psychologist, 27*(1), 65.

Lave, J., & Wenger, E. (1991). *Situated learning: Legitimate peripheral participation*. Cambridge: Cambridge University Press.

Lead States, N. G. S. S. (2013). *Next generation science standards: For states, by states*. Washington, DC: The National Academies Press.

Lederman, N. G. (1992). Students' and teachers' conceptions of the nature of science: A review of the research. *Journal of Research in Science Teaching, 29*(4), 331–359.

Lederman, N. G. (1999). Teachers' understanding of the nature of science and classroom practice: Factors that facilitate or impede the relationship. *Journal of Research in Science Teaching, 36*(8), 916–929.

Lee, O. (1995). Subject matter knowledge, classroom management, and instructional practices in middle school science classrooms. *Journal of Research in Science Teaching, 32*, 423–440.

Llewellyn, D. (2005). *Teaching high school science through inquiry: A case study approach*. Thousand Oaks: SAGE Publications.

Lotter, C., Harwood, W. S., & Bonner, J. J. (2006). Overcoming a learning bottleneck: Inquiry professional development for secondary science teachers. *Journal of Science Teacher Education, 17*, 185–216.

Ministry of Education of Singapore. (2007). Primary science syllabus. Ministry of Education of Singapore: Curriculum Planning & Development Division.

National Research Council. (1996). *National science education standards*. Washington, DC: National Academy Press.

National Research Council. (2000). *Inquiry and the National Science Education Standards*. Washington, DC: National Academy Press.

National Research Council. (2007). *Taking science to school: Learning and teaching science in grades K-8*. Washington, DC: National Academies Press.
National Research Council. (2012). *A framework for K-12 science education: Practices crosscutting concepts, and core ideas*. Washington, DC: National Academy Press.
Newton, D. P., & Newton, L. D. (2001). Subject content knoweldge and teacher talk in the primary science classroom. *European Journal of Teacher Education, 24*, 369–379.
Pajares, M. F. (1992). Teachers' beliefs and educational research: Cleaning up a messy construct. *Review of Educational Research, 62*(3), 307–332.
Rutherford, F. J. (1964). The role of inquiry in science teaching. *Journal of Research in Science Teaching, 2*, 80–84.
Schon, D. (1987). *Educating the reflective practitioner: Toward a new design for teaching and learning in the professions*. San Francisco: Jossey-Bass.
Schwartz, R. S., Lederman, N. G., & Crawford, B. A. (2004). Developing views of nature of science in an authentic context: An explicit approach to bridging the gap between nature of science and scientific inquiry. *Science Education, 88*(4), 610–645.
Shulman, L. (1986). Those who understand: Knowledge growth in teaching. *Educational Researcher, 15*(2), 4–14.
Shulman, L. S. (1987). Knowledge and teaching: Foundations of the new reform. *Harvard Educational Review, 57*(1), 1–22.
Smith, D. M., & Cooper, B. (1967). A study of the use of various techniques in teaching science in the elementary school. *School Science and Mathematics, 67*(6), 559–566.
Tamir, P. (1991). Professional and personal knowledge of teachers and teacher educators. *Teaching and Teacher Education, 7*, 263–268.
Usak, M., Ozden, M., & Eilks, I. (2011). A case study of beginning teachers' subject matter (SMK) and pedagogical content knowledge (PCK) of teaching chemical reaction in Turkey. *European Journal of Teacher Education, 34*, 407–429.
van Driel, J., Berry, A., & Meirink. (2014). Science teacher knowledge. In N. Lederman & S. K. Abell (Eds.), *Handbook of research on science education* (Vol. II). New York: Routledge.
Wenger, E. (1998). *Communities of practice: Meaning, learning, and identity*. Cambridge: Cambridge University Press.
Wheeler, G. F. (2000). In J. Minstrell & E. van Zee (Eds.), *Inquiring into inquiry learning and teaching in science* (pp. 471–496). Washington, DC: American Association for the Advancement of Science.
Yerrick, R., Radosta, M., & Greene, K. (2018). Technology, culture and young science teachers—A promise unfulfilled and proposals for change. In Y. J. Dori, Z. Mevarech, & D. Baker (Eds.), *Cognition, metacognition, and culture in STEM education* (pp. 117–138). Springer.
Zohar, A. (2004). *Higher-order thinking in science classrooms: Students' learning and teacher' professional development*. Dordrecht: Kluwer Academic Press.

Chapter 3
Students' Metacognition and Metacognitive Strategies in Science Education

Shirly Avargil, Rea Lavi, and Yehudit Judy Dori

3.1 Introduction

Scientific literacy depends, among other things, on cognitive and metacognitive skills as well as on motivation (Herscovitz et al. 2012). Specifically, it is dependent on cognitive and metacognitive abilities for locating, selecting, reading, monitoring, and critiquing various information sources (Wang et al. 2014; Yore and Treagust 2006). For this reason, researchers argue that metacognition is a central feature in life-long learning in general and science education in particular, and that metacognitive engagement is key for developing deeper conceptual understanding of scientific ideas (e.g., Anderson and Nashon 2007; Blank 2000; Choi et al. 2011; Georghiades 2004a; Koch 2001; Nielsen et al. 2009; Wang et al. 2014).

Adaptive and life-long learning are gaining central importance, with the ability to regulate and control one's thinking, or 'think about thinking', being an essential part of such learning (Chiu and Duit 2011; Choi et al. 2011). Therefore, in order to effectively seize the opportunities and tackle the challenges of the twenty-first century citizens need to have sufficient levels of scientific literacy and metacognitive skills (Choi et al. 2011; Yore and Treagust 2006).

This chapter presents a critical review of studies in science education as part of Science, Technology, Engineering, and Mathematics (STEM) education, focusing

S. Avargil (✉)
Faculty of Social Studies, School of Education, Bar-Ilan University, Ramat Gan, Israel
e-mail: shirly.avargil@biu.ac.il

R. Lavi
Faculty of Education in Science and Technology, Technion, Haifa, Israel

Y.J. Dori
Faculty of Education in Science and Technology, Technion, Haifa, Israel

The Samuel Neaman Institute for National Policy Research, Haifa, Israel

Y.J. Dori et al. (eds.), *Cognition, Metacognition, and Culture in STEM Education*, Innovations in Science Education and Technology 24,
https://doi.org/10.1007/978-3-319-66659-4_3

on metacognitive strategies, training, and assessment of students. This chapter focuses on science education in the following scientific disciplines: biology, chemistry, earth science, environmental science, general science, and physics. Our review was based on a search in three leading journals in science education, using the following key words: metacognition, metacognitive strategies, assessment and student thinking, and science education. Next, we narrowed our search to publications from the year 2000 and onwards. At the same time, we conducted the same search in the National Research Council (NRC) archive, for documents related to students' metacognition and metacognitive strategies in science education. Finally, we selected several chapters from the edited book *Metacognition in science education: Trends in current research* (Zohar and Dori 2012), since it is, to the best of our knowledge, the only book concerned with science education specifically. Although we are aware that there are other books related to metacognition (e.g., Mevarech and Kramarski 2014), we were not able to include them in this review, since these books concerned science education as a peripheral and not as a main topic.

The review presented herein was preceded by another review of research on metacognition in science education (Zohar and Barzilay 2013). This previous review was limited to one database (ERIC) and included only peer-reviewed journal articles, while our review spanned multiple databases and also included published book chapters. Another contribution of our review is that it includes papers on metacognition in science education published after 2013 (e.g., Wagaba et al. 2016; Wang and Chen 2014).

The first section of this chapter provides theoretical background and is concerned with the definition(s), importance and assessment of metacognition in science education. The second section provides information about the literature search of relevant resources on metacognition that are included in the review. The next section details the findings of the aforementioned literature search. The final section contains a discussion on whether, to what extent, and in what ways have metacognition and its assessment in science education been implemented, and what aspects of metacognition and metacognition research in science education are still lacking.

3.2 Theoretical Background

This section contains a brief overview of research on metacognition, in education in general and in science education in particular.

3.2.1 Definition of Metacognition

Numerous researchers have attempted to define metacognition (Sandi-Urena et al. 2011). If by *cognition* we mean the variety of learning skills students apply to complete a task, then *metacognition* can be defined as awareness of, and reflection upon, one's own cognitive process (Flavell 1976, 1981). Flavell (1979) described metacognition in more detail as (a) knowledge about peoples' cognition, (b) knowledge about cognitive tasks, (c) knowledge about strategies applied to the solution of different tasks, and (d) skills for monitoring one's cognitive activities. According to him, metacognition refers to the awareness of cognitive processes and the self-regulation and management of those processes in relation to the learning task, including conscious selection of strategies and matching the suitable strategy to task demands. Other researchers (Brown 1987; Veenman et al. 2006) made similar distinctions between knowledge of cognitive activities and regulation of such activities as two components of metacognition.

Jacobs and Paris (1987) noted that researchers have generally circumnavigated the problem of defining metacognition by referring to two broad classes of metacognition: knowledge about cognition, which includes declarative ('about') knowledge, procedural ('how to') knowledge, and conditional ('why' and 'when') knowledge, and regulation of cognition, which includes planning, evaluating, and monitoring. Their description will be used throughout this work when referring to elements or components of metacognition and these two terms are used interchangeably throughout the chapter. According to Brown (1978), knowledge of cognition is relatively stable, often can be stated, can be fallible and is age dependent, while regulation of cognition is relatively unstable and age independent.

Students' thoughts about their own capability influence cognitive performance, improve self-evaluation, and regulate their learning (Bandura 2000; Bouffard-Bouchard 1990; Jacobs and Paris 1987). Schraw and Moshman (1995) defined declarative knowledge as "knowledge about oneself as a learner and about which factors influence one's performance" (p. 352), clarifying that it is part of the knowledge of cognition. They described *good* learners as ones who have more knowledge about their own memory and are more likely to use what they know than *poor* learners. According to Schraw and Dennison (1994), knowledge of cognition is part of metacognitive knowledge, and it consists of declarative knowledge, procedural knowledge, and conditional knowledge. They defined declarative knowledge as knowledge of one's skills and abilities as a learner, procedural knowledge as knowledge of how to use different strategies, and conditional knowledge as knowledge of when and why to use the different strategies. They also defined regulation of cognition as the ability to plan learning strategies (e.g., goal setting), to manage information (e.g., strategies for processing information), to monitor, to identify mistakes, and to evaluate the learning (e.g., assessing the learning strategies, correcting performance errors and analyzing performance).

Bandura's (1997) social-cognitive learning theory as applied to school learning eventually led to the development of self-regulated learning theory, which stipulates that learning is governed by interacting cognitive, metacognitive, and motivational components.

According to Veenman (2011), reappearing major problems with metacognition research is its fuzzy definition, which is not only due to proliferation of terminologies, but also disagreement about the ingredients of metacognition and their interrelationships.

3.2.2 *Metacognition in General Education and in Science Education*

Ackerman and Goldsmith (2011) claimed that students' perception of their performance influences their ability to monitor the learning process. Thus, when there is no time limitation, a high perception of performance reflects positively on the actual performance. Researchers have found that when students' metacognition was improved, it was possible to also improve their learning outcomes (Thomas and McRobbie 2001).

Metacognition may enhance students' ability for contextual use of scientific conceptions, improve science reading comprehension and the ability to monitor the reading of popular scientific press (Georghiades 2004a; Michalsky 2013; Norris and Phillips 2012; Wang and Chen 2014; Wang et al. 2014). In disciplinary science learning in particular, studies have shown that in physics, for example: students who use metacognition are more likely to (a) develop conceptual understanding, (b) go through a process of knowledge construction and meaningful understanding (Anderson and Nashon 2007; Nielsen et al. 2009), (c) comprehend physics texts better (Koch 2001), and (d) have higher motivation and accurate views of what it means to understand physics (Taasoobshirazi and Farley 2013; Thomas 2013). Similar findings have been reported in chemistry and biology education research. In chemistry, a metacognition intervention that included reflection on the learning process and reduction of poor learning strategies (e.g., memorization) benefited students' standardized achievements (Thomas and McRobbie 2001). When Herscovitz et al. (2012) exposed high school chemistry students to metacognitive tools and strategies involving question posing skills, students were able to pose more complex questions, indicating that they developed a sophisticated understanding of concepts and processes. In biology education, awareness of the learning process and a stronger ability to monitor, regulate and control the learning contributed to meaningful understanding of various biology concepts like genetics and ecosystems and were found to improve scientific inquiry skills (Eilam and Reiter 2014; Martin et al. 2000; Zion et al. 2005). Several research based documents published by the NRC (2000; 2007), stated that it is critical for learners in general and students who study science in particular to develop their ability to think in a metacognitive way. In summary, researchers argue that metacognition is essential for science education. One of the questions that are of importance considers how instructional methods and assessments were used in the past 15 years to improve students' learning in this aspect.

Since it is an internal process rather than an overt behavior, metacognition is inherently difficult to measure, and individuals themselves are often unaware of their own metacognitive process (Desoete 2008; Georghiades 2004a). This has naturally led to difficulties with identifying and assessing metacognition and its related processes. However, if one considers metacognition to be an understanding of knowledge, then one can detect it in the learner in an indirect manner, either through effective use of this understanding, as witnessed by the learner's behavior, or by asking the learner to provide an overt description of it (Georghiades 2004a; NRC 2000).

In order to make assessment more effective, researchers have made various suggestions for carrying out assessment of metacognition in science education: (a) employ multiple methods of assessment and collect data from the same subjects by different means (Georghiades 2004a); (b) carry out assessment of the same learner across different times; (c) carry out assessment concurrently with the task rather than retroactively or prospectively, as it may be more effective (Cooper et al. 2008); (d) carry out assessment using real-life problems, questions and statements rather than abstract or textbook problems (Choi et al. 2011); and (e) any self-report sought from students regarding their learning processes should relate specifically to their science learning (Thomas et al. 2008).

With respect to assessment of metacognition in children, Garner and Alexander (1989) proposed three ways for carrying this out: (a) asking children directly about their metacognition; (b) having children think aloud while performing a task; and (c) asking children to teach a younger child a good solution to a problem. However, this assessment has several limitations: (a) children lack verbal fluency; (b) adult-child use of language is highly varied; (c) young children find discussing general cognitive events difficult; and (d) children have a tendency for describing specific just-experienced events.

3.3 Literature Search Procedure

Next we describe how we conducted the literature search, searching for studies that focus on metacognition use and assessment of students' learning outcomes in science education studies.

Our literature search concerned metacognition, assessment, and science education as archived in the three leading journals: *Journal of Research in Science Teaching*, *Science Education*, and *International Journal of Science Education*. We also searched documents published by organizations such as the NRC. We focused our search on studies that included K-12 students as well as at the college level. Finally, we also relied on the book *Metacognition in Science education* (Zohar and Dori 2012), since it is directly related to the topic of this chapter. Research on science teachers' metacognition and meta-strategic knowledge (Eldar et al. 2012; Zohar 2006, 2012) is beyond the scope of the current chapter.

We chose to focus on these three journals since they have the highest impact factor in research concerned with science education in the last 15 years. When searching within a specific journal, we used the words 'metacognition and assessment' in order to reduce the number of papers found and find the most relevant articles for our purpose. Search in the *Journal of Research in Science Teaching* revealed 104 papers, in *Science Education* 115 papers, and in the *International Journal of Science Education* 109 articles.

Our review included papers in the context of science education, with metacognition as a primary focus. We excluded papers in which the primary focus was on metacognition in other STEM domains (e.g., mathematics or technology) or papers in which metacognition was of secondary focus, even if their primary focus was on related topics such as self-regulated learning (e.g., Azevedo 2010; Greene and Azevedo 2009; Dignath and Buttner 2008). Another criterion for selection was that we chose papers from the last 15 years and those which include students, rather than teachers, as the subject of the investigation. The sites were last accessed in January 2015. Following the above process of selection, we were left with 23 papers from the three journals and with six chapters that fulfilled the above criteria. We refer to these articles as representative of the research on students' metacognition in science education.

The next section describes these papers and chapters, providing comprehensive description and characterization of students' metacognition assessment in science education.

3.4 Literature Search Findings

This section provides an overview of studies concerning metacognition in science education. Most studies of metacognition in science education contain an expectation of improving the outcomes of learning through practice of metacognition (Georghiades 2004a). Accordingly, the vast majority of the studies summarized herein are intervening studies, where metacognition-based pedagogical intervention was implemented on one or more groups of students. Each paper was analyzed and classified for its type: empirical research, theory, position, or review and critique. In empirical studies we also characterized the papers by classifying them to either describing a tool developed for assessing metacognition, describing an existing state or describing a pedagogical intervention. We classified these papers into three different categories: (a) research on assessment tools for metacognition – studies concerned exclusively with developing a method or tool for assessment of metacognition; (b) research on metacognitive learning processes – studies for which the researcher or researchers probed into students' metacognitive processes in order to procure information about these processes, without an explicit aim to improve learning outcomes; and (c) research on metacognition-based pedagogical intervention – studies with explicit research objective(s) to change students' metacognitive processes through training or pedagogical intervention, in order to improve learning outcomes. Furthermore, we examined in each paper the population description, the tools that were used and the metacognition components that were under investigation.

3.4.1 Review and Theoretical Papers

Our classification of papers into empirical and non-empirical revealed that from the 23 papers and six chapters chosen for this chapter, only two were reviews and one was theoretical. In the first review, the author discussed the literature on metacognition spanning the past three decades and identified the different definitions of the term and diverse origins of metacognitive processes (Georghiades 2004a). One of the concluding remark of Georghiades' review was that more research is needed in order to enhance understanding of metacognition and its aspects, specifically in science education. The author raised the questions of how metacognition can be identified, whether it can be taught, and if so, how. He also argued that research in metacognition pertaining to science education is in its infancy. Since this claim was made more than 10 years ago, in this chapter we review the representative studies that were conducted in the last 15 years as published in the leading journals of science education and the edited book (Zohar and Dori 2012). Veenman (2012) in the book *Metacognition is Science Education* (Zohar and Dori 2012) provided a review that emphasized the difficulty in establishing a consensus regarding metacognition investigation. Nevertheless, he provided a clear and concise review of many of the main approaches that appear in the literature regarding metacognition in general. At the end of his chapter, Veenman showed how metacognitive skills were integrated in science education, specifically in (a) scientific reading (in contrast to general reading skills), (b) science problem-solving, (c) scientific inquiry, and (d) scientific writing.

The paper which we classified as theoretical and non-empirical presented a framework for scientific literacy for South Korea that included five dimensions, one of which was metacognition (Choi et al. 2011). The authors' aim was to fill a gap they perceived to exist in present frameworks for scientific literacy, which was principally a lack of emphasis on (a) metacognition, (b) problem-solving skills for real-life (rather than conceptual or textbook) problems, and (c) global context for scientific issues. The authors based their proposal on a literature review and an online survey administered to 222 secondary school science teachers, 126 from the US and 96 from South Korea. This framework was subsequently reviewed by a team of five science educators from the US and Australia. The proposed framework contains five dimensions, the central one being (a) metacognition and self-direction and the rest being (b) content knowledge, (c) habits of mind, (d) character and values, and (e) science as human endeavor. The authors of this framework considered metacognition and self-direction to tie together the other four dimensions through the learner's reflection and management of cognition and learning. They considered the dimension of metacognition to include three elements, namely (a) self-directed planning (b) self-directed monitoring, and (c) self-directed evaluating. This proposed framework is another example of educators in science education calling for metacognition to be an integral part of science education. However, the fact that only a few reviews and theoretical papers were written in the context of metacognition in leading journals of science education and the book emphasizes the need to pay more attention to this issue. In the next section, we go beyond the categorization that Veenman (2012) had suggested and classify the empirical papers that investigate students' metacognition in science education into the three categories noted above and elaborated next.

3.4.2 *Empirical Papers*

We divided the empirical papers into three categories: (a) research on assessment tools for metacognition, (b) research on metacognitive learning processes, and (c) research on metacognition-based pedagogical intervention (details to follow on next page). Table 3.1 describes our categorization as well as the investigated population in each paper.

As can be seen in Table 3.1, we classified most papers as belonging to category (c), while only a couple of papers described the development of tools to specifically assess metacognition in science education.

This is not to say that there are no tools that assess metacognition in general, but rather that there is a need to develop or adjust specific tools to enhance the development and assessment of science students' metacognition. Moreover, each science discipline has its own discipline-based features. As described in *Discipline-Based Education Research (DBER): Understanding and Improving Learning in Undergraduate Science and Engineering* (NRC 2012b): "Metacognition is a necessary skill for meaningful learning and thus merits continued study in the context of DBER. Further research could clarify which metacognitive skills are useful to science and engineering because the skills may not be the same for each discipline, additional DBER could examine these similarities and differences" (p. 157). Additional examples exist for developing discipline-based assessment tools for metacognition (e.g., Cooper et al. 2008, to improve problem-solving skills), however, more validated tools are needed to address different discipline-based metacognition science skills. Thus, there is a need to examine and further investigate discipline-based assessment tool for metacognition. Furthermore, metacognition should be emphasized through K-12 and college level science education. As can be seen from Table 3.1, research is needed in all levels of education, but more research is needed for investigating metacognition in science education in kindergarten and at the early stages of elementary school and higher education levels. Next, we describe in more detail the papers in each category presented in Table 3.1, including what metacognition components were addressed and what scientific skills were promoted in each study.

3.4.2.1 Assessment Tools for Metacognition

Papers describing quantitative assessment tools for investigating students' metacognition (category *a* in Table 3.1) seem to be rare. Taasoobshirazi and Farley (2013) claimed that the majority of research on metacognition in this field has involved interviews or other qualitative methods and there is a need "to develop a valid, reliable, objective, and convenient tool that researchers and instructors can use to assess students' metacognition for solving physics problems" (p. 448). The Physics Metacognition Inventory (PMCI) self-reporting instrument included separate items for assessing declarative, procedural, and conditional knowledge – elements of

Table 3.1 Categorization of the empirical papers and chapters

Category and # of articles	Author(s) and year	Domain(s)/ discipline(s)	N students	Population
(a) Assessment tools for metacognition	Taasoobshirazi and Farley (2013)	Physics	505	Post-secondary school
2 papers	Thomas et al. (2008)	General science	465	Middle and high school
(b) Metacognitive learning processes 8 papers	Anderson and Nashon (2007)[a] and Nielsen et al. (2009)	Physics	50 + 14	High school
	Martin et al. (2000)	Biology	77	Post-secondary school
	Norris and Phillips (2012)	Biology, chemistry, and physics	91	High school and post-secondary school
	Schraw et al. (2012)	Environmental science	134	Elementary school
	Shin et al. (2003)	Physics	124	Middle school
	Wang and Chen (2014)[a] and Wang et al. (2014)	Biology, Earth science, and physics	556	Elementary school and middle school
(c) Metacognition-based pedagogical intervention	Ben-David and Zohar (2009)	Biology	119	Middle school
16 papers	Blank (2000)	Biology	92	Middle school
	Chiu and Linn 2012[b]	Chemistry	173 + 249	High school
	Conner and Gunstone (2004)	Biology and environmental science	16	High school
	Georghiades (2004b)	Physics	60	Elementary school
	Grotzer and Mittlefehldt (2012)	Physics	182	Middle school
	Hand et al. (2004)	Biology	93	Middle school
	Herscovitz et al. (2012)[b]	Chemistry	700 + 400	High school
	Koch (2001)	Physics	64	Post-secondary school
	Michalsky (2013)	Biology	198	High school
	Sandi-Urena et al. (2011)	Chemistry	Approx. 1000	Post-secondary school
	Thomas (2013)	Physics	29	High school
	Thomas and McRobbie (2001)	Chemistry	24	High school
	Wang (2015)	Biology	173	Middle school
	Ward and Wandersee (2002)	General science	17	Middle school
	Zion et al. (2005)	Biology	407	High school

[a]Different aspects of the same scientific domain and investigated population
[b]Two studies

knowledge of cognition, and separate items for monitoring, evaluation, debugging, and information management – elements of regulation of cognition, where the last two seem to replace the more commonly used 'planning' element. In addition to PMCI scores, the authors also collected the undergraduate students' physics course grades.

The students' total scores on the PMCI were found to be reliable and valid, relating to students' course grade and physics motivation. Men outperformed women in PMCI scores in (a) knowledge of cognition, while women outperformed men in (b) information management and (c) debugging. Taasoobshirazi and Farley (2013) concluded that men are more likely to understand their own problem-solving strengths and weaknesses, how to apply strategies, and when and why to apply them, while women were more likely to integrate free-body diagrams into their problem-solving and seek help when having difficulty with problem-solving. The authors suggested this tool can be used, with minor adjustments, in other disciplines. This tool could be valuable for assessing metacognition in physics education, however, it needs to be validated for use in other disciplines as well as other levels of education, like secondary and even more so elementary science education. As the authors acknowledge, there is also a need to investigate how the various components of metacognition interact with and impact problem-solving to be able to ascertain the relative contributions of each of these components to problem-solving success.

Thomas et al. (2008) sought to broadly assess aspects of science students' metacognition, self-efficacy and learning processes. The Self-Efficacy Metacognition Learning Inventory-Science (SEMLI-S) is a self-reporting instrument concerned with general science originally written in English and translated into Chinese. The final tool included several sub-scales, including (a) monitoring, evaluation, and planning, (b) science learning self-efficacy, (c) learning risks awareness, and (d) control of concentration. The tool can be used to collect students' pre and post data for investigating whether an intervention enhances metacognition.

These two tools, PMCI and SEMLI-S, found in our representative sample of articles on the topic of assessing science students' metacognition, represent a quantitative way to assess metacognition. Another example of a tool specifically designed for the field of science education is a multi-method assessment of metacognitive skillfulness in college chemistry problem-solving (Cooper et al. 2008) and metacognition in scientific reading, which was published in a paper prior to the 15-year time window of our review (Yore et al. 1998). In summary, there is a need to develop tools that specifically assess metacognition in relation to the context of science learning. As also noted by Thomas et al. (2008): "Most existing empirical self-report instruments that explore students' learning and metacognition … do not account for the classroom context or help students locate their self-report in relation to the learning of specific subjects such as science" (p. 1703). We stress this idea and suggest that tools should be discipline-based as well as scientific-practice based. Developing metacognitive skills for the scientific practice of *developing and using models* (NRC 2012a) might be different than the knowledge and regulation of cognition for the scientific practice of *obtaining, evaluating, and communicating information.* Another aspect that should be considered is the educational level for which these assessment tools are constructed. Furthermore, researchers should emphasize the different components of metacognition they assessed rather than general metacognitive skills.

3.4.2.2 Metacognitive Learning Process

Research on assessing students' metacognition learning processes in different science education domains and settings (category *b* in Table 3.1) and their connections to other scientific abilities includes various metacognition components and assessment tools. In most of the studies we found, researchers used both quantitative as well as qualitative tools to assess metacognition. Wang et al. (2014) sought to measure the level of science reading comprehension and metacognition of Taiwanese students from fourth to eighth grades and compare them to those of Canadian counterparts. They used the Reading Comprehension of Science Test (RCST), including items from biology, physics, and Earth science domains. The metacognition components that were assessed in this study, while the authors measured science reading comprehension, were knowledge of cognition, namely declarative, procedural, and conditional knowledge. Wang et al. (2014) found no growth in either group on science reading in middle school. However, they reported that higher metacognition level correlated with better science reading comprehension. The authors suggested that metacognitive skills may not transfer across domains without providing discipline-specific training and that science reading requires understanding of science related features in the text like evidence-based claims and counter-claims and evidence-based arguments. Thus, they claimed, science teachers may need to shift their teaching toward reading science materials with the intention of raising students' metacognitive reading awareness. Similar findings were reported in Wang and Chen (2014): in this study, the authors also concluded that prior science knowledge affects science reading completion and is mediated by metacognitive awareness, defined by the authors as "declarative, procedural, and conditional knowledge about reading" (Wang and Chen 2014, p. 176). Both studies focused on scientific reading in general and on middle school students. Research on scientific reading was also been conducted by Norris and Phillips (2012): in their chapter, they described research they conducted with high school and undergraduate students. They investigated students' metacognitive judgments on popular science texts, specifically judgments about the difficulty of the text and about the effect of students' prior beliefs on what they had read. Students systematically overestimated the degree of certainty in their report: while they were able to identify observation and method statements, they were generally unsuccessful in interpreting the role of statements in the text's reasoning. Students confused evidence statements as conclusions, and underestimated dramatically the demands of the text and the cognitive difficulty they had experienced with the interpretative tasks. Thus, Student performance on the reading tasks were reported to have only a very weak correlation with their perceived difficulty in reading the texts. Norris and Phillips (2012) recommended a view of reading which emphasizes strategies for interpreting scientific text over the simple view of reading as word recognition and information location.

Anderson and Nashon (2007) and Nielsen et al. (2009) conducted respective studies in physics education in the context of experiencing physics kinematics problem-solving while visiting an amusement park. Anderson and Nashon (2007) sought to identify the metacognitive characteristics evident in individuals and groups who participated in an amusement park physics program, and explore how these characteristics were involved in knowledge construction. The study was based on four groups with three or four high school students in each of them. The authors developed and administered the Metacognition Baseline Questionnaire (MBQ) to these high school students. Students were given novel kinematics problems concerning various cycles to try and solve. Assigning students into groups encouraged them to verbalize their thinking in order to present and discuss their ideas. Qualitative data was collected in order to probe deeper into students' metacognitive processes. The authors showed that the key dimensions of awareness, monitoring, and evaluation are critical to the resilience and sustainability of individual capacity to engage in meaningful learning. Thus, developing these capacities can contribute to empower students' meaningful understanding. Nielsen et al. (2009) also investigated high school students' metacognition in the context of an amusement park. The authors argued that as a result of the various learning activities undertaken by the participating students, they were able to develop deeper understanding of the kinematic concepts they encountered, enriching their prior conceptions. Individual combinations of metacognitive dimensions as represented by the obtained MBQ profiles seemed to dictate the student's approach to work within the group and as individual learner. The authors' approach regarded the qualitative assessment as an intervention tool intended to improve learning outcomes rather than as a tool for assessment of metacognition. They concluded that the problem-solving activities in the field and in-class activities enabled the students to develop further understanding of the kinematics concepts, enriching their prior comprehension, and allowed them to learn about themselves as learners. They suggested that if teachers had the option to receive students' individual metacognitive profiles, they could potentially utilize this information to improve learning in terms of group configuration and problems development for various classroom activities. Also in the domain of physics, Shin et al. (2003) examined aspects that predict success in solving ill-structured problems by ninth grade science students within the domain of scientific inquiry in astronomy. They evaluated the aspects of metacognition by an instrument called How Do You Solve Problems? The instrument included metacognitive statements related to reflection, planning and monitoring, problem-solving strategies, and information-selection. The authors found that knowledge of cognition, including information-selection and problem-solving strategies, was not a significant predictor in students' success to solve ill-structured problems. However, regulation of cognition, including planning and monitoring skills, did predict problem-solving level in unfamiliar contexts. They summarized that solving ill-structured problems requires that students not only have the necessary knowledge but also regulate their cognition, which includes (a) modifications of plans, (b) reevaluation of goals, and (c) monitoring one's own efforts. They also concluded that if the problem is not structurally complicated enough, the students may not use their regulation of cognition abilities even though they possess them.

Martin et al. (2000) investigated, among other things, differences in the metacognitive reflections of students employing diverse learning strategies. They audio-recorded clinical interviews with students of marine biology in post-secondary education. The authors aimed at probing metacognitive knowledge and relate this knowledge to students' predominant learning modes. They argued that many students lack the fundamental learning skills and metacognitive abilities essential for success in the 'information age' and that students are often unaware of the limitations posed by their learning style. They recommended conducting large-scale studies to demonstrate the promise of these approaches in a variety of science disciplines and ages. Schraw et al. (2012) investigated the relationship between self-reported metacognition, attitudes about an outdoor learning program, and field-based learning in an environmental education program. Their main research question concerned whether knowledge and regulation of cognition scores were related to attitudes and learning before and after completing a half-day field-based science curriculum. Students' attitudes and knowledge relating to the intervention were assessed before and after the intervention. Schraw and colleagues made use of the Junior Metacognitive Awareness Inventory (Jr. MAI), created by Sperling et al. (2002), to assess metacognition. The Jr. MAI was based on Schraw and Dennison's (1994) MAI, which was also used to assess metacognitive awareness. The Jr. MAI was intended for students in third through eighth grades and was used specifically to assess incoming metacognitive knowledge or changes in knowledge after an intervention to improve metacognitive skills. Schraw and colleagues removed some of the MAI's items due to irrelevance to younger populations and modified others by rewording of certain phrases to make them simpler or in order to provide a more familiar context.

Schraw and colleagues found two factors, knowledge and regulation of cognition, accounted together for 35% of variance, and reported that knowledge of cognition and regulation of cognition factors were moderately correlated. Knowledge of cognition correlated with attitudes and post-intervention knowledge scores. However, regulation of cognition scores did not correlate with these measures at the fourth grade level. The authors concluded that the Jr. MAI can serve to assess the knowledge and regulation of cognition in a valid and reliable manner and that metacognitive knowledge is related positively to increased learning and attitude change. Schraw and his colleagues suggested that future research should compare the role of metacognitive knowledge inside and outside the classroom. Lastly, the authors outlined several instructional strategies to promote metacognitive awareness, such as helping students to develop and refine their metacognitive knowledge and regulatory skills, and promoting metacognitive knowledge and regulation through active reflection and dialogue.

In summary, the above studies showed that metacognition is correlated with (a) robust and profound scientific understanding, (b) the ability to read scientific texts,

(c) effective learning strategies, and (d) problem-solving skills. Currently, it is known that metacognition can contribute to different aspects of scientific learning. Therefore, valuable research from this point forward should include research on specific components of metacognition that can be addressed to enhance different scientific skills, e.g., teaching students metacognitive strategies for reading to develop their question posing skills (Herscovitz et al. 2012). This goal was partially addressed by Shin et al. (2003) and Anderson and Nashon (2007). Identifying different components of metacognition and correlating them to different scientific skills might contribute to teachers and students understanding of metacognition and thus to the implementation of related strategies in the classroom (see also Nielsen et al. 2009). The next section contains studies from the last 15 years concerned with metacognition-based pedagogical intervention.

3.4.2.3 Metacognition-Based Pedagogical Intervention

The papers and chapters that described a pedagogical intervention are presented in Table 3.2 in more detail than in Table 3.1, category *c*. Some of these studies present a metacognitive intervention aimed at enhancing specific scientific skills or knowledge (e.g., Ben-David and Zohar 2009; Herscovitz et al. 2012; Koch 2001; Wang 2015), while others were aimed at enhancing metacognitive skills specifically and assessing them (e.g., Sandi-Urena et al. 2011; Thomas 2013; Thomas and McRobbie 2001). Studies of metacognitive pedagogical intervention in domains other than science education or those not concerned directly with metacognition, but with self-regulated learning, were not included in this review. For example: the IMRPOVE method, which aims to enhance mathematics learning through metacognitive intervention, was excluded from the present review (e.g., Mevarech and Kramarski 1997; see also Chap. 12 in this book).

In the case of papers concerning metacognitive skill assessment, the tools that were used before, after or during the intervention were either qualitative, quantitative, or both. For example: Sandi-Urena et al. (2011) used a multi-method assessment that combined two instruments, a prospective self-report named the Metacognitive Activities Inventory—MCAI (Cooper and Sandi-Urena 2009) and an online concurrent automated instrument (software) named Interactive Multimedia Exercises—IMMEX (Cooper et al. 2008). The intervention was aimed at enhancing students' metacognition awareness and was used in a problem-solving scenario. The authors engaged students in small groups collaboration and individual work that promoted reflection about the processes and the products in a problem-solving environment. Thomas (2013) used the Metacognitive Orientation Learning Environment Scale-Science, classroom observation, interviews with students, and the SEMLI-S (Thomas et al. 2008), to assess the metacognitive orientation of the classroom learning environment, students' views of what it means to learn physics, and how students considered they knew they had learnt physics. The intervention included a change in the teacher's pedagogy and explicit teaching of a triarchic model of representations altering the metacognitive orientation of a physics class-

Table 3.2 Summary of research on metacognition-based pedagogical intervention in science education

Author(s) and year	Scientific skills or concepts being assessed	Metacognitive skills being assessed	Aspects of metacognition in the intervention	Time of intervention
Ben-David and Zohar (2009)	Inquiry skills: (a) defining research questions (b) formulate research hypotheses	None	Awareness of the type of thinking strategies being used in specific instances	10 lessons
Blank (2000)	Students' understanding of targeted ecology concepts	None	Metacognitive classroom where students asked to reveal their science ideas and to discuss the status of their conceptions throughout the instruction	3 months
Chiu and Linn (2012)	Learning from and understanding of scientific visualizations	Study 1 – Self-assessment of learning: (a) generating explanations and (b) identifying difficulties with their understanding	Study 1 – Monitoring one's own progress while learning chemical concepts and processes	1 week + 1 week
		Study 2 – Students' revisiting of visualizations they were previously exposed to – Monitoring their level of understanding	Study 2 – Some students were given a multiple choice question immediately following the visualization in order to focus them on a specific idea, where an incorrect answer would refer them back to the visualization with added explanation	

(continued)

Table 3.2 (continued)

Author(s) and year	Scientific skills or concepts being assessed	Metacognitive skills being assessed	Aspects of metacognition in the intervention	Time of intervention
Conner and Gunstone (2004)	Essay writing about biological issues and social, ethical, or environmental implications	Declarative and procedural knowledge of cognition, and awareness and control of cognition	Prompts for reflection by teacher to tap into students' prior knowledge of learning strategies so they could use it to develop more independent and self-regulating learning	4.5 weeks
Georghiades (2004b)	Conceptual understanding and retention of current electricity ideas	None	Metacognitive reflection	4 weeks
Grotzer and Mittlefehldt (2012)	Conceptual understanding, identifying underlying relational causality and transferring this understanding of causal structures between topics	Metacognitive comments	Encourage greater monitoring and evaluation in students	16 weeks
Hand et al. (2004)	Addressing concepts, structures, functions and processes, creating analogies, developing arguments, and explaining processes	General metacognition	Intervention was aimed at promoting conceptual understanding and metacognition by using science writing heuristics and a textbook writing activity	8 weeks
Herscovitz et al. (2012)	Posing questions as part of reading comprehension	Study 1 – Use of reading strategies Study 2 – Use of chemistry understanding levels	Posing quetions after reading scientific text – knowledge of cognition; assessing the complexity of questions posed – regulation of cognition	4–5 months

(continued)

Table 3.2 (continued)

Author(s) and year	Scientific skills or concepts being assessed	Metacognitive skills being assessed	Aspects of metacognition in the intervention	Time of intervention
Koch (2001)	Text reading comprehension	None	Training in self-awareness by self-assessment of reading comprehension and ranking abilities and disabilities hierarchically	3 months
Michalsky (2013)	General scientific literacy	Cognitive or metacognitive regulation	Different self-addressable questions instructional method for reading of scientific texts – cognitive-metacognitive alone, motivational alone, or combined cognitive-metacognitive and motivational	12 weeks
Sandi-Urena et al. (2011)	Developing participants' awareness and use of regulatory metacognitive skills in domain-specific context	Metacognitive awareness and use	Engage in small group collaboration and individual work that promoted reflection about processes and products in a problem-solving situation	2.5 weeks
Thomas (2013)	Views of what it meant to understand physics and how they might learn and understand physics concepts	Metacognition awareness	Explicit teaching of a triarchic model of representations on the induction of metacognitive reflection in students considering physics phenomena	6 weeks
Thomas and McRobbie (2001)	Students' metacognition and learning processes in chemistry education	Self-concept and metacognition	The intervention served as a catalytic metacognitive experience that informed students about what was for some an alternative conception of learning	12 weeks

(continued)

Table 3.2 (continued)

Author(s) and year	Scientific skills or concepts being assessed	Metacognitive skills being assessed	Aspects of metacognition in the intervention	Time of intervention
Wang (2015)	Content knowledge and construction of scientific explanations in five inquiry-based biology activities	None	Metacognitive evaluation instruction to resolve inadequate self-evaluation using idea-unit standards during peer evaluation	10 weeks
Ward and Wandersee (2002)	Textual and visual explanations of abstract science concepts and principles	Questioning, reflecting on their learning, and creating visuals in dyads and explanations on their own	Using a metacognition-based visual learning model – The roundhouse diagram strategy	9 weeks
Zion et al. (2005)	General scientific ability and domain-specific inquiry skills	None	Metacognitive consciousness questions, concerning knowledge about (a) problem-solvers, (b) the goals of assignment, and (c) problem-solving strategies, and	12 weeks
			Executive questions, concerning (a) regulating, (b) controlling and (c) criticizing cognitive processes and products	

room. Students were engaged in metacognitive reflection related to the use of (a) macroscopic, (b) molecular/sub-micro, and (c) symbolic representations when considering physics phenomena. In Thomas and McRobbie's (2001) study, the authors conducted interviews with students, collected students' journals, formal assessment documents and grades, and videotaped classroom sessions. They used questionnaires to assess student's self-concepts, metacognition, and students' self-reported use of surface, deep, and achieving approaches to learning. The teacher in this study used the 'learning is constructing' metaphor while teaching chemistry in order to enhance students' metacognition. The use of the metaphor revealed students' conceptions of learning while considering it in the process of learning.

In these three papers (Sandi-Urena et al. 2011; Thomas 2013; Thomas and McRobbie 2001) that explicitly assessed metacognition, all the authors reported an increase in several components of metacognition. For example: Sandi-Urena et al. (2011) reported that the treatment group showed significant increase in metacognition awareness and increased ability in solving non-algorithmic chemistry problems. Although they reported a lower score of the treatment group in the MCAI, the authors argued that raising the awareness about metacognition developed within students a more critical view of their learning processes. They concluded that the intervention enhanced students' metacognitive skills and was a factor in the students' ability to solve ill-structured problems. The authors call for a clear differentiation that "needs to be made between instruction that fosters the use of processes associated with metacognition – reflection for instance – and the evidence for the actual development of metacognition" (p. 325). Ward and Wandersee (2002) used a graphical metacognitive technique that introduces both visual and textual modalities to instruction of abstract scientific concepts. The authors claimed that sixth grade students had become more aware of their learning as a result of using their metacognition. The authors argued that students who articulated their understanding through the bimodal tool, using both icon drawings and short sentences, were capable of asking more questions, better self-regulated their learning, and were more independent learners. Thomas (2013) argued that explicit representational frameworks (e.g., using a triarchic model comprised of macroscopic, molecular, and symbolic representations for science phenomena) can help students to use metacognitive skills in their learning processes. Other researchers that included an intervention to enhance metacognitions skills as well as assessing them were Chiu and Linn (2012) and Herscovitz et al. (2012). Both groups of researchers as well as Thomas (2013, mentioned earlier) described a discipline-based (i.e., chemistry) metacognitive intervention. Herscovitz and colleagues described the use of a discipline-based metacognitive tool relying on the four chemistry understanding levels – macroscopic, microscopic, symbol and process (see also Avargil, Herscovitz, and Dori 2012; Kaberman and Dori 2009). In their study they promoted the use of metacognition in chemical education based on the knowledge structure of chemistry and specific scientific practices, especially posing complex questions. Chiu and Linn (2012) described learning chemistry in a technology-rich environment and investigated how students monitored their own progress and the effect of this process on their performance. The authors investigated whether dynamic visualizations in chemistry impact students' judgments of their learning with and without prompting explanations to mediate students' understanding of visualizations and the chemical phenomenon. The authors used a technology-based visualization of atomic interactions during chemical reactions. Students were divided into dyads, where each dyad was placed into one of two conditions: 'explanation first,' where they were given prompts immediately following visualizations and then rated their understanding, or 'rating first,' where they rated their understanding first, then used the visualizations and were prompted to provide explanations. The students were later asked to rate their understanding once more. The authors administered pre- and post-tests before and after the intervention, in which they asked the students to rate their individual understanding of various chemistry concepts. A follow-up study included an investigation of students' activities when they realized they did not understand a concept. The authors reported that students in the 'rating first' group consistently rate themselves as more knowledgeable than those in the 'explanation first'. One possible explanation is that students in the 'explanation first' group had more time and specific instruction to reflect on their knowledge before they had to rate their understanding, and were given an opportunity to reflect on their understanding while identifying gaps in their knowledge. The authors established these findings reveal the importance of self-monitoring for learning with dynamic visualizations and the need to foster students' self-regulatory behavior.

Grotzer and Mittlefehldt (2012) conducted a pedagogical intervention that included introducing what they named 'metacognitive moves' into instruction, aimed at helping students to reflect upon and revise their underlying causal assumptions about density and pressure and develop meaningful learning. The authors explained the concept 'metacognitive move' as a set of questions students ask themselves in order to examine their cognition. The intervention, which was both material-based and teacher-facilitated, was intended to encourage greater monitoring and evaluation abilities in the students and explicit classroom discussion of these causal structures. Students were assessed pre and post-intervention for understanding science content with embedded casual complexity, for both learning units. Selected students were also interviewed to assess their conceptual understanding and metacognitive behavior. Students' ability to identify underlying relational causality was improved post-intervention when compared to pre-intervention. A high correlation was reported between the number of metacognitive comments students made during their interviews and higher science assessment post-test scores. Moreover, students who made more metacognitive comments were more likely to offer relational causal responses on their post-test and were also more likely to transfer their understanding from density to the context of the pressure unit. This study also raised the importance of metacognition in affecting other higher order thinking skills like transfer.

Zion et al. (2005) implemented the Metacognitive-guided Inquiry within Networked Technology (MINT) learning environment, which was based on a learner-centered approach and comprised cognitive, metacognitive, social and technological elements. They compared four different groups of high school participants who studied microbiology in an inquiry-based learning environment. They found that students who studied science in the MINT environment had better domain-specific (microbiology) scientific ability and inquiry skills than students who studied in other groups. Zion and colleagues also reported that the metacognitive guidance provided to the participants did not enhance general science ability like it did for domain-specific inquiry skills, meaning that the effect on the latter was greater than the effect on the former. This strengthened previous findings that 'far transfer,' i.e., improvement in general science knowledge and skills following a domain-specific learning process, is more difficult to attain than 'near transfer,' i.e., improvement in domain-specific performance following the learning process in the same scientific domain (Dori and Sasson 2013; Sasson and Dori 2015; Zohar 2004).

Not all the investigations that included a metacognition-based pedagogical intervention actually assessed students' metacognition skills. Moreover, when reviewing the papers described in Table 3.2, it was not clear in some of the papers how to identify specific elements of metacognition that were addressed, such as specific theoretical constructs of regulation of cognition or knowledge/awareness of cognition. Time of intervention varied between few weeks to 3 and 4 months. Another aspect, not specified Table 3.2, is the assessment of retention regarding scientific concepts and skills being assessed. Only three studies evaluated the retention of skills and knowledge (Ben-David and Zohar 2009; Blank 2000; Georghiades 2004b).

Analyzing the papers concerning metacognitive intervention, we found a couple of studies that showed gains in learning, but not in metacognition. Sandi-Urena et al. (2011) reported that scores on the IMMEX problem, which served as an indication of chemistry learning and metacognitive skills, increased significantly for both treatment and control groups from pre- to post-test. However, they also reported that MCA-I scores, representing regulation of cognition elements, such as planning, evaluating, and monitoring, decreased significantly for the treatment group from pre- to post-test, and did not change for the control group. The authors suggested that a possible explanation of this finding might be the effect of a change in students' self-report. Unlike an attitude inventory, the MCA-I does not assess the importance that students placed on the construct, but rather their use of it. Therefore, they argued, raising awareness of metacognition and increasing its students' perceived importance increased their critical self-view and their tendency to self-score more strictly. A similar finding was obtained in another metacognition-enhancing intervention assessed by the same instrument (Cooper and Sandi-Urena 2009). The authors explained that after students were given the correct answer to the problem, they realized that they had used superfluous information in their previous attempt to solve the problem and that therefore, the students overestimated their knowledge in the first stage relative to what was expected from them in order to solve the scientific problem. These studies show that gains in one component of metacognition do not guarantee gains in another metacognitive component, however this aspect needs further investigation.

Thomas and McRobbie (2001) reported that some students showed an increase in their metacognitive skills following the intervention, while others showed no such increase. The authors' explanation to this finding was qualitative and related to students' learning processes or styles rather than to age.

3.5 Discussion

The review presented in this chapter helped to identify gaps between what researchers in the field of metacognition in science education strive to achieve and what they have achieved in practice, based on the literature. This review should help researchers determine whether, to what extent, and in what ways metacognition and assessment in science education have been implemented, and what aspects of metacognition and metacognition research in science education are still require more work. We shall now discuss (a) evaluation criteria for metacognition assessment tools, (b) research gaps we have identified, and (c) recommendations based on our findings and conclusions.

3.5.1 Evaluation Criteria for Metacognition Assessment Tools

In light of the theoretical background and studies reviewed in previous sections, we specified optional requirements that effective tools for assessment of metacognition in science education should meet (see Table 3.3). All the tools listed in Table 3.3 were reported by their respective authors to be reliable and valid. Fulfilling as many of these requirements as possible would improve future tools and their effectiveness for assessing metacognition in science education.

3.5.2 Research Gaps

One of the objectives of this work was to review and compare relevant literature to provide readers with potential guidelines for further research on metacognition in science education and different methods for assessing metacognition in science education. Our literature search findings highlight the need for further research on metacognitive assessment of science students. Indeed, as Veenman (2012) noted, research on metacognition in science education is still a work in progress. Since each STEM discipline has its own body of knowledge, more research to define and investigate metacognitive pedagogical interventions and metacognitive skills that are unique to each discipline is required. Two such examples – in mathematics education – are described in Chaps. 12 (Mevarech and Fan 2018) and 13 (Kohen and Kramarski 2018) of this volume. Another example for domain-specific metacognitive assessment is described in Chap. 9 of this book by authors Wengrowicz et al. (2018), where the authors describe their newly developed method for meta-assessment in systems engineering education. Students were required to compare and contrast different conceptual models across various model quality criteria, which the authors claimed fostered students' metacognitive skills. The authors specifically mentioned planning, monitoring, and evaluating, all belonging to

Table 3.3 Requirements for an effective tool for assessment of metacognition in science education

Requirement criterion	Requirement description	Examples of tools	Development and research suggestions
Adaptability	Be administrable, or adapted to age ranging from young students to undergraduate and graduate students	Two theoretical examples:	Integrate into the assessment concrete items taken from students' learning experiences
		A. Guidelines for assessing metacognition in children (Georghiades 2004a)	
		B. Suggestions for implementation into science classrooms in the document: *Taking science to school – grade K-8* (NRC 2007)	
		An empirical example: Junior Metacognitive Awareness Inventory (Jr. MAI) – developed and used to assess elementary school students. Suitable also to different student populations – MAI (Schraw et al. 2012)	
		Other empirical examples: Georghiades (2004b) and Wang et al. (2014)	
Comprehen-siveness	Contain items that cover various elements of metacognition	Metacognition Baseline Questionnaire (MBQ) – includes multiple elements of recognition: awareness, control, evaluation, planning, monitoring, and self-efficacy (Anderson and Nashon 2007)	Assessment and intervention should explicitly address several elements of metacognition based on the literature

(continued)

Table 3.3 (continued)

Requirement criterion	Requirement description	Examples of tools	Development and research suggestions
Concreteness	Include items that address concrete real-life problems, questions or statements, rather than abstract or 'textbook' issues	Interactive Multimedia Exercises (IMMEX) – presents students with real-world scenarios and concrete problem cases as part of a metacognitive assessment tool (Cooper et al. 2008; Sandi-Urena et al. 2011)	Assessment and intervention should address specific discipline-based knowledge constructs
In-action assessment	Be administrable concurrently with the task used for assessment, rather than being anticipatory or retroactive	IMMEX – used to conduct concurrent assessment of students (Cooper et al. 2008)	
		Think-aloud protocol assessment tool while reading a scientific text and answering questions (Michalsky 2013)	
Multi-contextualization	Include items that address various contexts: personal, social and global, or otherwise can be effectively adapted for this purpose	A theoretical example: Choi et al. (2011)	Investigate metacognitive components under the umbrella of SRL to include motivation and resource management aspects that promote life-long learning
		An empirical example: IMMEX – used to create detailed scenarios and multiple problem cases, enables creating scenarios in different contexts – personal, social or global (Cooper et al. 2008; Sandi-Urena et al. 2011)	
		Another empirical example: investigating self-reported metacognition and attitudes about an outdoor learning program (Schraw et al. 2012)	

(continued)

Table 3.3 (continued)

Requirement criterion	Requirement description	Examples of tools	Development and research suggestions
Bimodality	The tool should include both visual and textual modalities	A visual strategy (a diagram) has been applied in science education for encouraging students to improve their science understanding of complex topics and their ability to demonstrate their mastery in both text and schemes (Ward and Wandersee 2002)	Assessment and intervention should be based on two concurrent modalities: visual and textual
Assessment repeatability and students' retention	Allow for repeated testing in time gaps for effective monitoring of students' retention	A theoretical example: Choi et al. (2011)	The tool should be suitable for multiple admissions to the same students without a reduction in reliability or substantial increase in cost
		Empirical examples:	
		A. Assessing students' physics understanding twice in 3 months with slight modification (Koch 2001)	
		B. Assessing students' question posing skill via a 4-month intervention in chemistry classes with pre- and post-tests (Herscovitz et al. 2012)	
		C. Assessment of retention in biology (Ben-David and Zohar 2009)	
STEM-orientation	Include items that test students' scientific thinking	Inventory of Science Reading Awareness (ISRA) is used to measure students' science reading awareness (Wang et al. 2014 based on Yore et al. 1998)	Assessment and intervention should include particular STEM domains as well as scientific practices or thinking skills
Disciplinary focus	Include items that are concerned with a particular STEM domain	Physics Metacognition Inventory (PMCI): Multiple items from physics (Taasoobshirazi and Farley 2013)	Metacognition needs to be intertwined with learning domain-specific science core ideas and scientific practices or thinking skills

regulation of cognition. They also mentioned students had to consider their own task-specific knowledge, which can be constituted as declarative knowledge of cognition, although this term was not mentioned explicitly regarding meta-assessment. For future work on this promising tool, we suggest to specify precisely what elements of metacognition are engaged by this meta-assessment, and assess each element using a self-report tool for assessment of metacognitive thinking. This could provide evidence for metacognition as well as indication of the level of importance of each element of metacognition to task performance.

If and when such disciplinary knowledge becomes more established, investigation of interdisciplinary metacognitive thinking will become feasible and its outcomes potentially valuable. Thomas (Thomas 2013, also based on Schraw 1998, and Thomas 2012) argued that teacher-led explanations regarding thinking and reasoning strategies are key for fostering metacognition in students. We suggest such teacher-led explanations should be subject-specific and take into consideration the science content being taught. Assessment tools for metacognition in science education should be adaptable to a wide age spectrum, ranging from elementary school to post-secondary education, and suitable for various scenarios, problems, contexts, societal and global situations. Moreover, Wang et al. (2014) and Schraw et al. (2012) showed that young students in the elementary and middle school levels should be able to assess their learning in STEM or science before making a career choice.

Science educators and teachers should be aware that metacognition is composed of different components and that gains in one component of metacognition does not ensure gains in another component (Cooper and Sandi-Urena 2009; Sandi-Urena et al. 2011). Further research is needed here. Moreover, significant gains in one age group do not mean that these gains will be significant or visible in another age group. For example: Schraw et al. (2012) reported that following a metacognitive-based intervention, fifth grade students improved their environmental science knowledge, knowledge of cognition, and regulation of cognition, and that these factors were correlated positively. However, examining fourth grade students, the authors found that the first two factors were not correlated with the third one, which was regulation of cognition. The authors explained this finding by explaining that knowledge about oneself as a learner relates better to attitudes and performance than to self-regulatory aspects of metacognition.

Some of our findings echo those of Zohar and Barzilay (2013), who also carried out a review of research on metacognition in science education: (a) the prevalence of research in specific scientific disciplines, rather than in general contexts; (b) studying of metacognition usually occurs along with or in relation to other constructs; and (c) metacognition is studied mainly amongst older students. Our review adds in the aspect of assessment tools for metacognition and their requirements and calls for further development of tools to assess metacognition.

3.5.3 Recommendations

We recommend teachers engage in explicit teaching of the different components from the theoretical metacognition construct viewpoint. This would require prior training of teachers in the knowledge and practice of metacognitive learning. They will be able to enhance their teaching strategies using different components of metacognition, while students will better understand their different meanings. Metacognition needs to be intertwined with learning science core ideas and scientific practices as an integral part of science education. The document published by the NRC (2012a) emphasized that learning science involves deep exploration of important concepts, allocating time for students to develop meaningful understanding, and the need to progress throughout K-12 education. For older, high school and university students, it is important that knowledge be anchored to specific science subjects (Thomas et al. 2008). As NRC reports (2012b; 2015) highlighted, effective instruction in science education should include student-centered approaches. The more advanced approaches advocate attention to students' metacognitive strategies, though the K-12 reports do not yet explicitly mention metacognition (NRC 2013, 2015).

Additionally, and as part of science teaching, the elements of metacognition that a metacognitive intervention targets should be made explicit and assessed using a tool that is designed to assess those specific metacognitive elements (Herscovitz et al. 2012; Wagaba et al. 2016). While using a metacognitive tool in the classroom, assessment of learning and assessment of metacognition should be carried out in periodically and in tandem. This would enable the evaluation of the relationship between students' learning and their metacognition. We recommend teachers conduct behavioral observations, interviews and questionnaires, and ask for reflections from students to identify individual students' metacognitive profiles, as students are usually unaware of the limitations posed by their learning style (Anderson and Nashon 2007). Martin et al. (2000) claimed that many students' learning skills and metacognitive abilities are not adapted to the 'information age', placing their long-term success in jeopardy. Teachers and educators should make individual metacognitive profiles transparent to students in order to advance their learning and metacognition skills in a technological setting (see also Chap. 14 in this book).

Adopting these recommendations would help to make metacognition as an inseparable part of science education, and enable the progress of students' metacognitive skills alongside science learning and its assessment.

Although the research community has made various attempts to define metacognition and its components in the science education body of knowledge (Zohar and Dori 2012), with respect to several components (e.g., reflection or transfer) there is no consensus in regard to whether they are part of cognition or metacognition. Nonetheless, we recommend that science education research relate to specific components of metacognition and that researchers define what metacognitive aspects are at the focus of their intervention or assessment. Doing so rather than relating to metacognition as a whole will, help advance research and development in this progressing field. For example: Wang and Chen (2014) found that students do not differentiate between various components of metacognitive awareness. In addition, we suggest that when conducting studies of metacognition in general science or in a particular scientific discipline, researchers account for study findings in the relevant discipline. For assessment of students' metacognition who study science, researchers should use at least one qualitative and one quantitative assessment tools; and if no adequate tool exists for the researcher's assessment of metacognition purpose, we recommend they design their own tool to fulfil at least part of the criteria in Table 3.3, such as concurrent, in different contexts, and cater to repeatability – pre, post, and retention.

In summary, this review chapter along with its discussion and recommendations contributes to STEM researchers, educators, and practitioners who seek to advance science and engineering education through metacognition. For researchers, we contribute by presenting what research is still lacking in the STEM field; for educators, this contribution lies in our recommendations for teachers' training and professional development with emphasis on metacognition-based pedagogical intervention; and for teachers, the contribution is practical by raising the awareness for the need to incorporate metacognition in the specific science or engineering topics they teach, with the goal of advancing their students' metacognitive skills.

References

Ackerman, R., & Goldsmith, M. (2011). Metacognitive regulation of text learning: On screen versus on paper. *Journal of Experimental Psychology Applied, 17*(1), 18.

Anderson, D., & Nashon, S. (2007). Predators of knowledge construction: Interpreting students' metacognition in an amusement park physics program. *Science Education, 91*, 298–320.

Avargil, S., Herscovitz, O., & Dori, Y. J. (2012). Teaching thinking skills in context-based learning: Teachers' challenges and assessment knowledge. *Journal of Science Education and Technology, 21*(2), 207–225.

Azevedo, R. (2010). Computer environments as metacognitive tools for enhanced learning. *Educational Psychologist, 40*(4), 193–197.

Bandura, A. (1997). *Self-efficacy: The exercise of control*. New York: Freeman.

Bandura, A. (2000). Exercise of human agency through collective efficacy. *Current Directions in Psychological Science, 9*(3), 75–78.

Ben-David, A., & Zohar, A. (2009). Contribution of meta-strategic knowledge to scientific inquiry learning. *International Journal of Science Education, 31*(12), 1657–1682.

Blank, L. M. (2000). A metacognitive learning cycle: A better warranty for student understanding? *Science Education, 84*(4), 486–506.

Bouffard-Bouchard, T. (1990). Influence of self-efficacy on performance in a cognitive task. *The Journal of Social Psychology, 130*(3), 353–363.

Brown, A. L. (1978). Knowing when, where, and how to remember: A problem of metacognition. In R. Glaser (Ed.), *Advances in instructional psychology* (pp. 77–165). Hillsdale: Erlbaum.

Brown, A. L. (1987). Metacognition, executive control, self-regulation, and other mysterious mechanisms. In F. E. Weinert & R. H. Kluwe (Eds.), *Metacognition, motivation, and understanding* (pp. 65–116). Hillsdale: Lawrence Erlbaum.

Chiu, M. H., & Duit, R. (2011). Globalization: Science education from an international perspective. *Journal of Research in Science Teaching, 48*(6), 553–566.

Chiu, J. L., & Linn, M. C. (2012). The role of self-monitoring in learning chemistry with dynamic visualizations. In A. Zohar & Y. J. Dori (Eds.), *Metacognition in science education: Trends in current research* (pp. 133–163). Dordrecht: Springer.

Choi, K., Lee, H., Shin, N., Kim, S. W., & Krajcik, J. (2011). Re-conceptualization of scientific literacy in South Korea for the 21st century. *Journal of Research in Science Teaching, 48*(6), 670–697.

Conner, L., & Gunstone, R. (2004). Conscious knowledge of learning: Accessing learning strategies in a final year high school biology class. *International Journal of Science Education, 26*(12), 1427–1443.

Cooper, M. M., & Sandi-Urena, S. (2009). Design and validation of an instrument to assess metacognitive skillfulness in chemistry problem solving. *Journal of Chemical Education, 86*(2), 240–245.

Cooper, M., Sandi-Urena, S., & Stevens, R. (2008). Reliable mutli-method assessment of metacognition use in chemistry problem-solving. *Chemistry Education Research and Practice, 9*, 18–24.

Desoete, A. (2008). Multi-method assessment of metacognitive skills in elementary school children: How you test is what you get. *Metacognition and Learning, 3*(3), 189–206.

Dignath, C., & Büttner, G. (2008). Components of fostering self-regulated learning among students. A meta-analysis on intervention studies at primary and secondary school level. *Metacognition Learning, 3*, 231–264.

Dori, Y. J., & Sasson, I. (2013). A three-attribute transfer skills framework–part I: Establishing the model and its relation to chemical education. *Chemistry Education Research and Practice, 14*(4), 363–375.

Eilam, B., & Reiter, S. (2014). Long-term self-regulation of biology learning using standard junior high school science curriculum. *Science Education, 98*(4), 705–737.

Eldar, O., Eylon, B., & Ronen, M. (2012). A metacognitive teaching strategy for preservice teachers: Collaborative diagnosis of conceptual understanding in science. In A. Zohar & Y. J. Dori (Eds.), *Metacognition in science education: Trends in current research* (pp. 225–250). Dordrecht: Springer.

Flavell, J. H. (1976). Metacognitive aspects of problem solving. In L. B. Resnick (Ed.), *The nature of intelligence* (pp. 231–235). Hillsdale: Erlbaum.

Flavell, J. H. (1979). Metacognition and cognitive monitoring: A new area of cognitive-developmental inquiry. *American Psychologist, 34*, 906–911.

Flavell, J. H. (1981). Cognitive monitoring. In W. P. Dickson (Ed.), *Children's oral communication skills*. New York: Academic Press.

Garner, R., & Alexander, P. A. (1989). Metacognition: Answered and unanswered questions. *Educational Psychologist, 24*(2), 143–158.

Georghiades, P. (2004a). From the general to the situated: Three decades of metacognition. *International Journal of Science Education, 26*(3), 365–383.

Georghiades, P. (2004b). Research report: Making pupils' conceptions of electricity more durable by means of situated metacognition. *International Journal of Science Education, 26*(1), 85–99.

Greene, J. A., & Azevedo, R. (2009). A macro-level analysis of SRL processes and their relations to the acquisition of a sophisticated mental model of a complex system. *Contemporary Educational Psychologist, 34*, 18–29.

Grotzer, T., & Mittlefehldt, S. (2012). The role of metacognition in students' understanding and transfer of explanatory structures in science. In A. Zohar & Y. J. Dori (Eds.), *Metacognition in science education: Trends in current research* (pp. 79–99). Dordrecht: Springer.

Hand, B., Wallace, C. W., & Yang, E.-M. (2004). Using a science writing heuristic to enhance learning outcomes from laboratory activities in seventh-grade science: Quantitative and qualitative aspects. *International Journal of Science Education, 26*(2), 131–149.

Herscovitz, O., Kaberman, Z., Saar, L., & Dori, Y. J. (2012). The relationship between metacognition and the ability to pose questions in chemical education. In A. Zohar & Y. J. Dori (Eds.), *Metacognition in science education: Trends in current research* (pp. 165–195). Dordrecht: Springer.

Jacobs, J. E., & Paris, S. G. (1987). Children's metacognition about reading: Issues in definition, measurement and instruction. *Educational Psychologist, 22*, 255–278.
Kaberman, Z., & Dori, Y. J. (2009). Metacognition in chemical education: Question posing in the case-based computerized learning environment. *Instructional Science, 37*, 403–436.
Koch, A. (2001). Training in metacognition and comprehension of physics text. *Science Education, 85*, 758–768.
Kohen, Z., & Kramarski, B. (2018). Promoting mathematics' teachers pedagogical metacognition – A theoretical practical model and case study. In Y. J. Dori, Z. Mevareach, & D. Bake (Eds.), *Cognition, metacognition and culture in STEM education*. Springer (Accepted).
Martin, B. L., Mintzes, J. J., & Clavijo, I. E. (2000). Restructuring knowledge in biology: Cognitive processes and metacognitive reflections. *International Journal of Science Education, 22*(3), 303–323.
Mevarech, Z., & Fan, L. (2018). Cognition, metacognition and mathematics literacy. In Y. J. Dori, Z. Mevareach, & D. Bake (Eds.), *Cognition, metacognition and culture in STEM education*. Springer.
Mevarech, Z. R., & Kramarski, B. (1997). IMPROVE: A multidimensional method for teaching mathematics in heterogeneous classrooms. *American Educational Research Journal, 34*(2), 365–394.
Mevarech, Z., & Kramarski, B. (2014). *Critical Maths for innovative societies: The role of metacognitive pedagogies, educational research and innovation*. Paris: OECD Publishing.
Michalsky, T. (2013). Integrating skills and wills instruction in self-regulated science text reading for secondary students. *International Journal of Science Education, 35*(11), 1846–1873.
National Research Council. (2000). How people learn: Brain, mind, experience, and school. In J. D. Bransford, A. L. Brown, & R. R. Cocking (Eds.), *Committee on developments in the science of learning*. Washington, DC: National Academy Press.
National Research Council. (2007). Taking science to school: Learning and teaching science in grades K-8. In R. A. Duschl, H. A. Schweingruber, & A. W. Shouse (Eds.), *Committee on science learning, kindergarten through eighth grade*. Washington, DC: The National Academy Press.
National Research Council. (2012a). *A framework for K-12 science education: Practices, crosscutting concepts, and core ideas*. Washington, DC: The National Academy of Sciences.
National Research Council. (2012b). *Discipline-based education research: Understanding and improving learning in undergraduate science and engineering*. Washington, DC: The National Academy Press.
National Research Council. (2013). *Monitoring progress toward successful K-12 STEM education: A nation advancing? Committee on the evaluation framework for successful K-12 STEM education, Board on science education and board on testing and assessment, division of behavioral and social sciences and education*. Washington, DC: The National Academies Press.
National Research Council. (2015). *What research says about effective instruction in undergraduate science and engineering?* Washington, DC: The National Academy Press.
Nielsen, W. S., Nashon, S., & Anderson, D. (2009). Metacognitive engagement during field-trip experiences: A case study of students in an amusement park physics program. *Journal of Research in Science Teaching, 46*(3), 265–288.
Norris, S. P., & Phillips, L. M. (2012). Reading science: How a naive view of reading hinders so much else. In A. Zohar & Y. J. Dori (Eds.), *Metacognition in science education* (pp. 37–56). Dordrecht: Springer.
Sandi-Urena, S., Cooper, M. M., & Stevens, R. H. (2011). Enhancement of metacognition use and awareness by means of a collaborative treatment. *International Journal of Science Education, 33*(3), 323–340.
Sasson, I., & Dori, Y. J. (2015). A three-attribute transfer skills framework–part II: Applying and assessing the model in chemical education. *Chemistry Education Research and Practice, 16*, 154–167.

Schraw, G. (1998). Promoting general metacognitive awareness. *Instructional Science, 26*(1/2), 113–125.

Schraw, G., & Dennison, R. S. (1994). Assessing metacognitive awareness. *Contemporary Educational Psychology, 19*(4), 460–475.

Schraw, G., & Moshman, D. (1995). Metacognitive theories. *Educational Psychology Review, 7*(4), 351–371.

Schraw, G., Olafson, L., Weibel, M., & Sewing, D. (2012). Metacognitive knowledge and field-based science learning in an outdoor environmental education program. In A. Zohar & Y. J. Dori (Eds.), *Metacognition in science education: Trends in current research* (pp. 57–77). Dordrecht: Springer.

Shin, N., Jonassen, D. H., & McGee, S. (2003). Predictors of well-structured and ill structured problem solving in an astronomy simulation. *Journal of Research in Science Teaching, 40*(1), 6–33.

Sperling, R. A., Howard, B. C., Miller, L. A., & Murphy, C. (2002). Measures of children's knowledge and regulation of cognition. *Contemporary Educational Psychology, 27*, 51–79.

Taasoobshirazi, G., & Farley, J. (2013). Construct validation of the physics metacognition inventory. *International Journal of Science Education, 35*(3), 447–459.

Thomas, G. P. (2012). Metacognition in science education: Past, present and future considerations. In B. J. Fraser, K. G. Tobin, & C. J. McRobbie (Eds.), *Second international handbook of science education* (pp. 131–144). Dordrecht: Springer.

Thomas, G. P. (2013). Changing the metacognitive orientation of a classroom environment to stimulate metacognitive reflection regarding the nature of physics learning. *International Journal of Science Education, 35*(7), 1183–1207.

Thomas, G. P., & McRobbie, C. J. (2001). Using a metaphor for learning to improve students' metacognition in the chemistry classroom. *Journal of Research in Science Teaching, 38*(2), 222–259.

Thomas, G., Anderson, D., & Nashon, S. (2008). Development of a tool designed to investigate elements of science students' metacognition, self-efficacy and learning processes: The SEMLI-S. *Internal Journal of Science Education, 30*(13), 1701–1724.

Veenman, M. V. (2011). Alternative assessment of strategy use with self-report instruments: A discussion. *Metacognition and Learning, 6*(2), 205–211.

Veenman, M. V. (2012). Metacognition in science education: Definitions, constituents, and their intricate relation with cognition. In A. Zohar & Y. J. Dori (Eds.), *Metacognition in science education: Trends in current research* (pp. 21–36). Dordrecht: Springer.

Veenman, M. V. J., Van Hout-Wolters, B. H. A. M., & Afflerbach, P. (2006). Metacognition and learning: Conceptual and methodological considerations. *Metacognition and Learning, 1*, 3–14.

Wagaba, F., Treagust, D. F., Chandrasegaran, A. L., & Won, M. (2016). Using metacognitive strategies in teaching to facilitate understanding of light concepts among year 9 students. *Research in Science and Technological Education*. doi:10.1080/02635143.2016.1144051.

Wang, C. Y. (2015). Scaffolding middle school students' construction of scientific explanations: Comparing a cognitive versus a metacognitive evaluation approach. *International Journal of Science Education, 37*(2), 237–271.

Wang, J. R., & Chen, S. F. (2014). Exploring mediating effect of metacognitive awareness on comprehension of science texts through structural equation modeling analysis. *Journal of Research in Science Teaching, 51*(2), 175–191.

Wang, J. R., Chen, S. F., Fang, I., & Chou, C. T. (2014). Comparison of Taiwanese and Canadian students' metacognitive awareness of science reading, text, and strategies. *International Journal of Science Education, 36*(4), 693–713.

Ward, R. E., & Wandersee, J. H. (2002). Struggling to understand abstract science topics: A roundhouse diagram-based study. *International Journal of Science Education, 24*(6), 575–591.

Wengrowicz, N., Dori, Y. J., & Dori, D. (2018). Metacognition and meta-assessment in engineering education. In Y. J. Dori, Z. Mevareach, & D. Bake (Eds.), *Cognition, metacognition and culture in STEM education*. Springer.

Yore, L. D., & Treagust, D. F. (2006). Current realities and future possibilities: Language and science literacy – empowering research and informing instruction. *International Journal of Science Education, 28*(2–3), 291–314.

Yore, L. D., Craig, M. T., & Maguire, T. O. (1998). Index of science reading awareness: An interactive-constructive model, test verification, and grades 4–8 results. *Journal of Research in Science Teaching, 35*(1), 27–51.

Zion, M., Michalsky, T., & Mevarech, Z. R. (2005). The effects of metacognitive instruction embedded within an asynchronous flearning network on scientific inquiry skills. *International Journal of Science Education, 27*(8), 957–983.

Zohar, A. (2004). *Higher order thinking in science classrooms: Students' learning and teachers' professional development*. Dordrecht: Kluwer Academic.

Zohar, A. (2006). The nature and development of teachers' metastrategic knowledge in the context of teaching higher order thinking. *The Journal of the Learning Sciences, 15*(3), 331–377.

Zohar, A. (2012). Explicit teaching of metastrategic knowledge: Definitions, students' learning, and teachers' professional development. In A. Zohar & Y. J. Dori (Eds.), *Metacognition in science education: Trends in current research* (pp. 197–224). Dordrecht: Springer.

Zohar, A., & Barzilay, S. (2013). A review of research on metacognition in science education: Current and future directions. *Studies in Science Education, 49*(2), 121–169.

Zohar, A., & Dori, Y. J. (Eds.). (2012). *Metacognition in science education: Trends in current research*. Dordrecht: Springer.

Chapter 4
Reconsidering Different Visions of Scientific Literacy and Science Education Based on the Concept of *Bildung*

Jesper Sjöström and Ingo Eilks

4.1 Introduction

Over the last 50 years, policy makers and STEM educators have argued for Scientific Literacy (SL) (Roberts 2007). SL has become a guiding framework in educational policy, for example, in the PISA studies (Sadler and Zeidler 2009). Laugksch (2000) has stated that SL has become a buzzword, conveying a rather vague notion of what the general public should know about science. However, there have been a number of attempts to systematically describe different elements of SL (e.g., Coll and Taylor 2009; Gräber and Bolte 1997). One example is Hodson (2009), who subdivided scientific and technological literacy into the following three elements:

1. Learning science and technology (e.g., conceptual understanding);
2. Doing science and technology (e.g., scientific inquiry); and
3. Learning *about* science and technology.

Roberts (2007, 2011) distinguished between two main orientations of SL: *Vision I*, which focuses mainly on learning about scientific content and scientific processes for later application, and *Vision II*, which focuses on understanding the usefulness

The original version of this chapter was revised: This chapter was previously published as non-open access. It has now been changed to open access under a CC BY 4.0 license. The correction to this chapter is available at https://doi.org/10.1007/978-3-319-66659-4_16

J. Sjöström (✉)
Department of Science-Environment-Society, Faculty of Education and Society, Malmö University, Malmö, Sweden
e-mail: jesper.sjostrom@mah.se

I. Eilks
Institute for Science Education, Department of Biology and Chemistry, University of Bremen, Bremen, Germany

Y.J. Dori et al. (eds.), *Cognition, Metacognition, and Culture in STEM Education*, Innovations in Science Education and Technology 24,
https://doi.org/10.1007/978-3-319-66659-4_4

of scientific knowledge in life and society by starting science learning from meaningful contexts. The tension between Vision I and II is related to the tension between "pipeline science – preparing future scientists" and "science for all" (Aikenhead 2006), but these two visions can also be seen as two different orientations of the science curriculum (Eilks et al. 2013).

Recently, a more advanced form of Vision II was suggested, called *Vision III,* which emphasizes scientific engagement (Liu 2013; Yore 2012) and "knowing-in-action" (Aikenhead 2007). As far as we know, Hodson (2003, 2009, 2011) did not use the term Vision III. But instead he used the term "critical scientific literacy", and added a forth element in addition to the three mentioned above: Engaging in socio-political action. Similarly, Santos (2009) has identified three types of SL, which can be described as: (a) practical view, (b) understanding human culture and (c) socio-political action.

Vision III or critical-SL can be understood based on the Central/Northern European educational tradition called *Bildung* (Hofstein et al. 2011; Sjöström 2013a). *Bildung* is a complex concept that can – as discussed further below – be explained in several ways, but typically it consists of two elements: an ideal picture (of desirable knowledge and cognitive skills) and free learning processes (Gustavsson 2014a), or in other words both "the process of personal development and the result of this development process" (Fischler 2011, p. 33). Schneider (2012) describes *Bildung* as a reflexive event and its function is to form the self in a complex meaning-making process that covers the whole range from early childhood to the advanced age. According to Wimmer (2003), *Bildung* encompasses all aims that are not covered by other concepts of pedagogical theory, such as socialisation, education, and instruction; it stands for them all and provides also something more. He describes it as "the central critical concept of modern pedagogy" (p. 185). Due to its both educational and political dimensions (Biesta 2002a), it allows us to say something different about science education and scientific literacy.

In general, one can say that *Bildung*-oriented science education is an example of humanized science education (Aikenhead 2006) that goes beyond many understandings of scientific literacy in the literature. However, it has many similarities with "science for [critical] citizenship" (e.g., Albe 2015), complex versions of socio-scientific issues (SSI) based science education (e.g., Bencze et al. 2012; Simonneaux and Simonneaux 2012; Zeidler 2015) and STSE (Science, Technology, Society, and Environment) education (e.g., Pedretti and Nazir 2011).

Similar to us, Wickman et al. (2012) connected scientific literacy in the European sense with *Bildung* (see also Fischler 2015) and Elmose and Roth (2005) tried to introduce the concept of *Bildung* to justify science teaching focusing on preparing students for political participation in a growing complex world. However, these papers are not explicitly discussing a Vision III of scientific literacy (i.e., critical-SL), and put no direct emphasis on educated socio-political action. But the definition by Wickman et al. emphasizes the importance of worldviews, values and ethics in science education. Similarly, such socio-cultural aspects were emphasized by Sadler and Zeidler (2009) in their SSI framework; regarding SL it is interesting that they explicitly placed themselves at the extreme of Vision II.

In this chapter it is suggested that SL and *Bildung* should be considered to be action-oriented – or even better, 'praxis-oriented'. *Bildung*-oriented education aims at making the student capable for a self-determined life in his/her socio-cultural environment, for participation in a democratic society, and for empathy and solidarity with others (e.g., Elmose and Roth 2005; Hofstein et al. 2011; Sjöström 2013a). In other words, Vision III of SL should imply a politicised science education aiming at emancipation and socio-ecojustice. This concept is also closely connected to more recent educational paradigms, for example, the ideas of Education for Sustainability (EfS) (Sjöström et al. 2015) and transformative learning (Mezirow 1997; Sterling 2011; Thomas 2009), where content and contexts should be considered from multifaceted perspectives. EfS aims on skills development for critical-democratic participation and for shaping society in a sustainable way. Simonneaux (2014a) emphasises participation and action as especially important parts of transformative science education: "when implementing post-normal education, it is not sufficient just to learn and to understand. Instead, the central purpose is to encourage participation and action in the scientific activity." (p. 51)

In other words, the different visions of SL have consequences for the teaching and learning in the STEM subjects. Within a *Bildung*-tradition there is awareness that our view of scientific content knowledge is dependent on our culture, for example our norms, values and worldviews, and it is dependent on the time we are living in (Sjöström 2013a). Furthermore, there is an awareness that learning (cognition) must be complemented with not only meta-learning (metacognition), but also with epistemic and transformative learning (Sterling 2011). Examples include scientific concepts and models, but also scientific processes (nature of science, NOS) and the embeddedness of science and technology in society (Sjöström and Talanquer 2014).

To summarize this chapter focuses on implications for science teaching and learning of *Vision III* of SL and its connection to a contemporary understanding of *Bildung*, EfS and transformative learning. We start with describing the concept of *Bildung*, focusing on the most complex type, which we call critical-reflexive *Bildung*. Thereafter, we first discuss implications of this version of *Bildung* on education in general and then its connection to different meanings of 'critical' in education. It is followed by in-depth discussions of implications of critical-reflexive *Bildung* on science education and scientific literacy, respectively.

4.2 The Concept of *Bildung*

In Central and Northern Europe (especially in German speaking countries and in Scandinavia) there is a philosophical-educational tradition called *Bildung*, which has been developed since the late eighteenth century by Johann Gottfried Herder, Wilhelm von Humboldt, Hans-Georg Gadamer, Erich Weniger, Wolfgang Klafki and others.[1] Because there is no precise English translation, the German term *Bildung* started to be used in the international educational literature (e.g., Elmose and Roth 2005; Hofstein et al. 2011; Sjöström 2013a).

---

[1] See (Westbury et al. 2000), for some translated original contributions from the history of *Bildung* and *Didaktik* in Central Europe.

Without doubt *Bildung* is a complex construct and a description of its genesis may help understanding the concept. According to Gustavsson (2012, 2014a) at least three versions of *Bildung* are well-established today and all of them have transformed over time from a national/European to a global focus. We call them (a) classical *Bildung*, (b) Anglo-American *Bildung*, or liberal education, and (e) critical-reflexive *Bildung*. In addition to these three versions Burman (2011) also identified two civic-oriented *Bildung*-traditions: (c) the Scandinavian folk-Bildung-tradition, and (d) Dewey's democratic education. In the following we will describe these five *Bildung*-traditions in a little more detail:

(a) *Classical Bildung:* this tradition is based on the German philosopher and educational politician Wilhelm von Humboldt (1767–1835) (2000, originally in German, 1793). Von Humboldt understood *Bildung* as "one in the tradition rooted process of individualization where humans through studies and reflections develop their personality in a diverse, harmonious and unique way, and thus become a human original rather than a copy of others" (Burman 2014, p. 127, our translation). However, today von Humboldt is often – at least at universities – more associated with free search for knowledge, free from both the state and the market. The works of von Humboldt are also sometimes misused. His idea that *Bildung* manifests itself mainly in language, led to a long time of devaluing the sciences for developing own worldviews in the individual. In some European countries, e.g. Sweden and Germany, this led to a long time of over-emphasizing the humanities to constitute classical *Bildung* against education in the STEM subjects.

(b) *Anglo-American Bildung:* the thoughts behind this tradition, which is called liberal education, can also be tracked back to von Humboldt (Løvlie and Standish 2002). The character-formation ideal is emphasized in the English version, whereas the canon was emphasized in the American version (Burman 2014). The latter has strong connections to American colleges. The liberal education tradition emphasizes humanism and generalization – in contrast to specialization – and also, that education must be free from short-term instrumental thinking. The thought of life-long learning, which for example is important in contemporary European policy debate, is related to this type of thinking. A famous representative for a more critical and cosmopolitical version of liberal education is Martha Nussbaum (born 1947). She argues for ethical self-reflection and critical approaches to the own culture and its traditions. This is needed to create enlightened citizens, rather than efficient workers and uncritical consumers. Nussbaum uses typical *Bildung*-type arguments for liberal education, however without explicitly using the term (Bohlin 2008).

(c) *The Scandinavian folk-Bildung-tradition*: from the late nineteenth century a unique tradition called *folkbildning* in Swedish (might be translated as '*Bildung* for the whole people') was developed in Scandinavia. It is a tradition that is less academically oriented than the classical German tradition. The German basic notion was combined with a pronounced benefit-approach. *Bildung* should be useful for the creation of a society with justice. The political dimension was much more explicit than in the classical German version, but it was not especially radical. An example of a famous Swedish pedagogue is Ellen Key (1849–1926). She emphasized *Bildung* as a relevant concept both on the individual and the societal level. Children should be educated to become civic citizens. School would encourage students to become free, responsible actors in society, with a developed individuality – cognitively, morally, as well as aesthetically (Burman 2014).

(d) *Dewey's democratic education*: the idea of a school for all was also developed in the USA by John Dewey (1859–1952). In the book *Democracy and Education: An Introduction to the Philosophy of Education* from 1916 he advocated that school has a crucial role to play in every democratic society. He suggested the basic mission of school is to prepare for citizenship. This requires that students can develop quite freely (Burman 2014). According to Väkevä (2012), Dewey's most important contribution to the *Bildung* tradition was his analysis of the social-ethical foundations of a society to promote democratic habits. Dewey used the term *Bildung* in his work, although not systematically (Bauer 2003). However, it is interesting that Kivelä et al. (2012) conclude that on a general level there is no significant difference between *Bildung* (in a growth-theoretical understanding) and the ideas of pragmatists such as Dewey, James, and Mead.

(e) *Critical-reflexive Bildung*: this understanding of *Bildung* is rooted in the work by Hans-Georg Gadamer (1900–2002) and Paul Ricoeur (1913–2005) and can be described with '*Bildung* as a journey' (Gustavsson 2012, 2014a). Especially during the 1950s and 1960s, and in interaction with the work of Gadamer and Ricoeur, the German educational philosophers Erich Weniger (1894–1961) and Wolfgang Klafki (1927–2016) developed a new understanding of *Bildung* connected to educational practice. They created the term *Allgemeinbildung*. Within this concept, part of the word, *Allgemein* (which can be translated as 'general') has two dimensions. The first dimension means to achieve *Bildung* for all persons (like in the Scandinavian approach of *folkbildning*). The second dimension aims at *Bildung* in all human capacities (e.g., Klafki 2000a). Klafki's thinking is based on the thought that responsible citizens in a democratic society need *Bildung*. This educational philosophy has a clear critical approach (see further below) and we regard critical-reflexive *Bildung* the most complex version of the five traditions. In the following, when the term *Bildung* is used, we mean this version, if not something else is specified.

4.3 Critical-Constructive *Didaktik* as an Educational Implication of *Bildung*

Bildung in a critical understanding is praxis-oriented, in addition to being oriented towards consciousness and critical literacy. In line with this, Mogensen and Schnack (2010, p. 60) argue that their concept of 'action competence' is "closely linked to democratic, political education and to […] the notion of '*Bildung*'." According to Marks et al. (2014, p. 286), *Bildung* "…inseparably bounds education to a democratic understanding of society. It defines all objectives of education under consideration of a societal perspective, for education in general, but also for all school educational domains in particular – among them science education."

For educational operation Klafki (2000b, originally in German, 1958; see also Fischler 2011) and others developed a tool called *Didactical Analysis* as being part of the so called *Critical-Constructive Didaktik*. At this point it is necessary to say that the *Bildung*-connected meaning of the term *Didaktik* in German and Scandinavian languages differs a lot from how the word *didactics* is used in English (Duit 2015). *Didaktik* in German and Scandinavian languages means the knowledge about teaching and learning and at the same time covers the research area about teaching and learning (Hopmann 2007; Kansanen 2009). According to Duit (2015, p. 325) *Didaktik* "stands for a multifaceted view of planning and performing instruction. It is based on the German concept of *Bildung* [… and] concerns the analytical process of transposing (and transforming) human knowledge (the cultural heritage) into knowledge for schooling that contributes to Bildung". Hopmann (2007, p. 109) has compared *Didaktik* and the Anglo-American concept of curriculum and instruction. He claims that "Didaktik is characterized as 'restrained teaching', based on (a) a commitment to Bildung, (b) the educative difference of matter and meaning, and (c) the autonomy of teaching and learning." Similarly, Kansanen (2009) compared subject-matter didactics with Lee Shulman's pedagogical content knowledge (PCK). The former is, according to Kansanen, a much broader idea also containing aspects of values and other characteristics related to the curriculum and pedagogy. The *Didaktik* tradition focuses predominantly on the *why*-question (and its implication on practice), while the pragmatic Anglo-American curriculum tradition focuses more on the *how*-question (Duit 2015).

Didactical Analysis in terms of Klafki reflects whether an issue or topic is relevant enough to be taught. It consists of a set of certain questions, e.g. what the general exemplary character of the topic is, or what meaning it has for the learner today and for his/her future (Klafki 2000b, published originally in German 1958). These questions try to identify epoch-typical relevant knowledge and key problems to learn about, which are of importance for the individuals and the society the students live in and operate today and in the future. Contemporary examples of science-related key problems, important for education, are among many others e.g. the questions of global warming (Selby 2014), alternative energy usages (Feierabend and Eilks 2011), or the *chemicalization* of our world (Sjöström and Stenborg 2014). Except learning the science behind such relevant issues, students also should get "the potential to learn about how such an issue is handled within society and one can learn about the interplay of science with ecology, economics, politics, cultural beliefs and values" (Marks et al. 2014, p. 287).

Classical *Bildung* (von Humboldt) already had a critical dimension because of its relationship to the critical philosophy by Immanuel Kant. However, in practice the critical dimension has not been particularly prominent in all the *Bildung*-versions. Especially the Anglo-American liberal education-tradition has traditionally had a relatively uncritical approach to the classical Greek and Latin culture (Gustavsson 2014b). However, through, for example, the work of Nussbaum in America and even more Klafki and also scholars of *Critical Theory* in Germany this has changed. For some of the latter a critical perspective "…is realised by reflection, by activating critique as a moral-philosophical-existential-political alternative" (Gur–ze'ev 2002, p. 404). An important concept in critical theory is *emancipation*, which can be defined as "eliminating oppression and creating conditions for effective agency" (Zembylas 2006, p. 665). Below this is called a 'critical-emancipatory approach'.

The concept of *Bildung* has itself been criticized and problematized, mainly by postmodern theorists (Løvlie et al. 2003, reviewed in Hansen 2008). Recently, Schaffar and Uljens (2015) identified the following two central points of criticism: (a) a logico-conceptual type of critique, where *Bildung* has been called a 'container word' and the meaning of emancipation has been questioned, and (b) a socio-cultural critique, whereby *Bildung* is reachable only for the elite and that it is thus serving and supporting existing cultural structures of power. However, Biesta (2002b) claims that *Bildung* still works as a critical concept in a postmodern world. But he has argued against "certain versions of the critical theory of *Bildung* and critical pedagogy" with the ambition "to 'read' power behind knowledge" (Biesta 2002b, p. 388). Instead, he referred to Latour's networks, in which knowledge and power are not separable. More recently, he discussed, based on writings by Freire, Foucault and Rancière, a dialogical approach to emancipation. In such an approach, doing things differently to show alternatives, are emphasized (Biesta 2012). To sum up we – just like Klafki, Kemp, Biesta and others (e.g. Kemp 2005) – claim that criticisms of the concept can be counteracted by arguing for a contemporary and complex version of *Bildung* (we call it critical-reflexive *Bildung*) and by emphasizing that *Bildung* is something for all citizens in our complex and globalized society. In the next section we discuss different meanings of the term 'critical' in an educational context.

4.4 *Bildung*-Oriented Education for Critical Thinking and Responsible Actions

The word critical is used in a variety of forms in curricula, for example as critical skills and critical thinking. Johnson and Morris (2010) have discussed how critical citizenship can be understood as the intersection between critical thinking and critical pedagogy. For them, critical thinking is associated to abstract and technical skills and has an individualistic focus, whereas critical pedagogy has a collective focus and is driven by a concern for socio-ecojustice.

The core explanation of critical thinking (CT) is that it is something cognitive, that is, for example logical reasoning. However, the term can also be understood in a broader sense. It has also to do with awareness of the own way of learning (meta-cognition) and philosophical-ideological awareness. Learning connected to the latter can be called epistemic and transformative learning (Sterling 2011). The core understanding of CT, that is cognitive and intellectual thinking, has been called the first wave, whereas a broader understanding of the term is called the second wave of CT (Walters 1994). Bohlin (2009, p. 190) has described it in the following way: "good thinking requires logical skills but is not exclusively defined by them; creative imagination, empathy, and self-reflective awareness of one's own presuppositions are equally important". Similarly, Hasslöf and Malmberg (2015) recently showed that critical thinking can have various meaning depending on different epistemological views; sometimes it is based on the educational aims of qualification and socialization, and sometimes subjectification. Especially the latter is related to the concept of *Bildung* (Biesta 2012; Schneider 2012; Straume 2015).

With reference to the moral philosopher Richard Hare (1919–2002), Vieira et al. (2011) described CT as one of the central ideas behind education and suggest that it forms the social basis for the achievement of equal rights and freedom within democratic societies. For Hare there are three justifications of CT: intellectual, pragmatic, and ethical. However, we think that the term *critical approach* better mirrors this broader meaning of CT and is more appropriate to be used in relation to *Bildung*. According to Gustavsson (2014b), a critical approach encompasses both to think and act critically, and to do so both in theory as well as in practice.

Another related term, already suggested above, is *critical praxis*. Critical praxis is an important goal of critical pedagogy. Critical approaches in education have followed two main lines: In Germany a critical-emancipatory approach was based on the early work of Habermas, and in North America a critical theory of education was developed based on writings by, e.g., Dewey and Freire (Biesta 2012). Freire's educational approach "…is essentially a humanistic pedagogy concerned with the real context of human conditions, particularly focused on the oppressive context" (Santos 2009, p. 364). The focus of critical pedagogy is the relationship between knowledge and power and its agenda is transformation of knowledge (e.g. curriculum) and pedagogy (e.g., teaching) (Cho 2010). With reference to Dewey and Freire, Reis (2014) claims that critical pedagogy suggests education as a democratizing force and in the same time being a catalyst for individual development and social transformation. More in detail, Shor (1992) in her book *Empowering Education* defined critical pedagogy as: "Habits of thought, reading, writing, and speaking which go beneath surface meaning, first impressions, dominant myths, official pronouncements, traditional clichés, received wisdom, and mere opinions, to understand the deep meaning, root causes, social context, ideology, and personal consequences of any action, event, object, process, organization, experience, text, subject matter, policy, mass media, or discourse" (p. 129). In other words, at the heart of critical pedagogy are the ideas of education for awareness, praxis and dialogue (Bader and Laberge 2014). It can be seen as the educational implication of *Bildung* (in its critical-reflexive version). The goal of *Bildung*-oriented education is transformation of both the subjects/individuals/citizens and the global society towards sustainability.

Transformative learning can be understood as a deep shift in perspective focusing on making the habits of mind in the young generation more open, more permeable and better justified (Cranton 2011). This is expected to occur when people start to critically reflect on their instrumental and communicative knowledge. Houwer (2014) supports this view by arguing that crises are opportunities for transformative practices. Transformative learning is also about addressing the critical dimensions of certain contexts (Sterling 2011) and focusing the transformation of attitudes, behaviors, values, beliefs, and corresponding action (Carter et al. 2014). According to Bohlin (2008, p. 8) transformative learning, although only seldom explicitly associated with the idea of *Bildung*, "indicates ways to implement the ideal of moral *Bildung* in educational practice" (see also Bohlin 2013). For us a core idea of critical-reflexive *Bildung* is to critically identify cultural presuppositions and to support alternative ways of thinking and acting in dialogue with the surrounding world.

4.5 Towards *Bildung*-Oriented Science Education

As we have pointed out previously, except for scientific concepts and models which are in focus in traditional science education, scientific processes and societal contexts need to be also emphasised in humanized, socio-critical and *Bildung*-oriented science education (Marks and Eilks 2009; Sjöström 2013a; Sjöström and Talanquer 2014). This means that without including ethical and socio-political perspectives into STEM teaching, science learning will miss essential aspects that contribute making it relevant education (Hofstein et al. 2011; Stuckey et al., 2013). This necessarily includes a focus on understanding uncertainties and balancing benefits and risks (Sjöström 2013a). It also is in line with the thinking of Albe (2013), who claims that we need to rethink our culture and the way science education is being thought. She argues for a shift from the almost exclusive focus on subject matter content to socio-educational aims and preparation for socio-political action. We agree that science education should go in a socio-scientific direction, but just like Klafki we also think that relevant subject matter content is important.

Santos (2009) discussed the implications of critical pedagogy (a Freirean perspective) on science education and teaching. It is a radical view of scientific literacy, where not only socio-political perspectives are incorporated; the focus is on the political aim of transforming society to overcome oppressive conditions. Freirean-oriented science education can, according to Santos, be characterized by the following three aspects: (1) discussions of socially relevant themes by SSIs, (2) establishment of a dialogical process in the classroom, and (3) engagement of students in socio-political actions. He writes: "Freirean science education ought to take

SSI as the goal of attaching social meaning to science content, and of helping students understand the oppressive context of modern society" (p. 374). Science teaching "should be developed with grounds on students' cultural context through socially relevant themes that incorporate issues of oppressive context in society, and that ought to be developed through a dialogical process in classrooms, engage students in sociopolitical action and thus make it possible to look forward to bringing equity and social justice into our world" (p. 377). More recently, Bader and Laberge (2014) also claimed that the general principles of critical pedagogy should be applied in science education. For them critical pedagogy is focusing on reflexivity on any ideologies that orientate our worldviews. They emphasize the importance of making school science meaningful for the students and claim that critical perspectives are still too often neglected in school science.

Hart (2012), who writes about what he calls a post-critical pedagogy for science education, focuses on the need to change the discourses in science teaching, rather than changing the students. He claims that traditional science education is based on a rationalist-objectivist foundation and that "serious consideration of how people learn implies changes […] to one that engages a range of personal sociocultural and political issues within a frame of multiple ways of knowing" (p. 104). In a way the tension can be understood as a conflict between modernism (including scientism) and postmodernism in science education (Blades 2008). It also mirrors a tension between views in traditional science education versus common views in the area of contemporary environmental education (Dillon 2014). The latter focusses much more on interactive relational production of knowledge. Similarly, Colucci-Gray and Camino (2014) write about 'science of relationships' and 'epistemic and reflexive knowledge'. On the other hand, contemporary science education (Bencze and Carter 2011), and actually also the field Education for Sustainable Development (ESD) (Jickling and Wals 2008), is sometimes framed in a neoliberal ideology. The discourse of ESD is partially focused on ecological modernisation (Sjöström et al. 2016). It is based on "assumptions about progress, a human-centered world, and individualism" (Bowers 2002, p. 28) and results in overvaluing the chances of technology compared to ethical and cultural values and politics (Bader and Laberge 2014).

Education for Sustainability (EfS) is a more critical alternative to a narrow focussed ESD (Simonneaux and Simonneaux 2012; Birdsall 2013; Thomas 2009). According to Albe (2013) it requires the individual to take the political dimension of any environmental issue and their intrinsic power relationships into consideration. The aim is to empower the individual for acting responsibly in terms of sustainability, which was also identified by Stuckey et al. (2013) as an essential justification in their model of relevant science education. Other related and critically oriented alternatives are called, e.g., ecojustice education (Bowers 2002; Mueller 2009), ecocritical pedagogy (Garrard 2010), and activist environmental education (Burns and Norris 2012). All these call for a much higher degree of transformation than it is normally the case in many ESD examples (Burmeister et al. 2012). In an abstract for a keynote speech at the 8th World Environmental Education Congress in Gothenburg, Sweden in the summer 2015 professor Arjen Wals wrote:" Perhaps a key lesson from the UN DESD [the United Nations World Decade of Education for Sustainable Development] that ended in 2014 is that we have come to realise that sustainability as such is not a destiny or a way of behaving that can be transferred or trained but rather represents our capacity for critical thinking, reflexivity and transformation."

To increase sustainability perspectives in science teaching, Littledyke (2008) argues for integrating cognitive and affective domains. For example, it was suggested to include politicisation of science education to address socio-scientific and environmental issues. In an illustrative figure he describes the difference between modern/traditional science and postmodern science, and also its consequences for pedagogy. According to him modern/traditional science is characterised by a stereotypical separation between cognitive and affective domains and it can be described with labels such as objectivism, reductionism-mechanistic and value-free. The corresponding pedagogy is described by him with labels such as transmission, non-contextual and facts-based. Instead he suggests constructive postmodern science that is characterised by integration between cognitive and affective domains, critically informed views of issues, systems thinking and uncertainty. The corresponding pedagogy Littledyke describes with labels such as active learning, interdisciplinary approach and real-life contexts. In line with this, Colucci-Gray et al. (2013) suggest that involvement of the learners is needed at a personal and emotional level to allow for finding ethical positions.

From a postmodern perspective on risks, the society cannot leave it to the experts to deal with them. According to Christensen (2009) postmodern risk-oriented science education has two challenges: (1) to work more with knowledge uncertainty, and (2) to work with both sides of science – the good and the bad, i.e. science as Janus-faced. Examples of teaching models, which takes these challenges in consideration, are the so called STEPWISE framework for activist science and technology education by Bencze and Carter (2011), a model of socio-scientific sustainability reasoning (S^3R) by Morin et al. (2014), and a framework for socio-critical and problem-oriented science teaching by Marks and Eilks (2009; see also e.g., Marks et al. 2014). In the latter Eilks and co-workers have conceptualised principles of socio-critical science teaching and corresponding evidence-based lesson plans. These start with current, authentic and controversial problems being debated in public, e.g., debates about alternative fuels, climate change, diets, or risk chemicals in consumer products (Eilks et al. 2013). All the lessons include learning of scientific content knowledge and experiments. However, by mimicking authentic non-scientific practices of information handling in society, all the lesson plans focus an understanding how science is used (and sometimes misused) by scientists and non-scientists in society. Examples included mimicking the work of e.g., politicians, representatives of pressure groups, journalists, consumer testers, or advertising experts. This approach was recently connected also to a further educational justification for critical science education. The suggested framework (see Fig. 4.1) is based on the socio-philosophical works of the Jewish-Polish philosopher Ludwik Fleck (1896–1961) (Stuckey et al. 2015).

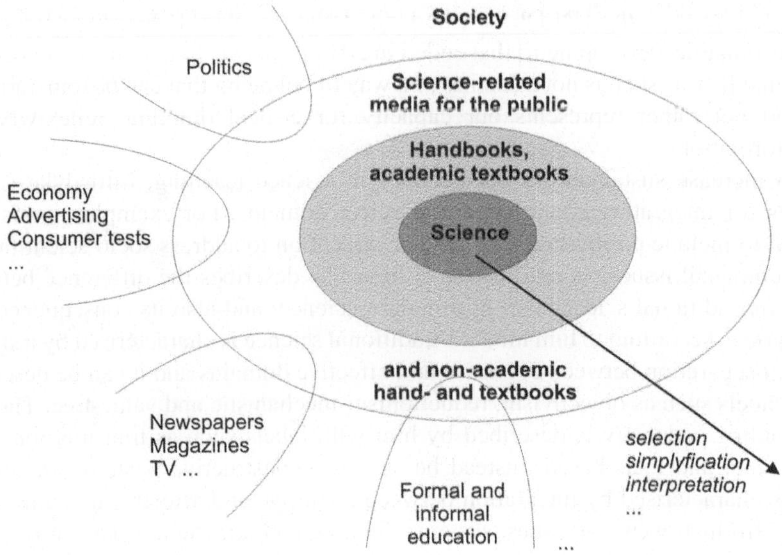

Fig. 4.1 A model for critically reflecting the science-to-society link (by Stuckey et al. 2015, based on Fleck 1935, and Bauer 2009)

The educational model based on Fleck justify reflective and critical learning about how information from the core domain of science is transferred into society, e.g., its presentation in the news media or its use in political debate. According to Stuckey et al. (2015) it is essential for understanding the often indirect and limited role science knowledge plays in societal decision making (see also Marks et al. 2014). The educational model, based on Fleck (1935) and Bauer (2009), illustrates how the core of real science endeavor is encircled by different media domains. It starts with journal and handbook science via scientific information for public understanding towards non-scientific practices of information use in society. With any further step away from the core of science, scientific facts or theories are purposely selected and presented; information is left out, intentionally or unintentionally biased, or used in suggestive ways. The model suggests that it is not only the understanding of science that allows for critically dealing with science-related media in everyday life. It is also necessary to understand the mechanism how science is transferred into and used within society, and at the same time selected, simplified and interpreted. It also suggests understanding and reflection about the skills and potential interests of all the persons involved in the information transfer processes.

4.6 Better Understanding the Different Visions of Scientific Literacy

In the beginning of this chapter we introduced a *Vision III* of scientific literacy (SL) that complements Roberts' (2007, 2011) Vision I, which focuses on scientific content and scientific processes for later application, and Vision II, which aims at understanding the usefulness of scientific knowledge in life and society. Vision III is about scientific "knowing-in-action" (Aikenhead 2007) and has also been called "critical scientific literacy" (Hodson 2009, 2011). It implies a politicised science education aiming at dialogical emancipation and socio-ecojustice, and emphasizes transdisciplinarity, philosophical values and praxis-oriented global citizenship.

To get a better understanding of this praxis-oriented vision of SL we have discussed the term 'critical' in relation to education (for example critical thinking and critical pedagogy) and also how it can be understood based on the Central/Northern European educational tradition called *Bildung*, which we above have described in detail. As we showed, it is a multifaceted tradition that has evolved over more than 200 years and we have paid most attention to the most complex version, which we call critical-reflexive *Bildung*. We have also discussed its educational implications and how it relates to other praxis-oriented educational paradigms such as Education for Sustainability and transformative learning.

The goal with this part of the chapter is to give an even better illustration of the different SL-visions, by comparing them in different ways. It is always difficult to categorize, but to describe it in a simplified way Vison I focuses on disciplinary scientific content knowledge, Vision II on usefulness of scientific knowledge in everyday life, and Vision III on critical praxis in relation to science and technology in society.

The tension between Vision I and II is already well described in the literature (e.g., Roberts 2007, 2011; Roberts and Bybee 2014; Wickman et al. 2012). Zoller (2012) makes a similar subdivision between something that can be called a 'traditional approach' versus an 'alternative approach'. Zoller's alternative approach is somewhere in between what we here call Vision II and Vision III. He recommends shifts from growth at any cost to sustainable development, from corrective responses to preventive actions, from disciplinarity to problem-solving orientation, from reductionist thinking to system thinking, and from lower-order cognitive skills to higher-order cognitive skills.

Wickman et al. (2012, p. 42) describe the rationalistic orientation and content focused character of Vision I in the following way: "we need to stay away from the non-cognitive" and "scientific reasoning are the cures for the irrational". To them, Vision I-thinking is characterized with a positivistic culture, scientific findings are often presented as objectively true or false, and values are seen as subjective. Smith and Gunstone (2009, p. 14) connect Vision I to a neoliberal ideology and write: "Science education's attempt to see educated citizens as 'mini-scientists' is both futile and self-defeating. [...] The dualistic thinking that separates the education of future scientist from that of future citizens itself draws from the dualism that sees

Table 4.1 Connections between the three visions of scientific literacy, different knowledge types/ideals, aims with scientific research and emphasis in science education

Vision	Knowledge types/ideals	Aim with scientific research (Sjöström 2013b)	Emphasis in science education
I: Pipe-line science	*Theoria/ episteme*	Development of scientific understanding (mode 1)	Epistemological
	Intellectual		
	Disciplinary rationality		
II: Science for all	*Techne*	Growth and wealth, including sustainable development (mode 2)	Everyday life and usefulness
	Pragmatic		
	Technical rationality		
III: Science for transformation	*Praxis/ phronesis*	Democracy and justice; critical sustainability (mode 3)	Ethics and transformation
	Emancipatory		
	Critical rationality		

science as separated from society". Roberts (2011) connects four (of the seven) curriculum emphasizes (solid foundation; structure of science; correct explanations; scientific skill development) to Vision I and the other three (self as explainer; everyday coping; science, technology, and decisions) to Vision II. According to Wickman et al. (2012) Vision II can be understood as learning about the various contexts in which students in their daily life are faced with problems involving science. However, there can be different complexity of contextualized science education (e.g., Sjöström and Talanquer 2014). Both Vision II and Vision III emphasize relevance, but if Vision II focuses on everyday-life relevance, Vision III focuses more on problematized relevance for critical citizenship and sustainability.

Lundqvist et al. (2013) have discussed Vision I and II based on the three types of knowledge identified by Aristotle: theoria, techne and praxis/phronesis (see also Roberts and Bybee 2014). They subdivided Vision II into two types: *Vision IIa* is based on the assumption that applying knowledge (Techne) is something different than only knowing (Vision I is, according to Lundquist et al. only focusing on Theoria, as a way of thinking and arguing). In *Vision IIb* (with similarities to what we here have called Vision III), Vision IIa is complemented with an emphasis of ethical and political values (Praxis/Phronesis).

Here we further highlight the tension between Vision II and III, with the risk of categorizing too hard. For example, the tension can be understood by help of terms such as: modernism (Vision II) and postmodernism (Vision III); neoliberalism (Vision II) and ideological awareness (Vision III); sustainable development (Vison II) and critical sustainability (Vision III); and cognition/metacognition (Vision II) and epistemic and transformative learning (Vision III). In Table 4.1 we further illustrate the differences between the three visions by connecting them to different knowledge types/ideals (for example Aristotle's three types of knowledge) and different emphasis in science education. However, regarding knowledge types we must emphasize that critical-reflexive *Bildung* (Vision III) does not only focus on phronesis, but on phronesis in addition to episteme and techne.

Table 4.1 also includes a connection to different aims with scientific research. These have been described with different modes (Sjöström 2013b), which correspond relatively well with the visons of SL (see also Wickman et al. 2012). Mode 1 emphasizes fundamental disciplinary knowledge, and Mode 2 collaboration and instrumental usefulness (Gibbons et al. 1994), or as Wickman et al. (2012, p. 41) writes: "Mode 1 is academic, scientist-initiated and discipline-based production of knowledge, whereas Mode 2 is context-driven research in the sense that it is more focused on solving specific problems, and invokes interdisciplinary knowledge as needed". Fuller (2002) has suggested a complementary Mode 3 which pays attention to what is useful for the public and the civil society. We think that this mode corresponds to Vision III of scientific literacy.

The three modes and visions are also in line with an analysis of traditional and alternative curriculum orientations in the historical development of science education curricula as suggested by Eilks et al. (2013). Traditional curricular approaches from the 1950s to the 1970s were described as mainly focusing the structure of the discipline, the history of science and the mimicking of the work of scientists. Curricula following a context-based science education paradigm emerging in the 1980s and 1990s were characterized as still focusing the learning of science concepts and processes as their main goal. However, they do so by embedding the learning of science in everyday-life, societal or technological contexts for promoting meaningfulness and applicability of the learned subject matter. The latter was put into contrast with SSI-based and EfS-driven curricula, which were suggested not only aiming on content learning via context, but from the beginning aiming at general educational skill development and transformative education via making authentic and controversial issues from everyday life and society the drivers for science education. Which approach is chosen needs to be decided by the objectives of the teaching and its target group (Stuckey et al. 2013).

Different actors in society seem to have differences in their views and interests on the different visions respective modes of science learning (Aikenhead 2006). The state and the industry seem to prefer – from somewhat different perspectives – more of Mode 2 science, whereas many academic researchers would like to go back to more of Mode 1 science (Sjöström 2013b). Mode 3 science, on the other hand, focuses on responsible research and innovations (Sjöström 2013b). The corresponding Vision III of SL focuses on developing critical citizenship.

As already mentioned above, from a simplified point of view SSI-education can be seen as typical for Vision III-driven science education. In a complex form this is true, but there are also many less complex forms of SSI-teaching, normally even more in practice than in theory. Recently, Simonneaux (2014a) discussed different curriculum orientations of SSI-education using continuums from 'cold' (mainly emphasizing, e.g., monodisciplinarity, scientific learning, and epistemic values) to 'hot' (also emphasizing transdisciplinarity, political citizenship, and philosophical values): "At the 'cold end' [...] knowledge mobilized in the classroom is single-disciplinary science. At the 'hot end', it is discussed in interdisciplinary

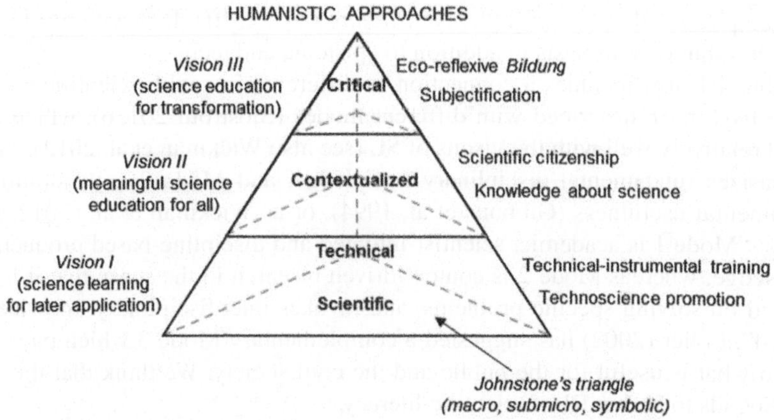

Fig. 4.2 Three levels of humanized science education

sessions in science and humanities" (Simonneaux 2014b, p. 106). In the middle of Simmonneaux's (2014a, b) model we find for example knowledge about science, critical thinking, social values, and scientific citizenship; STSE-contextualization is emphasized, but focus is on cognition and evidence-based argumentation. This is problematized at the hot end, which also contains e.g., ethical reflection.

Wickman et al. (2012, p. 17) writes: "to make meaningful actions possible, both knowledge and values are necessary […] it is crucial that not only cognitive dimensions but also values more generally are included." The continuum goes from 'techno-scientific rationality' – based on a belief that techno-scientific progress will resolve current problems – to a 'critical rationality', involving reflexivity towards the techno-sciences (Simonneaux 2014b, p. 107). Similarly, Pedretti and Nazir (2011) have discussed different orientations of STSE education – from application-oriented via socio-cultural-oriented to socio-ecojustice-oriented. Comparing this with the three visions of SL we would claim that Vision I is at the cold end, Vision II in the middle, and Vision III at the hot end of Simonneaux's continuum.

Another, and final, way to illustrate the increasing complexity from Vision I to Vision III is to use a tetrahedron model for *Bildung*-oriented chemistry education suggested in Sjöström (2013a) and Sjöström and Talanquer (2014). The top of the tetrahedron symbolises the human element and can be subdivided into three levels. These three levels are called: (1) applied chemistry, (2) socio-chemistry, and (3) critical-reflexive chemistry (Sjöström and Talanquer 2014). Figure 4.2 illustrates, based on the model, different orientations in humanized science education. The triangular bottom and level 1 corresponds to Vision I, and level 2 to Vision II. It is suggested that a politicised and eco-reflexive (Sjöström et al. 2016) science education aiming at critical-reflexive *Bildung,* subjectification and transformation, i.e., Vision III-driven science education, should be placed in the top of the tetrahedron.

4.7 Concluding Remarks

This chapter discusses the Central/Northern European educational theory of *Bildung* with respect to different visions of scientific literacy and science education. *Bildung* has a tradition of more than 200 years and forms the central socio-cultural theory of education in German speaking countries and Scandinavia (Westbury et al. 2000). Because of this history and the large influence of *Bildung* on societies in many Central/Northern European countries, *Bildung* needs to be considered as an inseparable part of culture in the corresponding countries, e.g. Germany and Sweden. Unfortunately, the unique concept of *Bildung* was largely neglected in the international discussion about goals and pedagogies in science education until quite recently (e.g., Sjöström 2013a).

Bildung is more a vision of development of a person in interaction with the surrounding society and the world, than it is a theory of the curriculum or a pedagogy. However, this vision has many implications for both fields. Since *Bildung* suggests that any kind of education should focus making the young generation capable for a self-determined life in society, for being able to participate and solve problems in it, as well as being empathic and to show solidarity, it suggests the development of certain skills (Hofstein et al. 2011). Similarly, Crippen and Antonenko (2018) in Chap. 5 of this volume discuss the need for science education to focus more on problem-solving skills by the individual in a societal context. We also agree with Avargil et al. (2018), who in Chap. 3 of this volume argue metacognitive skills are important for scientific literacy. However, we add that learning must be complemented with not only metacognition, but also with epistemic and transformative learning components (Sterling 2011).

Since *Bildung*, in the means of *Allgemeinbildung*, focuses on all learners and on all domains of personality development, science education has to contribute to corresponding educational skill development and to broaden its focus to all learners (also to those that will not embark in a later career in STEM professions). Relevant science education needs to recognize more thoroughly its societal dimension (Stuckey et al. 2013). It has to focus not only on science as an academic and industrial endeavor, but also to help understand science as a sociological construct embedded within society (Stuckey et al. 2015) and to learn about its relations to technology, culture and values as discussed from a different perspective by Waight and Abd-El-Khalick in Chap. 7 of this book. It needs to accept its responsibility for promoting critical scientific and technological literacy by promoting societal-oriented problem-solving and participation skills in the means of *Bildung*/Vision III, as outlined here.

This chapter suggests a stronger recommendation of concepts such as Education for Sustainability, transformative learning and complex SSI-based STEM education to focus on both the cognitive and the affective domains in the learner, when it comes to deal with information and issues stemming from science and technology. Many cases suggest the motivating character of SSI-based science education and in the meantime provide indication of potential for the development of *Bildung*/Vision

III-oriented skills (Sadler 2011; Marks and Eilks 2009; Marks et al. 2014). With a growing complex world, which Elmose and Roth (2005) describe as the risk society, such skills are needed to allow the younger generation to become critical-responsible citizens.

While learning about *Bildung* is an essential point in teacher education in all pre-service teacher education programs in the German-speaking and Scandinavian countries, this is not the case in the international literature, and it may even be unknown in teacher education in most countries. We suggest that teacher education also in these countries can benefit from a discussion of and reflection on *Bildung*. It might be discussed in comparison to international traditions and theories of the curriculum, education and teaching in science and technology education. Considering the basic philosophy of *Bildung* in science education might help teaching knowledge and skills in the young generation to transform our world and societies in a sustainable way. The goal of *Bildung*-oriented education is transformation of both the individuals/citizens/subjects and the society towards sustainability and development.

We conclude the chapter with the following three summarizing bullets:

- *Bildung,* in a critical understanding, is "the central critical concept of modern pedagogy" (Wimmer 2003, p. 185). *Bildung* has both educational and political dimensions. For over 200 years now, it became an essential and influential part of middle and northern European culture and educational policy. It should find better recognition and broader reception also in other countries and the international literature.
- Connecting *Bildung* with reflecting the goals of science education suggests that there should be a third vison of scientific literacy beyond the two visions described by Roberts (2007). *Bildung*-oriented STEM education needs to focus at a critical vision of scientific literacy, action competence, and critical praxis. This third vision (Vision III) of scientific literacy, inspired by a critical-reflexive understanding of *Bildung,* goes beyond contextualization of science learning. It describes a politicised vision of science education aiming at dialogical emancipation, critical global citizenship, and socio-ecojustice. This has consequences for the science curriculum that needs to incorporate more thoroughly a societal perspective and needs to incorporate stronger socio-scientific issues based science education (hot-type) and corresponding pedagogies.
- Vision III of scientific literacy asks for both reconsidering the contents and contexts of science education. Controversial, relevant and authentic socio-scientific issues, e.g., from the sustainability debate, shall become the drivers for the curriculum. Corresponding research, curriculum development, and teacher continuous professional development needs to be intensified.

References

Aikenhead, G. S. (2006). *Science education for everyday life: Evidence-based practice*. New York: Teachers College Press.

Aikenhead, G. S. (2007). Expanding the research agenda for scientific literacy. In C. Linder et al. (Eds.), *Promoting scientific literacy: Science education research in transaction*. Uppsala: Geotryckeriet.

Albe, V. (2013). On the road to science education for sustainability? *Cultural Studies of Science Education, 8*(1), 185–192.

Albe, V. (2015). Science for citizenship. In R. Gunstone (Ed.), *Encyclopedia of science education* (pp. 904–905). Dordrecht: Springer.

Avargil, S., Lavi, R., & Dori, Y. J. (2018). Students' metacognition and metacognitive strategies in science education. In Y. J. Dori, Z. Mevareach, & D. Bake (Eds.), *Cognition, metacognition and culture in STEM education* (pp. 33–64). Springer.

Bader, B., & Laberge, Y. (2014). Activism in science and environmental education: Renewing conceptions about science among students when considering socioscientific issues. In L. Bencze & S. Alsop (Eds.), *Activist science and technology education* (pp. 419–433). Dordrecht: Springer.

Bauer, W. (2003). On the relevance of *Bildung* for democracy. *Educational Philosophy and Theory, 35*(2), 212–225.

Bauer, M. W. (2009). The evolution of public understanding of science – discourse and comparative evidence. *Science and Technology in Society, 14*(2), 221–240.

Bencze, L., & Carter, L. (2011). Globalizing students acting for the common good. *Journal of Research in Science Teaching, 48*(6), 648–669.

Bencze, L., Sperling, E., & Carter, L. (2012). Students' research-informed socio-scientific activism: Re/visions for a sustainable future. *Research in Science Education, 42*(1), 129–148.

Biesta, G. (2002a). *Bildung* and modernity: The future of *Bildung* in a world of difference. *Studies in Philosophy and Education, 21*(4–5), 343–351.

Biesta, G. (2002b). How general can *Bildung* be? Reflections on the future of a modern educational ideal. *Journal of Philosophy of Education, 36*(3), 377–390.

Biesta, G. (2012). Have lifelong learning and emancipation still something to say to each other? *Studies in the Education of Adults, 44*(1), 5–20.

Birdsall, S. (2013). Reconstructing the relationship between science and education for sustainability: A proposed framework for learning. *International Journal of Environmental and Science Education, 8*(3), 451–478.

Blades, D. (2008). Positive growth: Developments in the philosophy of science education. *Curriculum Inquiry, 38*(4), 387–399.

Bohlin, H. (2008). *Bildung* and moral self-cultivation in higher education: What does it mean and how can it be achieved? *Forum on Public Policy Online*, no 2/2008, Retrieved January 10, 2015 from http://forumonpublicpolicy.com/summer08papers/archivesummer08/bohlin.pdf.

Bohlin, H. (2009). Perspective-dependence and critical thinking. *Argumentation, 23*(2), 189–203.

Bohlin, H. (2013). *Bildung* and intercultural understanding. *Intercultural Education, 24*(5), 391–400.

Bowers, C. A. (2002). Toward an eco-justice pedagogy. *Environmental Education Research, 8*, 21–34.

Burman, A. (2011). Svar på frågan: Vad är medborgerlig bildning? [Answer on the question: What is civic-*Bildung*?]. In A. Burman (Ed.), *Våga veta! Om bildningens möjligheter i massutbildningens tidevarv* (pp. 9–31). Huddinge: Södertörns högskola. (in Swedish).

Burman, A. (2014). *Pedagogikens idéhistoria: Uppfostringsidéer och bildningsideal under 2500 år [History of pedagogy: Ideas of education and ideals of Bildung during 2500 years]*. Lund: Studentlitteratur. (in Swedish).

Burmeister, M., Rauch, F., & Eilks, I. (2012). Education for sustainable development (ESD) and secondary chemistry education. *Chemistry Education Research and Practice, 13*(2), 59–68.

Burns, D. P., & Norris, S. P. (2012). Activist environmental education and moral philosophy. *Canadian Journal of Science, Mathematics and Technology Education, 12*(4), 380–393.

Carter, L., Rodriguez, C. C., & Jones, M. (2014). Transformative learning in science education: Investigating pedagogy for action. In L. Bencze & S. Alsop (Eds.), *Activist science and technology education* (pp. 531–545). Dordrecht: Springer.

Cho, S. (2010). Politics of critical pedagogy and new social movements. *Educational Philosophy and Theory, 42*(3), 310–325.

Christensen, C. (2009). Risk and school science education. *Studies in Science Education, 45*, 205–223.

Coll, R. K., & Taylor, N. (2009). Special issue on scientific literacy. *International Journal of Enviromental and Science Education, 4*, 197–349.

Colucci-Gray, L., & Camino, E. (2014). From knowledge to action? Re-embedding science learning within the planet's web. In L. Bencze & S. Alsop (Eds.), *Activist science and technology education* (pp. 149–164). Dordrecht: Springer.

Colucci-Gray, L., Perazzone, A., Dodman, M., & Camino, E. (2013). Science education for sustainability, epistemological reflections and educational practices: From natural sciences to trans-disciplinarity. *Cultural Studies of Science Education, 8*(1), 127–183.

Cranton, P. (2011). A transformative perspective on the scholarship of teaching and learning. *Higher Education Research & Development, 30*(1), 75–86.

Crippen, K., & Antonenko, P. D. (2018). Designing for collaborative problem solving in STEM. In Y. J. Dori, Z. Mevarech, & D. Baker (Eds.), *Cognition, metacognition, and culture in stem education* (pp. 89–116). Springer.

Dillon, J. (2014). Environmental education. In N. G. Lederman & S. K. Abell (Eds.), *Handbook of research on science education* (2nd ed., pp. 497–514). New York: Routledge.

Duit, R. (2015). Didaktik. In R. Gunstone (Ed.), *Encyclopedia of science education* (pp. 325–327). The Netherlands: Springer.

Eilks, I., Rauch, F., Ralle, B., & Hofstein, A. (2013). How to allocate the chemistry curriculum between science and society. In I. Eilks & A. Hofstein (Eds.), *Teaching chemistry – a studybook* (pp. 1–36). Rotterdam: Sense.

Elmose, S., & Roth, W.-M. (2005). *Allgemeinbildung*: readiness for living in risk society. *Journal of Curriculum Studies, 37*(1), 11–34.

Feierabend, T., & Eilks, I. (2011). Teaching the societal dimension of chemistry using a socio-critical, problem-oriented lesson plan based on bioethanol usage. *Journal of Chemical Education, 88*(9), 1250–1256.

Fischler, H. (2011). Didaktik – an appropriate framework for the professional work of science teachers? In D. Corrigan, J. Dillon, & R. Gunstone (Eds.), *The professional knowledge base of science teaching* (pp. 31–50). Dordrecht: Springer.

Fischler, H. (2015). Bildung. In R. Gunstone (Ed.), *Encyclopedia of science education* (pp. 118–122). Dordrecht: Springer.

Fleck, L. (1935; in English 1979). Entstehung und Entwicklung einer wissenschaftlichen Tatsache [The genesis and development of a scientific fact]. Frankfurt, Germany: Suhrkamp (in German).

Fuller, S. (2002). *Knowledge management foundations*. Boston: Butterworth Heinemann.

Garrard, G. (2010). Problems and prospects in ecocritical pedagogy. *Environmental Education Research, 16*(2), 233–245.

Gibbons, M., Limoges, C., Nowotny, H., Schwartzman, S., Scott, P., & Trow, M. (1994). *The new production of knowledge – the dynamics of science and research in contemporary societies*. London: SAGE.

Gräber, W., & Bolte, C. (Eds.). (1997). *Scientific literacy*. Kiel: IPN.

Gur–ze'ev, I. (2002). *Bildung* and critical theory facing postmodern education. *Journal of Philosophy of Education, 36*(3), 391–408.

Gustavsson, B. (2012). Bildningens traditioner i transformation [The traditions of *Bildung* in transformation]. In A. Burman & P. Sundgren (Eds.), *Svenska bildningstraditioner* (pp. 309–327). Göteborg: Daidalos. (in Swedish).

Gustavsson, B. (2014a). Bildung and the road from a classical into a global and postcolonial concept. *Confero: Essays on Education, Philosophy and Politics, 2*(1), 109–131.

Gustavsson, B. (2014b). Bildning och kritiskt tänkande i teori och praktik [*Bildung* and critical thinking in theory and practice]. In A. Burman (Ed.), *Att växa som människa – Om bildningens traditioner och praktiker* (pp. 183–206). Huddinge: Södertörns högskola. (in Swedish).

Hansen, K.-H. (2008). Rewriting *Bildung* for postmodernity: Books on educational philosophy, classroom practice, and reflective teaching. *Curriculum Inquiry, 38*, 93–115.

Hart, P. (2012). Creating spaces for rethinking school science: Perspectives from subjective and social-relational ways of knowing. In A. Zeyer & R. Kyburz-Graber (Eds.), *Science|environment|health: Towards a renewed pedagogy for science education* (pp. 103–125). Dordrecht: Springer.

Hasslöf, H., & Malmberg, C. (2015). Critical thinking as room for subjectification in education for sustainable development. *Environmental Education Research, 21*(2), 239–255.

Hodson, D. (2003). Time for action: Science education for an alternative future. *International Journal of Science Education, 25*(6), 645–670.

Hodson, D. (2009). *Teaching and learning about science: Language, theories, methods, history, traditions and values*. Rotterdam: Sense.

Hodson, D. (2011). *Looking to the future: Building a curriculum for social activism*. Rotterdam: Sense.

Hofstein, A., Eilks, I., & Bybee, R. (2011). Societal issues and their importance for contemporary science education: A pedagogical justification and the state of the art in Israel, Germany and the USA. *International Journal of Science and Mathematics Education, 9*(6), 1459–1483.

Hopmann, S. (2007). Restrained teaching: The common core of Didaktik. *European Educational Research Journal, 6*(2), 109–124.

Houwer, R. (2014). Hopeful practices: Activating and enacting the pedagogical and political potential in crises. In L. Bencze & S. Alsop (Eds.), *Activist science and technology education* (pp. 113–125). Dordrecht: Springer.

Jickling, B., & Wals, A. E. J. (2008). Globalization and environmental education: Looking beyond sustainable development. *Journal of Curriculum Studies, 40*(1), 1–21.

Johnson, L., & Morris, P. (2010). Towards a framework for critical citizenship education. *The Curriculum Journal, 21*(1), 77–96.

Kansanen, P. (2009). Subject-matter didactics as a central knowledge base for teachers, or should it be called pedagogical content knowledge? *Pedagogy Culture & Society, 17*(1), 29–39.

Kemp, P. (2005; in English 2010). Världsmedborgaren: Politisk och pedagogisk filosofi för det 21 århundradet [Citizen of the world: Cosmopolitan ideals for the 21st century]. Göteborg, Sweden: Daidalos (in Swedish).

Kivelä, A., Siljander, P., & Sutinen, A. (2012). Between *Bildung* and growth: Connections and controversies. In P. Siljander, A. Kivelä, & A. Sutinen (Eds.), *Theories of Bildung and growth* (pp. 303–312). Rotterdam: Sense.

Klafki, W. (2000a). The significance of classical theories of Bildung for a contemporary concept of Allgemeinbildung. In I. Westbury, S. Hopmann, & K. Riquarts (Eds.), *Teaching as a reflective practice: The German Didaktik tradition* (pp. 85–108). Mahwah: Lawrence Erlbaum.

Klafki, W. (2000b). Didaktik analysis as the core for preparation of instruction. In I. Westbury, S. Hopmann, & K. Riquarts (Eds.), *Teaching as a reflective practice: The German Didaktik tradition* (pp. 139–160). Mahwah: Lawrence Erlbaum.

Laugksch, R. C. (2000). Scientific literacy: A conceptual overview. *Science Education, 84*(1), 71–94.

Littledyke, M. (2008). Science education for environmental awareness: Approaches to integrating cognitive and affective domains. *Environmental Education Research, 14*(1), 1–17.

Liu, X. (2013). Expanding notions of scientific literacy: A reconceptualization of aims of science education in the knowledge society. In N. Mansour & R. Wegerif (Eds.), *Science education for diversity – Theory and practice* (pp. 23–39). Dordrecht: Springer.

Løvlie, L., & Standish, P. (2002). Introduction: *Bildung* and the idea of a liberal education. *Journal of Philosophy of Education, 36*(3), 317–340.
Løvlie, L., Mortensen, K. P., & Nordenbo, S. E. (eds.) (2003). *Educating humanity – Bildung in postmodernity*. Malden, MA: Blackwell Publishing (first published in 2002 as volume 36, issue 3 of the *Journal of Philosophy of Education*).
Lundqvist, E., Säljö, R., & Östman, L. (Eds.). (2013). *Scientific literacy – teori och praktik [scientific literacy – theory and practice]*. Malmö: Gleerups. (in Swedish).
Marks, R., & Eilks, I. (2009). Promoting scientific literacy using a socio-critical and problem-oriented approach to chemistry teaching: Concept, examples, experiences. *International Journal of Environmental and Science Education, 4*(3), 231–245.
Marks, R., Stuckey, M., Belova, N., & Eilks, I. (2014). The societal dimension in German science education – from tradition towards selected cases and recent developments. *Eurasia Journal of Mathematics, Science & Technology Education, 10*, 285–296.
Mezirow, J. (1997). Transformative learning: Theory to practice. *New Directions for Adult and Continuing Education, 74*, 5–12.
Mogensen, F., & Schnack, K. (2010). The action competence approach and the 'new' discourses of education for sustainable development, competence and quality criteria. *Environmental Education Research, 16*(1), 59–74.
Morin, O., Simonneaux, L., Simonneaux, J., Tytler, R., & Barraza, L. (2014). Developing and using an S3R model to analyze reasoning in web-based cross-national exchanges on sustainability. *Science Education, 98*(3), 517–542.
Mueller, M. P. (2009). Educational reflections on the "ecological crisis": Ecojustice, environmentalism, and sustainability. *Science & Education, 18*(8), 1031–1056.
Pedretti, E., & Nazir, J. (2011). Currents in STSE education: Mapping a complex field, 40 years on. *Science Education, 95*(4), 601–626.
Reis, P. (2014). Promoting students' collective socio-scientific activism: Teachers' perspectives. In L. Bencze & S. Alsop (Eds.), *Activist science and technology education* (pp. 547–574). Dordrecht: Springer.
Roberts, D. A. (2007). Scientific literacy/science literacy. In S. K. Abell & N. G. Lederman (Eds.), *Handbook of research on science education* (pp. 729–780). Mahwah: Lawrence Erlbaum.
Roberts, D. A. (2011). Competing visions of scientific literacy: The influence of a science curriculum policy image. In C. Linder, L. Östman, D. A. Roberts, P.-O. Wickman, G. Erickson, & A. MacKinnon (Eds.), *Exploring the landscape of scientific literacy* (pp. 11–27). London: Routledge.
Roberts, D. A., & Bybee, R. W. (2014). Scientific literacy, science literacy, and science education. In N. G. Lederman & S. K. Abell (Eds.), *Handbook of research on science education* (2nd ed., pp. 545–558). New York: Routledge.
Sadler, T. D. (2011). *Socio-scientific issues in the classroom*. Dordrecht: Springer.
Sadler, T. D., & Zeidler, D. L. (2009). Scientific literacy, PISA, and socioscientific discourse: Assessment for progressive aims of science education. *Journal of Research in Science Teaching, 46*(8), 909–921.
Santos, W. L. P. (2009). Scientific literacy: A Freirean perspective as a radical view of humanistic science education. *Science Education, 93*(2), 361–382.
Schaffar, B., & Uljens, M. (2015). Paradoxical tensions between Bildung and Ausbildung in academia: Moving within or beyond the modern continental tradition. In E. Westergaard & J. S. Wiewiura (Eds.), *On the facilitation of the academy* (pp. 1–15). Rotterdam: Sense.
Schneider, K. (2012). The subject-object transformations and 'Bildung'. *Educational Philosophy and Theory, 44*(3), 302–311.
Selby, D. E. (2014). Education for sustainable contraction as appropriate response to global heating. In L. Bencze & S. Alsop (Eds.), *Activist science and technology education* (pp. 165–182). Dordrecht: Springer.
Shor, I. (1992). *Empowering education: Critical teaching for social change*. Chicago: University of Chicago Press.

Simonneaux, L. (2014a). *Questions socialement vives* and socio-scientific issues: New trends of research to meet the training needs of postmodern society. In C. Bruguère, A. Tiberghien, & P. Clement (Eds.), *Topics and trends in current science education* (pp. 37–54). Dordrecht: Springer.

Simonneaux, L. (2014b). From promoting the techno-sciences to activism – a variety of objectives involved in the teaching of SSIs. In L. Bencze & S. Alsop (Eds.), *Activist science and technology education* (pp. 99–111). Dordrecht: Springer.

Simonneaux, J., & Simonneaux, L. (2012). Educational configurations for teaching environmental socioscientific issues within the perspective of sustainability. *Research in Science Education, 42*(1), 75–94.

Sjöström, J. (2013a). Towards *Bildung*-oriented chemistry education. *Science & Education, 22*(7), 1873–1890.

Sjöström, J. (2013b). Eco-driven chemical research in the boundary between academia and industry – PhD students' views on science and society. *Science & Education, 22*(10), 2427–2441.

Sjöström, J., & Stenborg, E. (2014). Teaching and learning for critical scientific literacy: Communicating knowledge uncertainties, actors interplay and various discourses about chemicals. In I. Eilks, S. Markic, & B. Ralle (Eds.), *Science education research and education for sustainable development* (pp. 37–48). Aachen: Shaker.

Sjöström, J., & Talanquer, V. (2014). Humanizing chemistry education: From simple contextualization to multifaceted problematization. *Journal of Chemical Education, 91*(8), 1125–1131.

Sjöström, J., Rauch, F., & Eilks, I. (2015). Chemistry education for sustainability. In I. Eilks & A. Hofstein (Eds.), *Relevant chemistry education – from theory to practice* (pp. 163–184). Rotterdam: Sense.

Sjöström, J., Eilks, I., & Zuin, V. (2016). Towards eco-reflexive science education – A critical reflection about educational implications of green chemistry. *Science & Education, 25*(3–4), 321–341.

Smith, D. V., & Gunstone, R. F. (2009). Science curriculum in the market liberal society of the twenty-first century: 're-visioning' the idea of science for all. *Research in Science Education, 39*(1), 1–16.

Sterling, S. (2011). Transformative learning and sustainability: Sketching the conceptual ground. *Learning and Teaching in Higher Education, 5*(11), 17–33.

Straume, I. S. (2015). The subject and the world: Educational challenges. *Educational Philosophy and Theory, 47*(13–14), 1465–1476.

Stuckey, M., Heering, P., Mamlok-Naaman, R., Hofstein, A., & Eilks, I. (2015). The philosophical works of Ludwik Fleck and their potential meaning for teaching and learning science. *Science & Education, 24*(3), 281–298.

Stuckey, M., Hofstein, A., Mamlok-Naaman, R., & Eilks, I. (2013). The meaning of ‚relevance' in science education and its implications for the science curriculum. *Studies in Science Education, 49*, 1–34.

Thomas, I. (2009). Critical thinking, transformative learning, sustainable education, and problem-based learning in universities. *Journal of Transformative Education, 7*(3), 245–264.

Väkevä, L. (2012). Experiencing growth as a natural phenomenon: John Dewey's philosophy and the Bildung tradition. In P. Siljander, A. Kivelä, & A. Sutinen (Eds.), *Theories of Bildung and growth* (pp. 261–279). Rotterdam: Sense.

Vieira, R. M., Tenreiro-Vieira, C., & Martins, I. P. (2011). Critical thinking: Conceptual clarification and its importance in science education. *Science Education International, 22*, 43–54.

Von Humboldt, W. (2000). Theory of Bildung. In I. Westbury, S. Hopmann, & K. Riquarts (Eds.), *Teaching as a reflective practice: The German Didaktik tradition* (pp. 57–62). Mahwah: Lawrence Erlbaum.

Wals, A. (2015). Transformative learning in vital coalitions for socio-ecological sustainability. Retrieved January 10, 2015, from http://weec2015.org/programme-2/keynote-speeches/.

Walters, K. S. (Ed.). (1994). *Re-thinking reason. New perspectives in critical thinking*. New York: State University of New York Press.

Westbury, I., Hopmann, S., & Riquarts, K. (Eds.). (2000). *Teaching as a reflective practice: The German Didaktik tradition*. Mahwah: Lawrence Erlbaum.

Wickman, P. O., Liberg, C., & Östman, L. (2012). Transcending science: Scientific literacy and *Bildung* for the 21st century. In D. Jorde & J. Dillon (Eds.), *Science education research and practice in Europe* (pp. 39–61). Rotterdam: Sense.

Wimmer, M. (2003). Ruins of *Bildung* in a knowledge society: Commenting on the debate about future of *Bildung*. *Educational Philosophy and Theory, 35*(2), 167–187.

Yore, L. D. (2012). Science literacy for all: More than a slogan, logo, or rally flag! In K. C. D. Tan & M. Kim (Eds.), *Issues and challenges in science education research* (pp. 5–23). Dordrecht: Springer.

Zeidler, D. (2015). Socioscientific issues. In R. Gunstone (Ed.), *Encyclopedia of science education* (pp. 998–1003). Dordrecht: Springer.

Zembylas, M. (2006). Science education as emancipatory: The case of Roy Bhaskar's philosophy of meta-reality. *Educational Philosophy and Theory, 38*(5), 665–676.

Zoller, U. (2012). Science education for global sustainability: What is necessary for teaching, learning, and assessment strategies? *Journal of Chemical Education, 89*, 297–300.

Chapter 5
Designing for Collaborative Problem Solving in STEM Cyberlearning

Kent J. Crippen and Pavlo D. Antonenko

5.1 Introduction

Whether we consider the implications of climate change, the global loss of biodiversity or the challenge of universal healthcare, the problems of the twenty-first century are increasingly complex and interdisciplinary. The nature of expertise that is required for tackling such problems is equally as complex and is further complicated by the exponential increase in scientific knowledge over the past decades. The focus on cyberinfrastructure as a systemic platform for communication, data sharing, data analysis, and rapid publication implies that the tools of science are also changing rapidly (Alberts 2011; Campbell 2008). Thus, the traditional ways of presenting content devoid of any meaningful context, disregarding the authentic practices of being a scientist or engineer, and requiring copious practice solving textbook problems are not sufficient for educating students that represent the next generation of professionals. In addition, these forms of learning turn away many students, particularly those from diverse backgrounds and send a strong message about the exclusivity of science, technology, engineering and mathematics (STEM). Building a diverse and capable workforce requires an approach that situates learning in the authentic, collaborative and problem-oriented work of current practice in STEM, thus allowing students to develop the requisite identity, efficacy and habits of mind as well as the domain knowledge and skills for success.

To address this situation, recent calls for reform of undergraduate STEM focuses on student-centered learning and research-based instructional strategies (National Academy of Engineering [NAE] 2005; National Research Council [NRC] 2011), including the widespread adoption of approaches such as case studies, problem-based

K.J. Crippen (✉) • P.D. Antonenko
School of Teaching and Learning, College of Education, University of Florida, Gainesville, FL, USA
e-mail: kcrippen@coe.ufl.edu

Y.J. Dori et al. (eds.), *Cognition, Metacognition, and Culture in STEM Education*, Innovations in Science Education and Technology 24,
https://doi.org/10.1007/978-3-319-66659-4_5

learning, peer instruction, and computer simulations as a means for improving the retention and recruitment of undergraduate students in the disciplines (President's Council of Advisors on Science and Technology [PCAST] 2012). Consistent with these calls for reform is the perspective of professionals, such as the engineers who cite problem solving and communication, ethics, lifelong learning, experiments, teams, engineering tools, and design as the most important competencies for professional practice (Passow 2007).

The continued evolution of our understanding of learning as well as our capacity for developing learning technologies provides a potential for using these resources to address the need for reform. Cyberlearning, or the use of networked learning technologies (National Science Foundation [NSF] 2008), can transform individual and collective cognition by providing affordances for learners to engage in such important cognitive and metacognitive processes as problem definition, information discrimination, reasoning and argumentation, solution negotiation, and evaluation. In addition, cyberlearning is recognized as a critical twenty-first century skill and core practice of STEM (NRC 2011). However, achieving these goals implies that cyberlearning is deployed in such a way that it serves the creation of adaptive expertise and involves the appropriate use of scaffolding—tools that help mediate and extend student capabilities (Belland and Drake 2013).

These circumstances suggest that design plays a potentially paramount role in the success of cyberlearning ventures for STEM. Thus, design frameworks are needed that fulfill the following criteria: (a) work within the constraints of a typical undergraduate course, (b) include authentic learning activities—those that more accurately portray the nature of work, knowledge and knowing within each discipline, (c) support the use of collaborative problem solving and the development of adaptive expertise, (d) match the affordances of networked computing technologies with the scaffolding needs of the learning activity, and (e) account for the practical needs of instructors without adding significantly to their workload. Indeed, this is not a small or straightforward task.

The purpose of this chapter is to present such a design framework, one that is grounded in existing theoretical frameworks and empirical research and to use our past and current research projects to illustrate the basis and utility of these ideas. This framework focuses on a learning environment for undergraduate STEM that involves cyberlearning and collaborative problem solving. We begin by describing how a situated perspective on learning grounds our design in a contemporary theory of learning. We then review the existing empirical literature that defines problem solving as a situated STEM practice and the role of authenticity and collaboration in learning via problem solving. Next, we overview the role of cyberlearning for supporting collaborative problem solving in STEM and finally, using an example learning environment from our work, we present a design framework for collaborative problem solving in STEM.

5.2 Learning as Authentic Practice

Our view of collaborative problem solving and cyberlearning for undergraduate STEM is rooted in a situated perspective on learning, consisting of three essential features: (1) a social and material context; (2) activities and interactions; and (3) participation and identity (Johri and Olds 2011). This perspective (a) recognizes context and apprenticeship as key components for explaining how a given environment affords opportunities for meaningful learning (Brown et al. 1989; Gee 2008), (b) emphasizes social processes and authentic participation over transmission and receipt of knowledge (Lave and Wenger 1991) and (c) postulates that learning occurs through problem solving, imitation, and engagement in authentic activities. For example, becoming a scientist involves much more than simply acquiring content knowledge and technical skill. It includes learning the instruments, strategies and heuristics of a professional, including how to communicate in a range of technical forms to a variety of audiences, how to work with others in teams, and how to mentor and apprentice less experienced others. Learning STEM can thus be viewed as the process of becoming a recognized participant in these practices, including the social dimensions that support and define them. With such activities, learning becomes more than an accumulation of knowledge; it is a transformation from novice to recognized expert and is accomplished through cognitive apprenticeship (Johri and Olds, 2011).

Designing a learning environment for cognitive apprenticeship implies that activities are analyzed, planned, carried out, evaluated and adjusted, so that the essence of the apprenticeship (observation, coaching and practice in a social and functional context) is consistent and supported over time. The cognitive aspect of the term implies that the thought processes as well as the skills to solve complex problems are made explicit so that they may be studied and acquired by students in an interactive and systematic fashion. In addition to declarative and procedural knowledge, these thought processes include heuristics, mental models and habits of mind. By including an expert's thought processes as explicit elements of the learning activities, an environment can be designed to support successful transfer of problem solving to unique situations and problems (Jonassen 2007).

The practice of problem solving—the complex set of cognitive and metacognitive skills that help learners understand and solve problems of various levels of difficulty—is shared across STEM, but is uniquely and often implicitly different within each discipline (Adelson 1981; Chi et al. 1981; Moss et al. 2006; Silver 1979). The different problem solving practices among the STEM disciplines originate in the epistemology of each domain. For example, problem solving in mathematics is similar to problem solving in engineering, but the epistemic practice of each discipline creates subtle differences in the process. The rational nature of mathematical knowledge assumes a single, absolute solution for a problem. The applied nature and design emphasis of engineering allows for a range of acceptable solutions that are defined in relation to contextual parameters; thus, multiple rationalized solutions for any given problem.

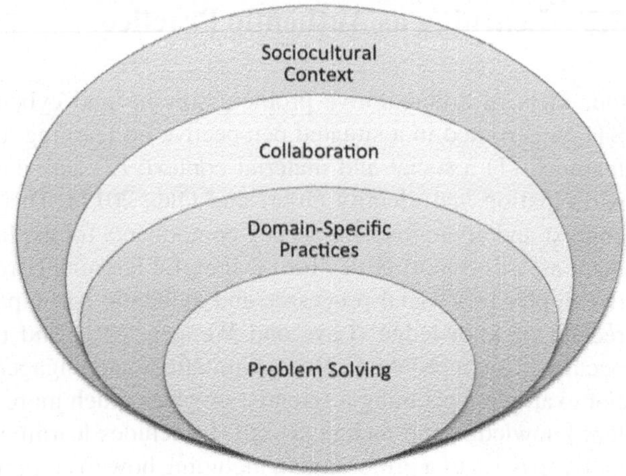

Fig. 5.1 The situated nature of problem solving in STEM

5.2.1 Problem Solving as Situated STEM Practice

We view problem solving in STEM as situated at the core of three distinct layers within an inter-related activity system (Fig. 5.1). These layers represent dimensions of authenticity that describe the situatedness of the problem and must be addressed in the design of a learning environment. The learning environment then serves the role of what Barab and Duffy (2012) describe as a practice field—"circumscribed activities or experiences for the learner" (p. 30) The goal of such a learning environment would be to provide a participatory experience that closely exemplifies that which exists in the world outside of formal school, in the context of society and professional work.

The core component of the system (i.e., model) is a general heuristic—the process of problem solving that transcends all domains. However, when a problem is situated within STEM, it interfaces with domain-specific practices such as inquiry, design and modeling. These practices emanate from and are related to the epistemic nature of each discipline (e.g., inquiry from science, design from engineering, or modeling from mathematics). While these practices and their relationship to the general attributes of problem solving are not mutually exclusive to a single domain, they do tend to associate in predictable ways. For example, modeling as a domain-specific practice in physics, which often involves the production of a mathematical equation from measurements of a physical system or the domain-specific practice of design in engineering, which involves specifying a set of attributes for a product under a set of specified conditions in order to meet the needs of a client. Each of these practices are enacted through collaboration or shared work among problem solvers, that involves specialized forms of discourse that reinforce the domain-specific practices as well as the problem solving process. This shared work emerges from and relates to the sociocultural context of the problem and its solvers. For problem solvers, this includes the everyday happenings and issues, ways of knowing, beliefs, values, and shared life experiences. This multidimensional perspective acknowledges and appreciates the potentially different cultural realities of people as problem solvers and allows for and leverages their context-rich ways of knowing and learning.

5.2.2 Problem Solving as Learning

Problem solving is at the center of many contemporary conceptualizations of learning including problem-based learning (Barrows 1996), inquiry learning (Hmelo-Silver et al. 2007), learning as conceptual change (diSessa 2006), case-based reasoning (Kolodner 2006), open-ended learning (Land and Hannafin 1996) and goal-based scenarios (Schank et al. 1994). These constructivist and sociocultural perspectives view learning as a process of active knowledge building and development of cognitive and metacognitive skills as students are enculturated into disciplinary norms and practices. According to Vygotsky (1978), this enculturation occurs when learning experiences are designed to incorporate common belief systems and habits of mind within the discipline externalized via discourse and the use of common tools. Cyberlearning problem-solving environments comprise one category of such common tools.

Problem-based learning is the approach that most explicitly addresses learning as problem solving and provides direct guidance for designing appropriate learning environments. The problem-based learning perspective, developed by Howard Barrows at McMaster University in Canada in the 1960s, is conceptualized as a process involving six core features (Barrows 1996):

1. Learning is student-centered.
2. Learning occurs in small student groups.
3. A tutor is present as a facilitator or guide.
4. Authentic problems are presented at the beginning of the learning sequence, before any preparation or study has occurred.
5. The problems encountered are used as tools to achieve the required knowledge and the problem-solving skills necessary to eventually solve the problems.
6. New information is acquired through self-directed learning.

While each of these six features is critically important for designing engaging and effective problem solving experiences for learners, this chapter will focus primarily on the aspects of problem authenticity, the notion of scaffolding and supporting collaboration in technology-based problem solving.

5.2.3 The Significance of Authenticity in Problem Solving in STEM

Authenticity is an overarching and critical point of emphasis in current conceptions of effective problem solving in STEM. Authentic problem solving involves "having students carry out tasks and solve problems in an environment that reflects the nature of such tasks in the world" (Collins 2006, p. 52). The use of authentic experiences for learning is based upon: (a) improving the transfer of learning by situating knowledge in the appropriate long-term context (i.e., real-world problems), (b) increasing student motivation for learning by including a meaningful context and building identity, and (c) improving the learning of content material by making the structural elements explicit (Edelson and Reiser 2006).

Knowledge and skills are learning outcomes that are tied to the situation and context in which they are experienced and learned (Sadler 2009). For example, learning that occurs in a school context is often found to be non-transferrable to new situations, even within the same school subject (van Merriënboer et al. 2006). This phenomenon is recognized as the transfer paradox and the difficulty is ascribed to the situated nature of knowledge; students know things best in the context in which they were learned. If this is the case, then learning should occur in contexts that are better representative of the ultimate educational goals.

For STEM, the term 'authentic' is most often ascribed to the context and practice of working professionals; it's what scientists, technologists, engineers and mathematicians do while at work. This form of problem solving, that based upon the practice of professionals is intended to avoid the transfer paradox by situating learning in the context and ways of knowing that represent one of the main goals of schooling and to support learning through collaboration (Bresnen et al. 2003).

Authenticity can also be applied along multiple dimensions that relate to the cultural backgrounds of students. Strobel et al. (2013) argue for *personal authenticity*–"projects are close to students' own life (i.e., life-stories of their neighborhood, biodiversity in the forest nearby)" (p. 144) and *value authenticity*–"personal questions get answered or projects satisfy personal or community needs" (p. 144). Such themes underlie the sociocultural context dimension of problem solving in STEM. Consistent with the theme of *personal authenticity* is the emerging practice of culturally responsive STEM, which is an approach that "teaches to and through [students'] personal and cultural strengths, their intellectual capabilities, and their prior accomplishments" (Gay 2010, p. 26). Researchers are currently documenting and exploring the efficacy and utility of culturally responsive STEM as a means for improving the practice of teachers as well as the retention and achievement of students (Brown and Crippen 2016; George 2013).

The authenticity of the social infrastructure of a learning environment is equally as important as the cognitive aspects. *Authentic collaboration* in lieu of competition has shown to improve the retention of women and underrepresented students (Qin et al. 1995), as have supportive, technology-enhanced experiences (Bennett et al. 1999). According to Bielaczyc (2006), the social infrastructure would include the beliefs of students and instructors, the classroom practices of instructors, the organization of people and tools across space and time and the interactions among students, instructors, and the world outside of the course.

Loss of interest is known to be one of the biggest reasons for undergraduates changing majors from STEM (Seymour and Hewitt 2000). Within the competitive and often unfriendly climate of undergraduate STEM (Krapp and Prenzel 2011), learning through authentic practice is theorized to use student interest in the process to build identity as well as the necessary efficacy for persisting with challenging coursework. With sustained success and the further development of skill, the situational interest derived from authentic problem solving should translate into personal interest and efficacy (Hidi and Renninger 2006; Schraw and Lehman 2001). Cognition aside, these represent key affective variables for impacting a student's choice to continue studying STEM (i.e., to persist).

5.2.4 *Problem Types and Authenticity*

In his design theory of problem solving, Jonassen (2000, 2007, 2011) argues that all scientific problems can be described and classified based upon four dimensions: (a) structuredness, (b) complexity, (c) dynamicity, and (d) domain specificity. Some problems are more authentic in the sense that they represent the complexity of real-world problem solving, or what Jonassen refers to as "workplace problem solving" (Jonassen 2007, p. viii), while other problems are simplified to be more accessible for novice learners with low prior knowledge in the domain and emerging problem-solving skills. All four of these must be considered when determining the nature and types of problems for undergraduate STEM students.

The most challenging problems are ill-structured. This typically means that the problem itself is not defined for the learner, the rules for solving the problem are not prescribed or are implied in the problem description and that the problem may not possess one correct, convergent solution. Ill-structured problems are the types of poorly-defined challenges that most people encounter in their professional lives when they have to analyze the main issue at hand, determine and prioritize sub problems, examine evidence for and against several possible approaches for solving the problem and select the most viable approach in light of the perspectives of the multiple stakeholders influenced by the issue and, potentially, by the solution. These types of open-ended challenges are rarely integrated into formal education because they require much time that instructors often do not have. They also require extensive cognitive, metacognitive, and emotional support as most students become confused and frustrated early in the process and experience major cognitive disequilibrium (Graesser et al. 2005) that they cannot overcome without carefully designed supports. Undergraduate STEM students need to be exposed to ill-structured problems, but this requires specially designed supports and scaffolds and not all of their problems need to be of this type. The degree and duration of this form of problem solving remains an open research question.

Difficult problems are also complex in the sense of the breadth and depth of prior knowledge required of the learner and the relational complexity of the problem elements. Cognitive load theorists describe this phenomenon of relational complexity as intrinsic cognitive load (Sweller 2010), which is a function of the number of interacting information units required to understand and solve the problem. For example, one problem-solving activity might require students to compare the dietary composition of two meal plans based on recommended dietary allowance values for protein and select one meal plan that achieves these values with a minimal calorie count. Another activity may ask students to consider five meal plans inspired by cuisines from around the world and prioritize them based on the most appropriate ratio of daily calorie intake and recommended dietary values for protein, vitamin B1, and iron. The second problem is more complex because it involves a larger number of interacting information elements and higher levels of prior knowledge resulting in increased relational complexity compared to the first problem.

Another dimension of problem complexity is dynamicity (Jonassen 2000). Dynamic problems are complex because the variables and relationships between variables change over time. Dynamic problems are often encountered in authentic scientific practices because most science domains study dynamic systems, which also often involve social spaces, such as weather and climate, water resource management, pollution, spread of pandemic and epidemic diseases, or ecosystem biodiversity. For example, the culling of gray wolves that occurred in Yellowstone National Park in the 1950s as a response to farmers' concerns about their livestock population, led to a disruption of the ecological balance and caused an increase in the population of elks, which in turn caused a decline in streamside vegetation, resulting in soil erosion, changes in the fish habitat and so on (Jonassen and Hung 2008). Dynamic problems like this are typically ill-structured, whereas static problems tend to be well-structured, as in the typical textbook.

Domain-specificity is also discussed as a key feature that can be used to describe and distinguish problems (Jonassen 2011). This claim is supported by research demonstrating that different domains rely on different schemas of scientific knowledge and cognitive strategies (Mayer 1992; Sternberg and Frensch 1991). For instance, one study showed that graduate students in probabilistic sciences of psychology and medicine perform better on statistical and conditional reasoning problems than students in chemistry or law, who do not rely on such analytical methods and procedures in their domain. The differences and similarities between the ways of thinking and reasoning in physics, biology, chemistry, engineering, geoscience, and astronomy are described in a recent National Research Council report—*Discipline-Based Education Research: Understanding and Improving Learning in Undergraduate science and Engineering*. This publication argues that students in different domains develop reasoning skills through solving situated, ill-structured problems that require forms of logic that are domain-specific. The cognitive operations and strategies are developed as a result of deep engagement with pragmatic reasoning schemas rather than exercises in formal logic, which supports the importance of integrating authentic problems in undergraduate STEM curricula.

The recent literature on designing problem-solving experiences in science education argues that educators should move away from using algorithmic, 'plug-n-chug' problems, towards achieving curricular goals and incorporating more complex (cf. authentic) problems (e.g., Antonenko et al. 2011; Edelson and Reiser 2006; Hmelo-Silver 2006). The logic behind this argument is that authentic problem solving encourages students to not only develop knowledge of discrete concepts within the domain, but also to understand and practice the multifaceted cognitive and metacognitive skills involved in solving problems of the real world. From a sociocultural perspective, engagement in authentic problem solving helps enculturate learners into the disciplinary norms and practices of STEM by providing them with opportunities to use the common tools, discourse, and conventions of professionals to address issues that are contemporary and culturally relevant.

5.2.5 Cyberlearning

Advances in cyberinfrastructure provide unique opportunities for authentic learning via access to data as well as software tools that support analysis, visualization and collaboration. With the emergence of large, easily accessible data sets (i.e., Big Data) students can engage in learning experiences that mirror the practices of STEM professionals who are addressing real world phenomena. Cyberlearning can transform the scaffolding of both individual and collective cognition because it provides affordances for supporting learners as they engage in such important cognitive and metacognitive processes as problem definition, information discrimination, reasoning and argumentation, solution negotiation and evaluation. In addition to supporting learners' cognitive processes, such as elaboration or problem categorization, cyberlearning can scaffold metacognition—that is, planning, differentiating forms of knowledge and information, as well as monitoring, and reflective activities. Designers can create supports for the cognition and metacognition of individual learners' (i.e., self-regulation) as well as for teams of students (i.e., socially shared regulation). Numerous examples of such cyberlearning tools have been successfully developed for middle and high school students, including the Web-Based Inquiry Science Environment (WISE) (Linn et al. 2005), the Biology Guided Inquiry Learning Environment (BGuILE) (Reiser et al. 2001), and the Physics Education Technology Project (PhET) (Moore et al. 2014).

As a learning strategy, cyberlearning is related to the concept of online learning as well as the more recent trend towards blended learning. Through meta-analysis, both of these strategies have demonstrated a small to moderate effect size when compared to traditional forms of learning across a number of studies (Means et al. 2009). In addition, a recent review by Donnelly, Linn and Ludvigsen (Donnelly et al. 2014) involving cyberlearning in a K-12 science context identified 30 unique environments that were found to support inquiry learning with an average effect size of 0.87. The authors cite teachable agents, collaboration and dynamic visualizations as key features of successful interventions. However, other research in this context has documented issues for struggling learners (Kern et al. 2014), especially for complex multi-step problems with limited supports (Villanueva and Hand 2011). This suggests a strong need for additional research on the general concept of cyberlearning in a STEM context, especially in relation to problem solving and scaffolding of process skills and collaboration. This represents a key design challenge for learning applications of cyberinfrastructure in STEM.

5.2.6 *Scaffolding Problem-Solving Process Skills*

A problematic, yet common, assumption among educational researchers and designers is that when instructors provide authentic, problem-based experiences, students will automatically be engaged (e.g., Belland et al. 2013; Blumenfeld et al. 2006). Evidence shows that this is not always the case (Antonenko et al. 2011, 2014; Dolmans and Schmidt 2006; Hung 2011). Since the simplification of ill-structured problems is not an acceptable solution, educational designers and learning scientists employ the notion of scaffolding as a metaphor for designing learner supports. Similar to how scaffolds are used in building construction to provide both adjustable and temporal support to the building under construction, instructional scaffolds are used by learning designers to create temporary supports for students who need help with certain skills. Scaffolding is intended to bridge the gap between what students can do on their own and what they can do with the help of a more knowledgeable other (Vygotsky 1978). Effective scaffolds help learners become successful at new tasks and extend their competencies into new areas. Scaffolding students as they solve authentic problems has been shown to lead to improved learning performance (Puntambekar and Kolodner 2005), metacognitive skills (Sandi-Urena et al. 2011), and argumentation ability (Belland et al. 2011).

Although discipline-based education researchers point out the benefits of domain-specific scaffolding, most cyberlearning environments that scaffold problem solving are designed to provide pedagogical and technological affordances (Kozma 1991) that support the development of STEM problem-solving skills that are invariant across problem-solving situations and can be transferred across diverse STEM problems. Support for domain-general scaffolding of the problem-solving process can be found in *Discipline-based Education Research: Understanding and Improving Learning in Undergraduate Science and Engineering*:

> What changes from one problem to another is the specific knowledge students need to bring to bear on their solution attempts, rather than the underlying cognitive processes. This general pattern of consistent results across disparate types of problems lends support to the committee's view that findings from cognitive science research on problem solving may be applicable in undergraduate science and engineering domains in which they have not yet been investigated. After all, humans have a single cognitive system, with specific operating parameters and constraints, that underlies their learning and problem solving regardless of the problem or discipline under investigation. (Singer et al. 2012, p. 77).

Because novice problem solvers frequently do not know what process to use, or employ a non-transferable, idiosyncratic process (Hmelo-Silver 2006; Krajcik et al. 1998), scaffolding of problem solving occurs most frequently at the core of the situated nature of STEM problem solving, at the level of *problem-solving process*. To scaffold the development of problem-solving skills, the problem-solving process is broken down into steps that reflect key cognitive and metacognitive processes, and each step is explicitly scaffolded for students. Such multi step problem-solving procedures have been developed across the STEM domains, including examples from mathematics (Polya 1957), chemistry (Bunce and Heikkinen 1986), biology (Hurst and Milkent 1996), engineering (Ryan et al. 2007) and physics (Reif et al. 1976).

This pedagogy can be taken a step further when a tangible space is provided for students to follow the explicit multistep procedure. For example, Problem Sheets (ALPS) devised by van Heuvelen (1991) contain separate, identified sections where students must represent the problem graphically and develop a qualitative analysis before working on the mathematical representation. Also included are sections on evaluation of units and magnitude of answers. This approach has been transformed to a computer environment, the Hierarchical Analysis Tool (HAT), in which students are required to first choose the principles involved in the problem from a pull-down menu, then choose the associated concepts, and only then select the equations needed to solve the problem (Dufresne et al. 1992; Leonard et al. 1996). This tool has been shown to help students both categorize problem types more effectively and improve their problem-solving performance. A similar environment is provided by the Story Problem-Solving Environment (SPSE, Jonassen 2004) where a story problem is presented to students who must then follow a series of tasks, ranging from identifying which scientific principles are involved, to qualitatively analyzing the problem, to building a quantitative representation of the problem.

5.2.7 *Scaffolding Collaboration in Problem-Solving Environments*

Scaffolding of collaboration in a cyberlearning environment is generally informed by the design guidelines generated by a learning science referred to as Computer-Supported Collaborative Learning (CSCL). In what was perhaps one of the first studies of computer-supported collaborative problem solving, Roschelle and Teasley (1995) define collaboration as a "coordinated, synchronous activity that is the result of a continued attempt to construct and maintain a shared conception of a problem" (p. 70). Computer-supported collaboration then involves the use of computer technology in the construction and maintenance of a shared conception of a problem. A shared conception of the problem is achieved by participating in a joint problem space, which is a shared knowledge structure that supports collaborative problem solving by integrating: (a) goals, (b) descriptions of the current state, (c) awareness of available problem solving actions; and (d) associations that relate goals, features of the current problem state, and available actions.

Joint problem space is defined as a "negotiated and shared conceptual space, constructed through the external mediational framework of shared language, situation, and activity—not merely inside the cognitive contents of each individual's head" (Roschelle and Teasley 1995, p. 70). The important question stemming from this definition is how can STEM educators design learning environments that provide affordances for such a sharing of language, situation, and activity? Simple placement of students into groups does not guarantee effective collaborative learning. Carefully designed instructional scaffolds are needed to ensure that students are supported in joint construction of knowledge as they work on solving a problem in a cyberlearning environment.

Moderation and scaffolding of collaborative problem solving require complex facilitation skills and highly depend on the knowledge and experience of the facilitator (Weinberger et al. 2009). A popular approach to scaffolding collaborative problem solving that circumvents this limitation is scripting (Fischer et al. 2007; Weinberger et al. 2005). Scripts can be defined as activity structures that recommend and sequence epistemic and social activities during a collaborative problem-solving task, supporting learners at both the cognitive and metacognitive level. In cognitive psychology scripts are often defined as a type of procedural schema that, when internalized, guides individuals' cognitive processing and behavior during particular sets of dynamic events such as typical interactions at a restaurant or at a library (e.g., Schank and Abelson 1977).

Epistemic scripts are perhaps the most common mechanism for scaffolding collaborative problem solving. Epistemic scripts focus on supporting collaborative learners by specifying the aspects of the *task*, or procedure, learners ought to take to solve the problem together. Many of the process frameworks discussed in the previous section such as the ALPS framework (Van Heuvelen 1991) can be defined as epistemic scripts, although many of them were not designed for collaborative problem-solving environments.

Unlike epistemic scripts, social scripts structure learners' *interactions* with one another in a learning situation. Social scripts typically accomplish this by specifying the roles that team members should assume when discussing the problem and possible solution strategies. As Palincsar and Herrenkohl (1999) noted, "... to deeply engage students with the cognitive content and with other participants in the classroom, they need to be given roles with concomitant rights and responsibilities" (p. 169). Examples of roles that have been used in social scripts include case analyst and critic (Weinberger et al. 2005), recaller and listener (O'Donnell 1999), and questioner and explainer (King 1997).

Scripts are also known to produce negative effects. Overscripting is the effect that is observed when scripts specify too much for the learners, to the point where they "micro-manage" the collaborative activities of learners (Weinberger et al. 2009) and suppress the natural patterns of interaction that learners have developed over years (Dillenbourg 2002). Overscripting is a controversial issue—how much scripting is too much? (e.g., Stegmann et al. 2011); however, educators generally agree that a carefully designed *external* collaboration script will build on *internal* scripts of the learners (Kollar et al. 2006) and help them assimilate, accommodate, or equilibrate new knowledge and skills.

Because epistemic and social scripts focus on scaffolding either the problem-solving procedures or the participant interactions, it seems that each type of script would be insufficient in a problem-based learning situation if used in isolation of the other script type. Effectively integrating epistemic and social scripting in a technology-enhanced problem solving environment represents a significant design challenge for designers as well as educators.

5.2.8 A Design Framework for Undergraduate STEM

Our perspective for developing a design framework emanates from that of design-based research (DBR), a tradition often used for constructing technology-enhanced learning environments (Brown 1992; Confrey 2006). A design framework is a core component of any design and serves to identify the fundamental characteristics for achieving a particular set of goals in a given context (Edelson 2002). Table 5.1 illustrates our design framework for a STEM learning environment that involves collaborative problem solving with cyberlearning by defining the features of the learning environment in relation to the four elements that have been defined by Sandoval (2013) as: (a) *Materials and Tools*—including software programs, instruments, manipulable materials, media, and other resources, (b) *Task Structures*—what learners are expected to do—their goals, criteria, standards, and so on, (c) *Participant Structures*—how participants (e.g., students and instructors) are expected to participate in tasks, the roles and responsibilities participants take on and (d) *Discursive Practices*—ways of communicating.

Figure 5.2 illustrates how the elements of the design framework relate to the mediating processes of situated problem solving in STEM. The arrows in the diagram represent the statements in the cells of Table 5.1. These relationships can be interpreted as individual design conjectures in the form of: *if learners engage in collaborative problem solving in STEM with these tools and structures, through this discursive practice, then this mediating process will emerge* (Sandoval 2013).

In the following section we use a software environment of our construction called ECLIPSE (Environment for Collaborative Learning Integrating Problem Solving Experiences) to illustrate the application of our design framework. Our recent research has involved using ECLIPSE in undergraduate science and engineering courses as means for providing authentic problem solving experiences while building our theoretical understanding of how such experiences afford meaningful learning and authentic participation.

Table 5.1 The design framework for a STEM learning environment that involves collaborative problem solving with cyberlearning

Materials and tools	Task structures	Participant structures	Discursive practices
A software system that supports problem-centered learning, a high degree of user-user interactivity, conventions of social media use, hyperlinking to outside media sources and robust forms of communication and assessment	Authentic, ill-structured, dynamic problems that emanate from the situated nature of problem solving in STEM and are associated with appropriate learning goals and standards	Structures that make explicit the roles and responsibilities of collaboration	Scaffolding to support collaborative social exchange during the processes of problem definition, information discrimination, reasoning and argumentation, solution negotiation and evaluation
Access to highly interactive media as well as valid and reliable data and reference material	Structures that make explicit the sociocultural context of the problem and any proposed solutions	Structures that allow instructors and peers to provide feedback as well as cognitive, metacognitive and emotional support	Scaffolding to support learners in developing a shared conception of a problem
Tools for formative and summative assessment that can be applied to individuals or teams that involve the appropriate use of multiple forms of representation	Structures that make explicit the epistemic practice(s) of the STEM discipline(s) for problem solving	Scaffolding to support the cognitive and metacognitive processes related to problem solving process skills	Scaffolding to support learners in differentiating and evaluating the forms of knowledge involved in successful problem solving

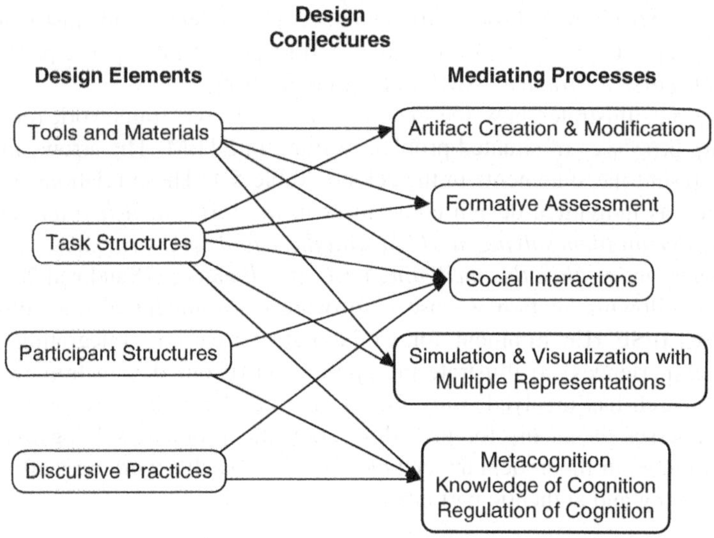

Fig. 5.2 The relationships among the elements of the design framework and mediating processes of situated problem solving in STEM

5.2.9 *Application of the Design Framework: Scaffolded Problem Solving with ECLIPSE*

ECLIPSE, or Environment for Collaborative Learning Integrating Problem Solving Experiences, is an example of a cyberlearning technology that is designed to scaffold collaborative problem solving and development of twenty-first century skills. ECLIPSE is not a general-purpose technology or Learning Management System like Blackboard or Sakai, which consist of instructor uploaded learning resources, discussion forums, and a grade book. Such general-purpose technologies provide content presentation, communication, and assessment tools but the important task of designing effective instructional scaffolding, activities, and structuring learning still rests on the shoulders of the instructor.

As a cyberlearning technology, ECLIPSE provides cognitive and metacognitive scaffolding that encourages and structures the problem-solving process for students. ECLIPSE is also not a mere collection of online forms, or electronic worksheets. Its design makes use of the unique affordances of media-rich dynamic web application technologies (e.g., Asynchronous Javascript and extensible markup language) and integrates social media metaphors and conventions (e.g., tagging and voting) to engage students in collaborative processes that are more akin to those they use when communicating, collaborating and learning with technology outside of a classroom. Because the design of authentic problem-solving activities is challenging, ECLIPSE also provides instructional design supports for instructors.

5.2.10 *Scaffolds for Instructors*

The scaffolds for instructors within ECLIPSE are designed to help educators create and manage classes and teams, to design new problem-solving activities, to adopt activities from a crowd sourced Activity Library, and to use rubrics and learning analytics to assess the performance of individual students and teams. ECLIPSE uses worked examples (Renkl and Atkinson 2007) and easy-to-use tools to provide epistemic guidance to instructors for: (a) designing an effective scenario for an authentic problem-solving experience, (b) identifying appropriate STEM learning resources (e.g., using existing cyberinfrastructure such as the National Science Digital Library), (c) monitoring the progress of individual students and teams and provide formative feedback, (d) assessing learning outcomes and (e) reflecting on the effectiveness of the activity in the teaching portfolio.

Since most of the research on problem solving in STEM has examined methods for improving well-structured problem solving our approach with ECLIPSE is to provide tools for designing and scaffolding more open-ended activities (i.e., ill-structured problems). Thus, in the instructor module, special attention is given to explicating the differences between well-structured (textbook) and ill-structured (authentic) problems in order to enable educators to evaluate and compare the complexity, structuredness, and dynamicity of various types of problems. To achieve this, ECLIPSE includes an interactive Problem Analyzer tool that explains the differences and highlights the features of a spectrum of typical STEM problems based on Jonassen's (2011) typology: story problems, rule-using problems, troubleshooting problems, diagnosis problems, strategic performance problems, policy analysis problems, design problems, and dilemmas.

Instructors have access to learning and collaboration analytics that help them effectively and efficiently assess progress, teamwork, flow of collaboration and provide formative and summative feedback. Because ECLIPSE is designed to support integration of authentic problem solving in large-enrollment courses, these tools allow an instructor to quickly determine where each student and team is in the problem-solving process, what they have contributed, and how effectively they are approaching a solution to the problem.

5.2.11 Scaffolds for Students

Scaffolds for students in ECLIPSE are provided so as to support both individual and collaborative cognition and metacognition during each step or 'challenge episode' (Hadwin et al. 2011) of the problem-solving process. As learners gain expertise, instructors can fade scaffolding by gradually deactivating some or all of these scaffolding features. This process of fading encourages transfer of problem-solving skills rather than deepening reliance on the scaffolds.

DEEPER, the process scaffolding framework (Azevedo et al. 2011) used in ECPLISE, scaffolds the problem solving process and skills using the following steps: Define, Explore, Explain, Present, Evaluate, and Reflect (Table 5.2). Unlike the previous epistemic scripting frameworks that focus primarily on planning and argumentation skills, DEEPER also provides supports for the development of stakeholder analysis (Define), information discrimination (Explore), and solution communication (Present), which are important twenty-first century skills of information and media literacy (p.21.org). Our research has shown that teams of students using DEEPER in a large enrollment entomology course had significantly higher problem-solving performance than a comparison group using a more traditional rationale-based approach (Antonenko et al. 2014).

Since understanding of collaborative processes requires attention to both individual and group-level information (Hadwin et al. 2011), each step in ECLIPSE provides scaffolding for students to engage in individual processing as well as negotiation of shared meaning with teammates. As part of the collaborative process, students work together to

1. Identify a specific problem to research and solve,
2. Explore information resources some of which are by design irrelevant to solving the problem (to increase authenticity),
3. Develop evidence-based arguments for the solution,

Table 5.2 Interaction of cognition and metacognition in collaborative problem solving in the DEEPER framework (Antonenko et al. 2014)

Define[a]	Analyze the situation and *think about whether you have read about or heard of similar issues in the past*
	Define one main problem that needs to be solved
	Identify the causes of this problem
	Identify the consequences of this problem
	Create a list of what you know about the problem and what you need to know to solve the problem
	Identify the stakeholders, which are people or groups influenced by the problem
Explore	*Determine which information resources are relevant to solving the problem*
	Analyze information resources and identify claims and evidence that are useful for solving the problem
Explain	Align information from the resources with your problem solving needs
	Determine whether claims for solving the problem are supported by evidence
	Build your claim for solving the problem based on the information you have analyzed
	Consider all possible impacts of the solution on stakeholders
	Discuss all solution proposals and select the best one
Present	*Determine what information to include in the presentation of the solution*
	Decide what solution presentation format to choose for the stakeholders
Evaluate	*Evaluate* your team's solution
	Evaluate others teams' solutions
Reflect[b]	*Reflect* on what has been learned in terms of problem-solving strategies
	Reflect on what has been learned in terms of new knowledge as part of the problem-solving experiences

[a]Cognitive task components are presented in normal font while the metacognitive task components are italicized. This differentiation is provided for illustrative purposes but is somewhat artificial as the cognitive and metacognitive components are highly interactive

[b]In addition to step-specific scripting that was provided to scaffold collaborative problem solving, the ECLIPSE software provided additional affordances for supporting social cognition and metacognition such as color-coding active, unexplored, and completed steps, dynamically arranging lists of learning resources based on teammates' relevance votes etc

4. Select appropriate media and structure solution presentation for the stakeholders,
5. Evaluate the quality of solution, and
6. Reflect on the experience of solving the problem and explore relevant STEM careers.

Technology-supported scaffolds are employed to assist with the teams' collective memory and shared metacognition (Iiskala et al. 2011). For example, the Explain step displays by relevance votes all evidence harvested by each teammate in the Explore step. This allows students to quickly see the most relevant information, collaboratively construct claims, link them to evidence, and then to develop the team's solution proposals. This scaffold is designed to assist with both metacognitive knowledge and metacognitive regulation. In terms of metacognitive knowledge, it helps learners become more aware of relevant task variables that are reflected in the evidence provided by each teammate. Importantly, it also helps increase the learners' awareness of their teammates' cognitive process and the approach they took to extracting relevant evidence and claims. Metacognitive regulation is supported by encouraging learners to engage in enhanced comprehension monitoring, debugging, and evaluative activities (e.g., check their comprehension against the ideas recorded by the teammates).

ECLIPSE makes effective use of the affordances of contemporary web development technology (c.f., Gibson 1966; Kozma 1991). For example, instead of asking STEM learners to check relevant concepts in what is essentially an electronic worksheet (a common practice in many technology-based environments), ECLIPSE will scaffold concept discrimination by having learners create tags, assign tags to information resources, and share them with the team. Each collection of tags in the form of a tag cloud is then discussed and voted on by teammates—similar to the way information is shared and evaluated in social media outside of school. Figure 5.3 illustrates a screenshot from the Team portion of the Explore step within ECLIPSE where students see each other's analyses of information resources, the tags provided by each teammate, as well as the claims and evidence and relevance votes for each resource. Each resource is analyzed and discussed by the team and the evidence and claims that they extract are sorted by relevance votes, which help the team in developing solution proposals in the Explain steps.

The learner module of ECLIPSE also provides important teamwork coordination tools, as well as learning and collaboration analytics. The philosophy underlying the

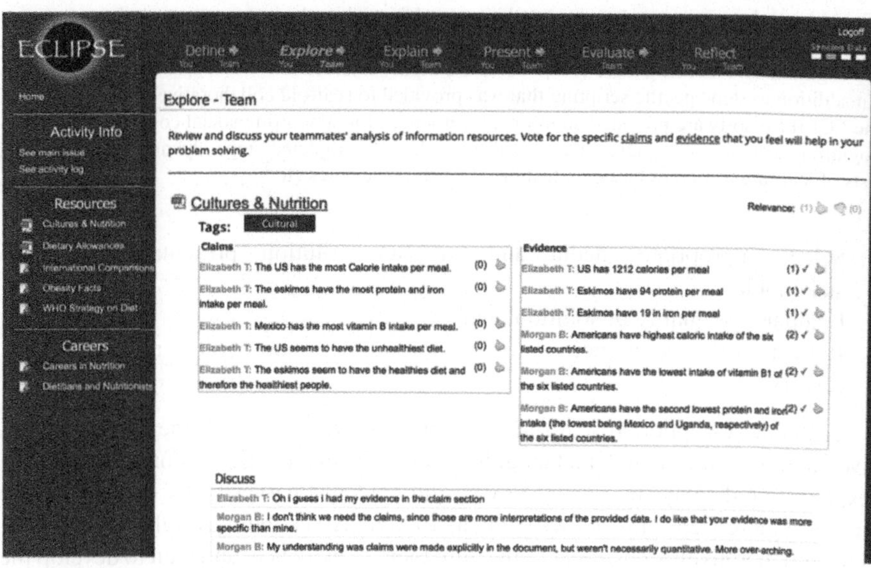

Fig. 5.3 Teammates tag, discuss, and vote on the relevance of claims and evidence in ECLIPSE's Explore step

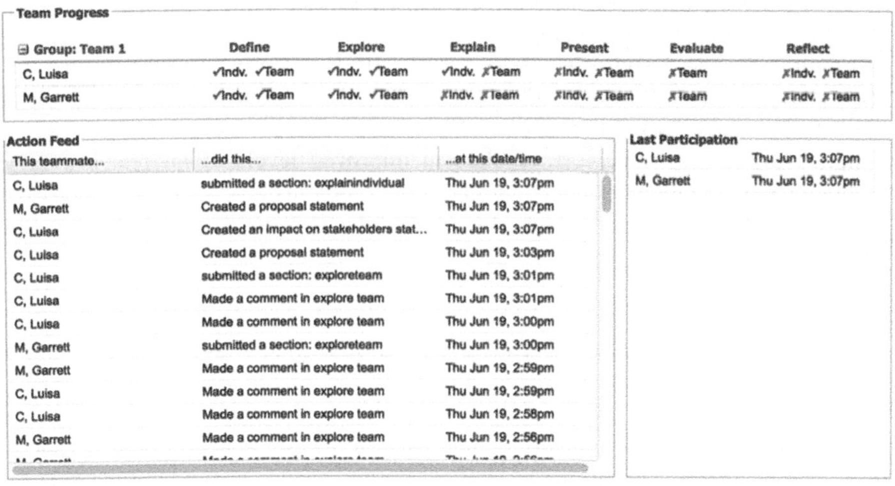

Fig. 5.4 ECLIPSE's activity feed tool that helps students coordinate collaborative learning

design of analytics tools is that they should be available not only to instructors and researchers (i.e., the common practice) but also to students, as these tools can serve as effective scaffolds for regulating individual and collaborative problem solving. ECLIPSE allows students to view a log of their teammates' actions, task submissions, and a history of comments (Fig. 5.4). Simple charts and graphs are generated to summarize data such as commenting activity and recipients of most votes, which will allow students to compare their work to their teammates' contributions. The empirical literature on technology-supported scaffolding is replete with examples of technologies and scaffolds that students *choose* not to access and use (e.g., Azevedo and Aleven 2014). Thus, a summary of the most important student and team activity will be shown to students upon the completion of each problem-solving step. Students also have access to teamwork coordination tools. For instance, using the popular Facebook™ metaphor, students can 'poke' a teammate to attract her or his attention and use a global messaging tool to coordinate activities.

To summarize, the ECLIPSE learning environment provides (a) Materials and Tools, (b) Task Structures, (c) Participant Structures, and (d) Discursive Practices (Fig. 5.2). It scaffolds collaborative problem solving by providing students with both materials (e.g., learning resources) and tools (e.g., technological affordances for collaborative resource tagging). ECLIPSE establishes a clear sequence of task structures using the DEEPER epistemic scripting framework of six evidence-based problem-solving steps (Table 5.2). Teammate interaction is facilitated using participant structures that include specific prompts (e.g., "discuss the viability of all solution proposals and select one") as well as collaborative tasks. Finally, discursive practices are supported on the synchronous and asynchronous level in a metacognitively meaningful way. Teammates plan, monitor and reflect on their learning individually and together. Their discourse is supported by synchronous commenting (cf., chat) and asynchronous voting, ranking, and discussion activities embedded in each step of the DEEPER process.

Empirically, the design framework for ECLIPSE has proven useful in the contexts of high school Biology and Physics as well as undergraduate Entomology (Antonenko et al. 2014; Antonenko and Nichols 2013). Problem-solving activities designed using this framework resulted in improved levels of interest in STEM problem solving, a greater understanding of the problems that were used, a broader and more informed range of solutions, and enhanced appreciation of authentic STEM practices.

5.3 Implications

Clearly, the forms of problem solving that we describe are not fully inclusive of all the learning experiences within any particular course and we are not advocating for them to be such. In fact, the type of problem solving that we have described is not even inclusive of all of the problem-solving experiences in a single course. Under certain circumstances, well-structured problem solving is still appropriate and can also be done effectively and is improved with cyberlearning (Biesinger and Crippen 2010; Crippen and Earl 2007). Finding the balance among all of the needed requirements, outcomes and forms of support for enacting relevant, twenty-first century STEM learning requires a contemporary, empirically grounded guiding framework that recognizes the situated nature of problem solving. From our perspective, cognitive apprenticeship (Johri and Olds 2011; Stewart and Lagowski 2003) meets that need and offers the potential for framing the entire undergraduate experience in any STEM domain, including the use of cyberlearning. To adopt such a perspective would change the current conversation on reform of undergraduate STEM from simply requiring courses to focusing on the situated nature of problem solving in STEM via the four dimensions of content, method, sequencing and sociology (Collins 2006). An example of that latter dimension can be found in Chap. 6 of this volume with Yerrick et al. (2018) description of digital native pre-service elementary teachers' technological culture, as well as Chap. 10 of this volume with Carberry and Baker's (2018) description of the cultural considerations in engineering and engineering education—particularly the sociology of over- and under-represented groups. *Thus, we view the exploration of situated frameworks, such as that of cognitive apprenticeship, for envisioning and enacting undergraduate STEM as a key implication of the ideas that we presented related to collaborative problem solving in STEM cyberlearning.*

From our experience, the construction of good ill-structured problems is the most difficult part of implementing collaborative, team-based problem solving. Fully embracing the situated nature of problem solving in STEM and doing so in a way that minimizes contrived, one-dimensional or unrealistic solutions requires multidisciplinary expertise. In addition to the different ways of knowing and perspectives that are provided by people working in different fields, including these elements as problem dimensions offer the potential for resonating with a greater diversity of students and informing them of the range of career opportunities that might be available. *Engaging expertise from multiple disciplines, including the social and applied sciences in the development of ill-structured problems enhances the situated nature of problem solving in STEM and thus affords collaboration* via *cyberlearning.*

Recovery of the discussion (i.e., recitation) component of large lecture courses offers one potential mechanism for delivering the kind of learning experiences for which we are advocating (Crippen et al. 2015), including the use of technology-enhanced learning environments such as ECLIPSE. This integration of out-of-class, online technology with face-to-face, in-class instructor-assisted collaborative work would result in a truly blended approach to learning, with the ascribed potential for positively influencing engagement and learning outcomes in large lectures courses (e.g., Gonzalez 2014). However, since common practice is to have graduate teaching assistants responsible for the instruction for these sections, training for them, as well as on-going support and coordination are critical structures for successful implementation. Thus, what we view as an obvious implication for STEM educators is to *explore existing course structures and available technologies and integrate them appropriately to provide a robust blended learning experience for students based upon the situated nature of problem solving in STEM.*

Our experience with the conceptual and empirical aspects of collaborative problem solving research indicates that neither epistemic, nor social scripts may be sufficient scaffolds when used on their own. Epistemic scripts serve to provide problem-solving teams with procedural support regarding task completion, whereas social scripts structure learners' interactions. Effective use of these scaffolds will require a balanced approach that is informed by new research. *Therefore, scaffolding of collaborative problem solving should involve an organic integration of epistemic and social scripts in the learning environment, as both task performance and quality of student interactions are critical to problem-solving success.*

Finally, problem solving is a practice that spans the student experience within a STEM discipline as well as overlapping across the disciplines. Both of these dimensions need to be further explored as a progression of learning that is coherent, developmental and comprehensive. The potential exists for a learning progression of problem solving in STEM that spans middle school through undergraduate and builds comprehensive understanding within a discipline as well as across STEM. Identifying and articulating such a progression represents a current grand challenge for STEM education. *Problem solving in a STEM domain should be viewed as a developmental progression along multiple dimensions that are distinct, yet complimentary to other domains where learning results in a coherent and comprehensive understanding of the domain as well as across STEM.*

5.4 Conclusions

This chapter argues for a reconceptualization of the problem-solving process in STEM education from an overly prescriptive, 'plug-n-chug' process to one that reflects the authentic practices and collaborative skills involved in real-world problem solving in STEM. To this end, collaborative problem solving is discussed as a core STEM practice that can be scaffolded using cyberlearning tools at the cognitive and metacognitive, as well as epistemic and social levels. Cyberlearning can transform the scaffolding of both individual and collective problem solving because it provides affordances, materials, tools and frameworks for learner engagement in such important processes as problem definition, information discrimination, reasoning and argumentation, solution negotiation and evaluation. In addition to supporting learners' cognitive processes, such as elaboration or problem categorization, cyberlearning environments scaffold metacognition—that is, planning, monitoring, and reflective activities. We have also described a cyberlearning environment entitled ECLIPSE: Environment for Collaborative Learning Integrating Problem Solving Experiences to illustrate the design of collaborative problem-solving technologies that follow our design framework. This chapter contributes to the growing body of scholarship on design-based research frameworks for scaffolding computer-supported collaborative learning and problem solving.

5.5 Recommendations

Based upon ideas presented, we offer the following recommendations to STEM educators:

- Explore the use of situated frameworks, such as cognitive apprenticeship for envisioning undergraduate STEM teaching and learning.
- Engage expertise from multiple disciplines, including the social and applied sciences, in the development of ill-structured problems.
- Explore existing course structures and available technologies and integrate them appropriately to provide a robust blended learning experiences for students based upon the situated nature of problem solving in STEM.
- Scaffolding of collaborative problem solving should involve an organic integration of epistemic and social scripts in the learning environment, as both task performance and quality of student interactions are critical to problem-solving success.
- View problem solving as a developmental progression along multiple dimensions that are distinct, yet complimentary to other domains where learning results in a coherent and comprehensive understanding of the domain as well as across STEM.

5.6 Key Sentences

The situated nature of problem solving in STEM involves a core, general heuristic that is embedded in domain-specific practices in a sociocultural context and enacted with collaboration.

Undergraduate learning should involve authentic, ill-structured, multidimensional problems that emanate from the situated nature of problem solving in STEM.

Problem authenticity, scaffolding, and supporting collaboration are key design goals for engaging and effective cyberlearning experiences.

A STEM cyberlearning environment that focuses on collaborative problem solving requires specific features to provide cognitive, metacognitive and emotional support in the form of materials and tools, task structures, participant structures, and discursive practices.

References

Adelson, B. (1981). Problem solving and the development of abstract categories in programming languages. *Memory and Cognition, 9*(4), 422–433.

Alberts, B. (2011). *Special issue: Dealing with data* (Vol. 332): AAAS.

Antonenko, P., Jackman, J., Kumsaikaew, P., Marathe, R., Niederhauser, D., Ogilvie, C., & Ryan, S. (2011). Understanding student pathways in context-rich problems. *Education and Information Technologies, 16*(4), 323–342.

Antonenko, P., Jahanzad, F., & Greenwood, C. (2014). Fostering collaborative problem solving and 21st century skills using the DEEPER scaffolding framework. *Journal of College Science Teaching, 43*(6), 78–87.

Antonenko, P., & Nichols, J. (2013). *ECLIPSE: Environment for collaborative learning integrating problem solving experiences, Proceedings of World Conference on E-Learning in Corporate, Government, Healthcare, and Higher Education 2013* (pp. 2280–2284). Chesapeake: AACE.

Azevedo, R., & Aleven, V. (Eds.). (2014). *International handbook of metacognition and learning technologies*. New York: Springer.

Azevedo, R., Cromley, J. G., Moos, D. C., Greene, J. A., & Winters, F. I. (2011). Adaptive content and process scaffolding: A key to facilitating students' self-regulated learning with hypermedia. *Psychological Testing and Assessment Modeling, 53*(1), 106–140.

Barab, S., & Duffy, T. (2012). From practice fields to communities of practice. In D. Jonassen & S. Land (Eds.), *Theoretical foundations of learning environments* (pp. 29–65). New York: Lawrence Erlbaum Associates.

Barrows, H. S. (1996). Problem-based learning in medicine and beyond. In L. Wilkerson & W. H. Gijselaers (Eds.), *New directions for teaching and learning. Bringing problem-based learning to higher education: Theory and practice* (pp. 3–13). San Francisco: Jossey-Bass.

Belland, B., & Drake, J. (2013). Toward a framework on how affordances and motives can drive different uses of scaffolds: Theory, evidence, and design implications. *Educational Technology Research and Development, 61*(6), 903–925. doi:10.1007/s11423-013-9313-6.

Belland, B. R., Glazewski, K. D., & Richardson, J. C. (2011). Problem-based learning and argumentation: Testing a scaffolding framework to support middle school students' creation of evidence-based arguments. *Instructional Science, 39*(5), 667–694.

Belland, B. R., Kim, C., & Hannafin, M. J. (2013). A framework for designing scaffolds that improve motivation and cognition. *Educational Psychologist, 48*(4), 243–270.

Bennett, D., Brunner, C., & Honey, M. (1999). *Gender and technology: Designing for diversity*. New York: Education Development Center.

Bielaczyc, K. (2006). Designing social infrastructure: Critical issues in creating learning environments with technology. *The Journal of the Learning Sciences, 15*(3), 301–329. doi:10.1207/s15327809jls1503_1.

Biesinger, K. D., & Crippen, K. J. (2010). The effects of feedback protocol on self-regulated learning in a web-based worked example learning environment. *Computers & Education, 55*(4), 1470–1482.

Blumenfeld, P. C., Kepler, T. M., & Krajcik, J. S. (2006). Motivation and cognitive engagement in learning environments. In R. K. Sawyer (Ed.), *The Cambridge handbook of the learning sciences* (pp. 475–488). New York: Cambridge University Press.

Bresnen, M., Edelman, L., Newell, S., Scarbrough, H., & Swan, J. (2003). Social practices and the management of knowledge in project environments. *International Journal of Project Management, 21*(3), 157–166. doi:10.1016/S0263-7863(02)00090-X.

Brown, A. L. (1992). Design experiments: Theoretical and methodological challenges in creating complex interventions in classroom settings. *The Journal of the Learning Sciences, 2*(2), 141–178. doi:10.1207/s15327809jls0202_2.

Brown, J. S., Collins, A., & Duguid, P. (1989). Situated cognition and the culture of learning. *Educational Researcher, 18*(1), 32–42.

Brown, J. C., & Crippen, K. J. (2016). Designing for culturally responsive science education through professional development. *International Journal of Science Education, 38*(3), 470–492. doi:10.1080/09500693.2015.1136756.

Bunce, D. M., & Heikkinen, H. (1986). The effects of an explicit problem-solving approach on mathematical chemistry achievement. *Journal of Research in Science Teaching, 23*, 11.

Campbell, P. (2008). Big data: Science in the petabyte era [Special Issue]. *Nature, 45*, 7209.

Carberry, A. R., & Baker, D. R. (2018). The impact of culture on engineering and engineering education. In Y. J. Dori, Z. Mevarech, & D. Baker (Eds.), *Cognition, metacognition, and culture in STEM education* (pp. 217–239). Springer.

Chi, M. T. H., Feltovich, P. J., & Glaser, R. (1981). Categorization and representation of physics problems by experts and novices. *Cognitive Science, 5*, 121–152.

Collins, A. (2006). Cognitive apprenticeship. In R. K. Sawyer (Ed.), *The Cambridge handbook of the learning sciences* (pp. 47–60). New York: Cambridge University Press.

Confrey, J. (2006). The evolution of design studies as methodology. *The Journal of the Learning Sciences, 11*, 105–121.

Crippen, K. J., & Earl, B. L. (2007). The impact of web-based worked examples and self-explanation on performance, problem solving, and self-efficacy. *Computers & Education, 49*(3), 809–821. doi:10.1016/J.Compedu.2005.11.018.

Crippen, K. J., Wu, C.-Y., Boyer, T., de Torres, T., Brucat, P., & Korolev, M. (2015). A design-based apprenticeship approach to transform freshman chemistry for engineering students. *QScience Proceedings (Engineering Leaders Conference 2014), elc2014*(71), 1 6. doi:10.5339/qproc.2015.elc2014.71.

Dillenbourg, P. (2002). Over-scripting CSCL. In P. A. Kirschner (Ed.), *Three worlds of CSCL: Can we support CSCL* (pp. 61–91). Heerlen: Open University of the Netherlands.

Dolmans, D. H. J. M., & Schmidt, H. G. (2006). What do we know about cognitive and motivational effects of small group tutorials in problem- based learning? *Advances in Health Sciences Education, 11*(4), 321–336.

Donnelly, D. F., Linn, M. C., & Ludvigsen, S. (2014). Impacts and characteristics of computer-based science inquiry learning environments for precollege students. *Review of Educational Research*. doi:10.3102/0034654314546954.

Dufresne, R. J., Gerace, W. J., Hardiman, P. T., & Mestre, J. P. (1992). Constraining novices to perform expert-like problem analyses: Effects on schema acquisition. *The Journal of the Learning Sciences, 2*(3), 307–331.

Edelson, D. C. (2002). Design research: What we learn when we engage in design. *The Journal of the Learning Sciences, 11*(1), 105–121. doi:10.1207/S15327809JLS1101_4.

Edelson, D. C., & Reiser, B. J. (2006). Making authentic practices accessible to learners: Design challenges and strategies. In K. Sawyer (Ed.), *The Cambridge handbook of the learning sciences* (pp. 335–354). Cambridge, MA: Cambridge University Press.

Fischer, F., Kollar, I., Mandl, H., & Haake, H. M. (2007). *Scripting computer-supported collaborative learning: Cognitive, computational and educational perspectives*. New York: Springer.

Gay, G. (2010). *Culturally responsive teaching: Theory, research, and practice* (2nd ed.). New York: Teachers College Press.

Gee, J. P. (2008). A sociocultural perspective on opportunity to learn. In P. A. Moss, D. C. Pullin, & J. P. Gee (Eds.), *Assessment, equity, and opportunities to learn* (pp. 76–108). New York: Cambridge University Press.

George, J. (2013). "Do you have to pack?" Preparing for culturally relevant science teaching in the Caribbean. *International Journal of Science Education, 35*(12), 2114–2131.

Gibson, J. J. (1966). *The senses considered as perceptual systems*. Boston: Houghton Mifflin.

Gonzalez, B. Y. (2014). A six-year review of student success in a biology course using lecture, blended, and hybrid methods. *Journal of College Science Teaching, 43*(6), 14–19.

Graesser, A. C., Lu, S., Olde, B. A., Cooper-Pye, E., & Whitten, S. (2005). Question asking and eye tracking during cognitive disequilibrium: Comprehending illustrated texts on devices when the devices break down. *Memory and Cognition, 33*(7), 1235–1247.

Hadwin, A. F., Järvelä, S., & Miller, M. (2011). Self-regulated, co-regulated, and socially shared regulation of learning. In B. J. Zimmerman & D. H. Schunk (Eds.), *Handbook of self-regulation of learning and performance* (pp. 65–84). New York: Routledge.

Hidi, S., & Renninger, K. A. (2006). The four-phase model of interest development. *Educational Psychologist, 41*(2), 111–127. doi:10.1207/s15326985ep4102_4.

Hmelo-Silver, C. E. (2006). Design principles for scaffolding technology based inquiry. In A. M. O'Donnell, C. E. Hmelo-Silver, & G. Erkens (Eds.), *Collaborative reasoning, learning and technology* (pp. 147–170). Mahwah: Erlbaum.

Hmelo-Silver, C. E., Duncan, R. G., & Chinn, C. A. (2007). Scaffolding and achievement in problem-based and inquiry learning. *Educational Psychologist, 42*(2), 99–107.

Hung, W. (2011). Theory to reality: A few issues in implementing problem-based learning. *Educational Technology Research and Development, 59*(4), 529–552.

Hurst, R. W., & Milkent, M. M. (1996). Facilitating successful prediction problem solving in biology through application of skill theory. *Journal of Research in Science Teaching, 33*(5), 541–552.

Iiskala, T., Vauras, M., Lehtinen, E., & Salonen, P. (2011). Socially shared metacognition of dyads of pupils in collaborative mathematical problem-solving processes. *Learning and Instruction, 21*(3), 379–393.

Johri, A., & Olds, B. M. (2011). Situated engineering learning: Bridging engineering education research and the learning sciences. *Journal of Engineering Education, 100*(1), 151–185.

Jonassen, D. H. (2000). Toward a design theory of problem solving. *Educational Technology Research & Development, 48*(4), 63–85.

Jonassen, D. H. (2004). *Learning to solve problems: An instructional design guide*. San Francisco: Pfeiffer Publishing.

Jonassen, D. H. (2007). *Learning to solve complex, scientific problems*. Mahwah: Lawrence Erlbaum Associates.

Jonassen, D. (2011). *Learning to solve problems: A handbook for designing problem-solving learning environments*. New York: Routledge.

Jonassen, D. H., & Hung, W. (2008). All problems are not equal: Implications for PBL. *Interdisciplinary Journal of Problem-Based Learning, 2*(2), 6–28.

Kern, C. L., Crippen, K. J., & Skaza, H. (2014). The fidelity and usability of 5-DIE: A design study of enacted cyberlearning. *Journal of Computers in Mathematics and Science Teaching, 33*(1), 313–336.

King, A. (1997). ASK to THINK – TEL WHY®®: A model of transactive peer tutoring for scaffolding higher-level complex learning. *Educational Psychologist, 32*(4), 221–235.

Kollar, I., Fischer, F., & Hesse, F. W. (2006). Collaboration scripts – a conceptual analysis. *Educational Psychology Review, 18*(2), 159–185.

Kolodner, J. L. (2006). Case-based reasoning. In K. L. Sawyer (Ed.), *The Cambridge handbook of the learning sciences* (pp. 225–242). Cambridge: Cambridge University Press.

Kozma, R. B. (1991). Learning with media. *Review of Educational Research, 61*(2), 179–212.

Krajcik, J. S., Blumenfeld, P. C., Marx, R. W., Bass, D. M., & Fredricks, J. (1998). Inquiry in project-based science classrooms: Initial attempts by middle school students. *The Journal of the Learning Sciences, 7*(3), 313–350.

Krapp, A., & Prenzel, M. (2011). Research on interest in science: Theories, methods, and findings. *International Journal of Science Education, 33*(1), 27–50. doi:10.1080/09500693.2010.518645.

Land, S., & Hannafin, M. J. (1996). A conceptual framework for the development of theories-in-action with open learning environments. *Educational Technology Research and Development, 44*(3), 37–53.

Lave, J., & Wenger, E. (1991). *Situated learning: Legitimate peripheral participation*. New York: Cambridge University Press.

Leonard, W. J., Dufresne, R. J., & Mestre, J. P. (1996). Using qualitative problem-solving strategies to highlight the role of conceptual knowledge in solving problems. *American Journal of Physics, 64*(12), 1495–1503.

Linn, M. C., Davis, E. A., & Bell, P. (2005). Inquiry and technology. In M. C. Linn, E. A. Davis, & P. Bell (Eds.), *Internet environments for science education*. Mahwah: Lawrence Erlbaum.

Mayer, R. E. (1992). Cognition and instruction: Their historic meeting within educational psychology. *Journal of Educational Psychology, 84*(4), 405–412.

Means, B., Toyama, Y., Murphy, R., Bakia, M., Jones, K. (2009). Evaluation of evidence-based practices in online learning: A meta-analysis and review of online learning studies. Retrieved from Washington, DC: http://files.eric.ed.gov/fulltext/ED505824.pdf

van Merriënboer, J. J. G., Kester, L., & Paas, F. (2006). Teaching complex rather than simple tasks: Balancing intrinsic and germane load to enhance transfer of learning. *Applied Cognitive Psychology, 20*(3), 343–352. doi:10.1002/acp.1250.

Moore, E. B., Chamberlain, J. M., Parson, R., & Perkins, K. K. (2014). PhET interactive simulations: Transformative tools for teaching chemistry. *Journal of Chemical Education, 91*(8), 1191–1197. doi:10.1021/ed4005084.

Moss, J., Kotovsky, K., & Cagan, J. (2006). The role of functionality in the mental representations of engineering students: Some differences in the early stages of expertise. *Cognitive Science, 30*(1), 65–93.

National Academy of Engineering. (2005). *Educating the engineer of 2020: Adapting engineering education to the new century*. Washington, DC: The National Academies Press.

National Research Council. (2011). *Successful K-12 STEM education: Identifying effective approaches in science, technology, engineering and mathematics*. Washington, DC: National Academies Press.

National Science Foundation [NSF]. (2008). *Fostering learning in the networked world: The cyberlearning opportunity and challenge*. Arlington: National Science Foundation Task Force on Cyberlearning.

O'Donnell, A. M. (1999). Structuring dyadic interaction through scripted cooperation. In A. M. O'Donnell & A. King (Eds.), *Cognitive perspectives on peer learning* (pp. 179–196). Mahwah: Erlbaum.

Palincsar, A. S., & Herrenkohl, L. R. (1999). Designing collaborative contexts: Lessons from three research programs. In A. M. O'Donnell & A. King (Eds.), *Cognitive perspectives on peer learning* (pp. 151–177). Mahwah: Erlbaum.

Passow, H. J. (2007). What competencies should engineering programs emphasize? A meta-analysis of practitioners opinions informs curricular design. Paper presented at the 3rd International CDIO Conference, Cambridge, MA.

Polya, G. (1957). *How to solve it*. Princeton: Princeton University Press.

President's Council of Advisors on Science and Technology [PCAST]. (2012). *Engage to excel: Producing one million additional college graduates with degrees in science, technology, engineering, and mathematics*. Washington, DC: Executive Office of the President; President's Council of Advisors on Science and Technology.

Puntambekar, S., & Kolodner, J. L. (2005). Toward implementing distributed scaffolding: Helping students learn science from design. *Journal of Research in Science Teaching, 42*(2), 185–217.

Qin, Z., Johnson, D. W., & Johnson, R. T. (1995). Cooperative versus competitive efforts and problem solving. *Review of Educational Research, 65*(2), 129–143. doi:10.3102/00346543065002129.

Reif, F., Larkin, J. H., & Brackett, G. C. (1976). Teaching general learning and problem-solving skills. *American Journal of Physics, 44*(3), 212–217.

Reiser, B. J., Tabak, I., Sandoval, W. A., Smith, B., Steinmuller, F., & Leone, T. J. (2001). BGuILE: Stategic and conceptual scaffolds for scientific inquiry in biology classrooms. In S. M. Carver & D. Klahr (Eds.), *Cognition and instruction: Twenty five years of progress*. Mahvah: Erlbaum.

Renkl, A., & Atkinson, R. K. (2007). An example order for cognitive skill acquisition. In F. E. Ritter, J. Nerb, E. Lehtinen, & T. M. O'Shea (Eds.), *In order to learn: How the sequence of topics influences learning* (pp. 95–105). New York: Oxford University Press.

Roschelle, J., & Teasley, S. (1995). The construction of shared knowledge in collaborative problem solving. In C. E. O'Malley (Ed.), *Computer supported collaborative learning* (pp. 69–97). Heidelberg: Springer-Verlag.

Ryan, S., Jackman, J., Antonenko, P., Kumsaikaew, P., Marathe, R., Niederhauser, D., Ogilvie, C., (2007). Student selection of information relevant to solving ill-structured engineering economic decision problems. In *Proceedings of the 114th Annual Conference of the American Society for Engineering Education*. Honolulu, HI.

Sadler, T. D. (2009). Situated learning in science education: Socio-scientific issues as contexts for practice. *Studies in Science Education, 45*(1), 1–42. doi:10.1080/03057260802681839.

Sandi-Urena, S., Cooper, M. M., & Stevens, R. H. (2011). Enhancement of metacognition use and awareness by means of a collaborative intervention. *International Journal of Science Education, 33*(3), 323–340.

Sandoval, W. (2013). Conjecture mapping: An approach to systematic educational design research. *The Journal of the Learning Sciences, 23*(1), 18–36. doi:10.1080/10508406.2013.778204.

Schank, R. C., & Abelson, R. P. (1977). *Scripts, plans, goals and understanding. An inquiry into human knowledge structures*. Hillsdale: Erlbaum.

Schank, R. C., Fano, A., Bell, B., & Jona, M. (1994). The design of goal-based scenarios. *The Journal of the Learning Sciences, 3*(4), 305–345.

Schraw, G., & Lehman, S. (2001). Situational interest: A review of the literature and directions for future research. *Educational Psychology Review, 13*(1), 23–52. doi:10.1023/a:1009004801455.

diSessa, A. A. (2006). A history of conceptual change research: Threads and fault lines. In K. L. Sawyer (Ed.), *Cambridge handbook of the learning sciences* (pp. 265–281). Cambridge: Cambridge University Press.

Seymour, E., & Hewitt, N. M. (2000). *Talking about leaving: Why undergraduates leave the sciences*. Boulder.: Westview Press.

Silver, E. A. (1979). Student perceptions of relatedness among verbal problems. *Journal of Research in Mathematics Education, 10*, 195–210.

Singer, S. R., Nielsen, N. R., & Schweingruber, H. A. (Eds.). (2012). *Discipline-based education research: Understanding and improving learning in undergraduate science and engineering*. Washington, DC: National Academies Press.

Stegmann, K., Mu, J., Gehlen-Baum, V., Fischer, F. (2011). The myth of over-scripting: Can novices be supported too much? In H. Spada, G. Stahl, N. Miyake, & N. Law (Eds.), Connecting computer-supported collaborative learning to policy and practice: CSCL conference proceedings, Vol. I International Society of the Learning Sciences, Long Papers.

Sternberg, R. J., & Frensch, P. A. (Eds.). (1991). *Complex problem solving: Principles and mechanisms*. Hillsdale: Erlbaum.

Stewart, K. K., & Lagowski, J. J. (2003). Cognitive apprenticeship theory and graduate chemistry education. *Journal of Chemical Education, 80*(12), 1362. doi:10.1021/ed080p1362.

Strobel, J., Wang, J., Weber, N. R., & Dyehouse, M. (2013). The role of authenticity in design-based learning environments: The case of engineering education. *Computers & Education, 64*(0), 143–152. doi:10.1016/j.compedu.2012.11.026.

Sweller, J. (2010). Element interactivity and intrinsic, extraneous and germane cognitive load. *Educational Psychology Review, 22*(2), 123–138.

Van Heuvelen, A. (1991). Overview, case study physics. *American Journal of Physics, 59*(10), 898–907.

Villanueva, M. G., & Hand, B. (2011). Science for all: Engaging students with special needs in and about science. *Learning Disabilities Research and Practice, 26*(4), 233–240.

Vygotsky, L. S. (1978). *Mind in society: The development of higher psychological processes*. Cambridge, MA: Harvard University Press.

Weinberger, A., Ertl, B., Fischer, F., & Mandl, H. (2005). Epistemic and social scripts in computer-supported collaborative learning. *Instructional Science, 33*(1), 1–30.

Weinberger, A., Kollar, I., Dimitriadis, Y., Mäkitalo-Siegl, K., & Fischer, F. (2009). Computer-supported collaboration scripts: Perspectives from educational psychology and computer science. In N. Balacheff, S. R. Ludvigsen, T. de Jong, S. Barnes, & A. W. Lazonder (Eds.), *Technology-enhanced learning: Principles and products* (pp. 155–173). Berlin: Springer.

Yerrick, R., Radosta, M., & Greene, K. (2018). Technology, culture and young science teachers—A promise unfulfilled and proposals for change. In Y. J. Dori, Z. Mevarech, & D. Baker (Eds.), *Cognition, metacognition, and culture in STEM education* (pp. 117–138). Springer.

Chapter 6
Technology, Culture, and Young Science Teachers: A Promise Unfulfilled and Proposals for Change

Randy Yerrick, Michael Radosta, and Kelsey Greene

6.1 Introduction

There is little doubt that the turn of the twenty-first century included a seismic technocultural (Shaw 2008) shift, with the advent of the Internet and a rapid adoption of ubiquitous computing devices. Sociologist Daniel Bell (1979) predicted the shift coming decades prior, recognizing nascent innovations in networked data systems as an emergent 'intellectual technology,' that would change commerce, society, and culture in profound ways. As computers became more powerful, and information resources more robust, information and communication technologies (ICTs) were championed as transformational tools across the social and economic spectrum, whose adoption was viewed as both beneficial and essential for a given enterprise. Education, and in particular science education, was one such enterprise in which ICT adoption would purportedly transform.

Educational futurist Marc Prensky (2001) described the emergence of the digital age as a "...singularity- an event which changes things so fundamentally that there is absolutely no going back." (p. 1). He coined the term 'digital natives' to describe students born after 1980 whose brains and social habits have changed as a result of frequent exposure to ICTs. These children would be profoundly different than their "digital immigrant" teachers, who have "one foot in the past." (p. 1). It was anticipated that once digital natives became teachers themselves, a technological transformation of education would occur. These young, technology-savvy teachers (Rideout, Foehr, and Roberts 2009) would bring engaging twenty-first-century learning into classrooms.

R. Yerrick (✉) • M. Radosta • K. Greene
Department of Learning and Instruction, Graduate School of Education, University at Buffalo, Buffalo, NY, USA
e-mail: ryerrick@buffalo.edu

Y.J. Dori et al. (eds.), *Cognition, Metacognition, and Culture in STEM Education*, Innovations in Science Education and Technology 24,
https://doi.org/10.1007/978-3-319-66659-4_6

While Prensky was hypothesizing differences between digital natives and immigrants, Larry Cuban (2009) published his research on the impacts of educational technology investments made in schools from kindergarten through college. He and his fellow researchers went into the heart of Silicon Valley to find schools where they should expect cutting-edge examples of technology-based learning. Instead, they found that ICTs were being used as basic organizational, publishing, and communication tools. Instruction and classroom practice were largely unaffected, despite hundreds-of-millions of dollars invested. The premise of young technologically savvy students and teachers remains popular even today despite the fact that scholarly research has come to a fairly critical consensus regarding its validity (Bennett et al. 2008).

More recently, Wang et al. (2014) provided an evidentiary basis to consider the mertits of Prensky and others' predictions. Investigating preservice elementary teachers (PETs) born after 1980, they found that, while students used ICTs regularly, their use of knowledge production tools, such as video editing applications, blogs, wikis, and simulations was minimal, and ICTs in generally were more likely to be used outside of school. In addition to the surveys of technology use, the teachers also provided their perceptions of obstacles towards implementing technology in the classroom. Many cited continued problems of access, lack of time to plan and prepare technology-based lessons, state mandated tests precluding technology-based instruction, unreliability of technology, and lack of knowledge regarding how to use it pedagogically. Their findings suggest little has changed in the decade since Cuban's (1986) book. Even when teachers and students are both regular and capable users of technology, the tools are not used for learning in innovative ways. The cultures of personal technology and school-based learning remain sharply divided.

As educators of future science teachers, we were interested and concerned about developing preservice teachers' use of learning technologies for elementary science education. Our own experiences and established research point to stubborn challenges infostering high quality science teaching at the elementary level (Abell et al. 1998; Yerrick et al. 2008; Yerrick et al. 1997). We felt that an up-close investigation was warranted and could be helpful for unpacking these long-standing challenges by documenting a concerted effort to developing digital natives into elementary science teachers. In this chapter, we consider the digital native proposition and the challenges of science teacher education with a group of preservice elementary teachers learning to teach science using technology for both pedagogy and their own learning. We will use a sociocultural lens to consider the literature and collect and interpret data (Lave and Wenger 1991; O'loughlin 1992; Wertsch 1998), in order to describe and analyze the participation and responses of digital native-aged preservice elementary teachers in a graduate program learning to use technology to teach science. The findings will discuss the cultural influences that might have informed the participation of the preservice teachers under consideration, with implications for teacher development in the digital age.

6.2 Tools and Actions: A Sociocultural Approach to Technology and Teaching

To investigate and discuss the state of how the culture of ICTs, digital natives, and science teaching interact necessitates the provision an epistemological framework. We adopt a sociocultural (Wertsch 1998) and situated (Lave and Wenger 1991) stance on teaching and learning. A sociocultural analysis assumes a dialectical and horizontal relationship between persons and their situated context; a person's thinking is an ongoing interaction with the material and symbolic aspects of present and past experiences (Hutchins 1995; McVee et al. 2005; Wertsch 1998). With this theoretical lens, digital native science teachers are regarded neither as independent agents, nor as subjects under the control of cultural technologies and institutionalized learning. Instead, they are individual participants from a variety of powerful social contexts, which have provided the source material for their thoughts and actions. Furthermore, as expressed by Lave and Wenger (1991), the intellectual conceptions of technology for teaching and learning are enmeshed with their personal uses of technology and past enactments of technology for learning.

As an analytical tool, sociocultural theory looks at "action" as the basic unit of analysis (Wertsch 1998, p. 12). As actions are described and interpreted in research, sociocultural theorists strive to be careful not to decontextualize their psychological, linguistic, social, and technological components. Within this chapter we are considering science teachers' words and actions with educational technology in the context of prior learning experiences with science, while avoiding suggestions of causal mechanisms or a priori claims about teacher learning. Instead, words and actions are used as a way to understand and interpret the meanings (which were simultaneously socially, emotionally, intellectually, and materially constructed) that could be informing the actions of a teacher or group of teachers in the research setting (Burke, as cited in Wertsch 1998).

For this particular analysis and discussion of digital natives, ICTs, and processes of science teaching, a second layer of sociocultural theory is needed, and that is the role meditational tools play in guiding actions (Wertsch 1998). In this use of the term, tools can be intellectual or physical. Drawing substantially on Vygostky (1987, as cited in Wertsch 1998), Wertsch argues that it is impossible to isolate thinking from the tools people use to think, such as language, numbers, text, and computers. In order to consider young science teachers and how they regard technology tools for learning, it is necessary to examine the cultural nature of those tools and affordances and how they interact with the culture of teaching and learning science, and education in general.

6.3 Addressing the Challenge: A Context to Foster Innovative Teaching

We have so far strived to establish that the promise of technology to transform education has yet to be fulfilled, and highlight the possible cultural and functional incompatibilities between personal technology practices and pedagogical uses. To provide a context to make further sense of the digital native generation, technology, and science teaching, this chapter will turn to qualitative data from a cohort of 23 preservice elementary science teachers [PETs] who meet the criteria of being digital natives. There is much reported on the enculturation, beliefs, needs, and competency of digital natives PETs. Researchers have addressed in this group the means and processes through which preservice elementary teachers acquire skill and science knowledge as they are socialized into the profession (Bell et al. 2011; Harlow et al. 2013; Yerrick et al. 2008) and the beliefs through which teachers filter their efforts to improve (Deniz and Akerson 2013; Shroyer et al. 2014; Yerrick, Parke, and Nugent 1999). Researchers have also identified the level of content knowledge, pedagogical, and pedagogical content knowledge accessible to this group (Adler 2012; del Pozo et al. 2011; Shim et al. 2011), as well as effective means in raising the collective knowledge and skill of this group (Boone et al. 2011; Schraw et al. 2006; Yerrick and Hoving 2003).

From this research-based assessment, we contend that supporting elementary teachers to transform classrooms into digital learning science contexts, teacher education must address at least these three areas: (a) teachers' knowledge of science concepts and understanding of children's prior experiences with and understanding of scientific concepts, (b) teachers' knowledge of a variety of pedagogical strategies for teaching science, and (c) teachers' understanding of their own past experiences as science learners and how these experiences influence their pedagogical choices Our approach was dedicated to the "content-specific technology integration strategies" of science education that Wang and colleges (2014, p. 657) proposed to address the problem of low-level implementations of pedagogical technology. Our approach also emphasized "meaningful connections between technology and teaching," and "understand[ing] the enabling conditions for technology use" proposed by similar research (Lei 2009, pp. 92–93).

Specifically, our approach to technology in teacher preparation focused largely on the iPad as a student-centered educational tool to mediate thinking in class in conjunction with an online flipped classroom environment (Sams and Bergmann 2012) of preparatory content. At the core of both in-class and online learning are reflective activities through electronic and in-person discussion to help construct a shared meaning about teaching and learning science with technology. Participants were provided access to iPads, video recording kits, digital cameras, peripheral tools like probeware and digital microscopes, and a variety of apps and online resources for teaching science. Class activities included evidence gathering and synthesizing data and information, taking pictures as evidence, gathering and analyzing measurement data with probeware, and composing videos and podcasts.

The software resources included electronic discussion boards, vodcasting (video-based pod casting), and social networking tools for accumulating and sharing timely resources for weekly assignments.

Included in the online electronic resources were more than three dozen video recorded science lessons that featured teacher instruction and interviews, interviews with children, and examples of technologically mediated student artifacts. These video recorded lessons where extracted directly from the learning contexts these PETs were targeting to function within—namely the elementary science classroom grades K-6th. As such, these artifacts were intentionally inserted weekly throughout the course and directed discussions were designed to have PETs recognize the innovative capabilities for inserted scientific inquiry tools as well as the landmarks for reflective change as observed in actual PETs' accounts from past years. The PETs engaged in discussions of these lessons asynchronously on a discussion board. From their live engagement in science lessons and their vicarious engagement in science lessons via technology, PETs were then asked to plan, facilitate, record, and reflect upon a lesson they taught in public schools. As a culminating event, PETs compiled and shared 10 min videos they edited with one another at the end of the semester to gain a deeper and broader representation of what it means to understand content and to assess content instruction.

For the purposes of examining the premise of digital native teachers, the group of PETs discussed who participated in this course were mostly born after 1980, and a few after 1990. The instructor was a certified teacher and technology specialist, as well as a professional developer and published researcher on inquiry-based pedagogy and using technology to teach science. To get ascertain how their science methods course approach compared to other teacher preparation coursework, we asked them to report on their use of tools throughout their time in their teacher preparation program. Figure 6.1 gives a brief summary of their reported classroom technology use. This data was collected from students reporting on what they viewed as actual use during observation period in two separate contexts—our inquiry approach and other educational courses they had taken or were concurrently enrolled.

The PETs responses demonstrate that our approach to teaching science with technology required them to use technology more frequently and towards an inquiry approach. Therefore, our data analysis will examine how this group of PETs responded to what would be a novel suite of technology-rich materials and activities that were intended to foster innovative approaches to teaching science. Through our analysis, we address the broader cultural implications of educational technology by describing and interpreting PETs actions and responses as they learned to use the tools.

We present our findings in three parts. First, we report on the experiences and perspectives on digital tools that the PETs had developed over the course of their own learning. We explore this participation in light of the expectations made of them as digital natives. Second, we look at how novice teachers regarded and enacted tools for specific learning outcomes in both the classroom and in the online environments. Finally, we examine particular digital tools that best supported their

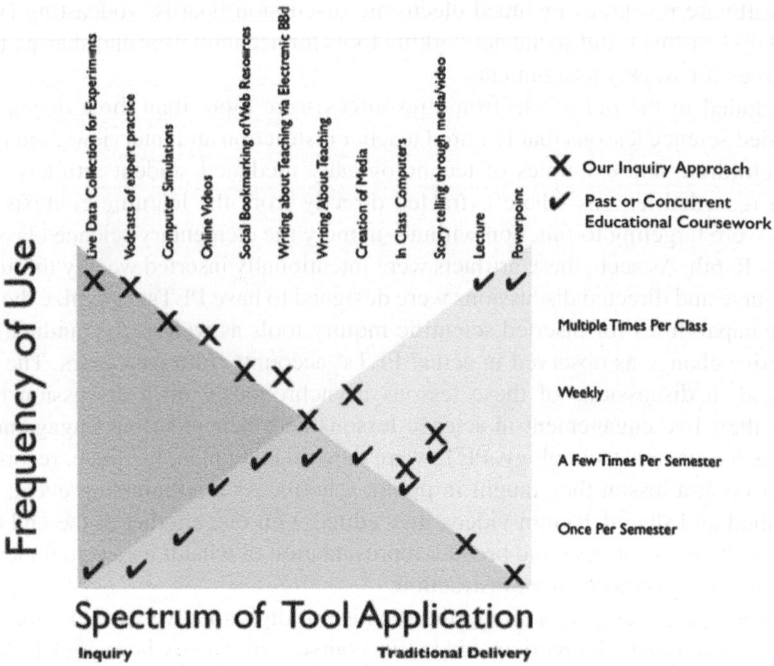

Fig. 6.1 Participants' assessment of preservice classroom technology experiences

goals and expectations as novice teachers in an age of ubiquitous digital technologies. From this analysis, we suggest implications for teacher educators looking to develop the next generation of science teachers in ways that are consistent with reform recommendations, and go beyond the false promises of technology by engaging cultural perspectives and needs of young, emerging teachers.

6.3.1 Prior Experiences and Attitudes: Familiar Tools, Unfamiliar Practices

To get at the idea they may or may not represent the digital native portrayed in literature, we asked the PETs to evaluate themselves as interested and capable users of technology. Most did not rate themselves as technological experts though they did show an affinity on the whole toward technology. A large majority, 87%, agreed they enjoyed technology and 96% said that technology was a major part of their life consistent with the evolving prevalence and frequency technology use (Pew 2015). All of the PETs carried Internet enabled phones, many had computers or laptops they brought to class, and nearly all of them reported having a computer at home to do homework. All students had some kind of Internet connection at home. The majority of reported use of their computer and phone included word processing, creating slideshows, browsing the Internet, using email texting, and using social media. Self-reported social media usage included posting selfie's and other photos, as well as coordinating social events and calendars through shared online websites and apps. A few students had experience using digital still camera, or video camera, or computer mobile device to create media representing their learning in education courses and other university learning contexts.

When asked how they characterized themselves as electronically savvy or advanced, most described themselves more as users than potential leaders of technology implementation. When asked the most innovative thing they had done with technology, there were some advanced uses like the creation and publishing of digital media projects, but they mostly referred to acquiring information, such as finding a creative app, or some other personal use of their smart phone. We pressed PETs for actual examples of their learning with technology and more importantly their examples of teaching (e.g., student teaching, volunteering, educational field work, and other school contexts). Their generated examples of teaching with technology was very sparse among this collection of digital natives (less than 15%). Some PETs had used blogs for teaching; others had created a web page, wiki, or an iMovie to teach a topic. None of them identified themselves as a potentially creative teacher through the implementation of technology.

A rarity among this group, Sarah, shares her self-characterization, which come closest to Prensky's representation of savvy learners:

> I think of myself as an "Old Fashioned Digital Native". I like technology, I see the necessity of it and love how easy it makes doing things. I use technology every day but part of me enjoys a more traditional method, like pen to paper. The teachers I work with are older with no experience [with technology] and the younger teachers I am around really have little knowledge in using technology outside of personal uses like their phone. I am definitely the 'go-to' person at my school when the computers are not working because I am not afraid to try and fix or make something work electronically.

This initial part of Sarah's response suggested a prospective teacher who was capable with technology, and disposed towards the problem-solving strategies that are often required to be a successful innovator. Her experiences with younger teachers also seem to align with the research previously cited. Yet, despite her positive attitude, she acknowledges with some emotion that she was wholly unfamiliar with the affordances of technology as tool for learning science:

> I thought I was all that and pretty good with technology before this class, but I was just using internet searches and online classroom blogs or discussion posts. I was overwhelmed quickly when I learned ICT was endless. I quickly was drowning and felt clueless. I now have ways to think about teaching science I would have never thought of. I was able to see good teaching in action through the different technology we used (e.g., videos, cameras, probeware, iPads). It all served as a good examples implementing strategies to teach science and reflecting on the teaching experience. It helped to demonstrate for me good examples of teaching and children's understanding when I have not had ANY experience student teaching prior to this class, certainly not science.

Despite her recent experience as a high school student, she identified a school culture that still had "technology time," separating using tools from learning the content and processes. As a result, she found our approach to be intellectually and emotionally challenging, which suggests that adoption of technology for inquiry-based science learning, is not a familiar cultural practice, even for a somewhat savvy digital native. As she had observed, digital technology has increased the possibilities of inquiry-based learning, but a vicarious affinity with digital technology did not increase her capacity to use it pedagogically.

Few students were more advanced than Sarah who had high levels of knowledge, experience, and skills using technology. One was Lori, who was a central part of assisting in the class and modeling for her peers. She described herself this way:

> I think I knew a lot entering the class. Certainly more than my peers. They complained a lot more than they tried to figure stuff out. Sometimes you just gotta' try it. As a digital native I think of myself as loving the way technology can excite you and create something that might not have been possible before... It still stresses me that I have to work with the administration to teach the way I want, the way that will reach my students the best. It's a totally different thing, though, to teach with technology. You've got to know what can go wrong and what kids are going to ask you before you teach with it. But just recently, I taught a lesson using inquiry and it really felt natural. I never would have predicted that.

Lori was an example of a true digital native for seeing technology and learning as an embedded practice. She even anticipated some friction between her ideas of modern learning and an aging traditional school culture. Yet, even she was surprised to discover a novel quality to teaching and learning with inquiry after having her own experience teaching science that way to children.

Since these two PETs were atypical, one might wonder if the remaining majority of PETs were simply unaware of their own practices as advanced technological users, like a fish unaware of the water they swim in. Response data revealed this to be not the case, as they tended to *disfavor* technology integration in their own coursework and learning. Also, we found from analyzing interviews, surveys, and the learning artifacts they produced that their comfort using social technology contrasted their use of technology for course integration in classrooms. Of all the enrolled elementary teachers, 53% agreed they had not had the used most of the inquiry technology in prior semesters and 37% exited the class strongly disagreeing that they could use technology to improve their teaching. More than 50% of the students disagreed that "Using technology in class is always a positive experience" and roughly a third of the class preferred courses, which *excluded* technology integration. A substantial portion of the class was not comfortable using technology use beyond Facebook, email, texting, browsing the Internet, and other social networking uses of their devices. Yet, while they expressed both aversion in their own coursework and challenges using technology, there was strong professed desire to be open to learning with new technologies, though this did not consistently manifest itself in their enacted teaching and learning in education courses nor in schools.

This contrast of professed interest and enacted initiative to learn and implement technology has been a critical, under-investigated dichotomy for quite some time (Pedersen and Yerrick 2001; Rakes and Casey 2002; Settlage et al. 2004). Also, in spite of seemingly contradictory responses, 78% expressed public interest in using technology in their future teaching. The practical challenges and experiences of using technology tools in the course were perhaps challenging commonly shared beliefs of technology's innate potential to support teaching and learning. As in other studies, we found that PETs were like other university students who favored technology use, but not necessarily when integrated into courses for professional learning (Yerrick et al. 2013). What this seemingly contradictory response pattern suggests is that our collective use and aspirations with technology and our culture of teaching and learning science are still disparate practices, and digital natives are not coming to teacher preparation with ideas or desires to change practice.

6.3.2 *Enactments in Class: Obstacles to Inquiry with Tools*

In this section we will explain how PETs interactions with ICTs were limited their understanding of inquiry and the use of the tools themselves. Their classroom activities had been designed so that the PETs could use tools to understand inquiry from the perspective of a learner, not the teacher. These were followed by recommendations for teachers hoping to use inquiry as a basis to guide their pedagogical choices. PETs experienced no less than [NUMBER OF] enacted inquiry lessons in chemistry, biology, physics, and earth science, modeling pedagogical practices and engaging PETs in inquiry activities so they could experience and learn science vis-a-vis an inquiry oriented pedagogy. Built into the enacted inquiry activities were technological tools and scaffolded tasks for students to practice data collection (probeware), scientific observation and inference (cameras), predictive power of models (simulations), and other tools for teaching science with inquiry.

However, when asked to employ tools for an inquiry task, many PETs' responses were typically characterized by fact finding, Google searches, over-simplistic answers, and truncating and subverting the inquiry task toward a thin representation of an intended rich knowledge construction task. During a key lesson and activity on inquiry, PETs were assigned to work in collaborative groups to use the technology tools in different ways to construct a collective understanding of Darwinian evolution. Using the classic teaching example of the peppered moth (Grant et al. 1998), paper "moths" were camouflaged and placed throughout the room and then hunted with certain selective restraints. As Cohen (1986) argued, this was a collaborative task intended to be sufficiently complex to warrant group cooperation and accountability and the centrality of each individual's role. This activity was designed so that one person was nominated as the designated content "expert", another was nominated as "photographer" documenting survival rates of camouflaged moths, and a "media" member was responsible for constructing a brief multimedia production (60 s) representing the group's learning from the activity. Performing as students, it was observed that many PETs engaged the learning task with limited enthusiasm or effort. "Experts" reported that they hadn't read the entire website given to them explaining the scientific concept, "Photographers" failed to photograph the most discrepant examples they found, and the "Media" experts took no notes or photos of the process.

In an attempt to animate the participation of the PETs, the instructor circulated throughout the room offering pointed advice regarding expert practices. Despite the effort, some PETs were apparently stalled and talking about non-coursework related events. Many challenges were seemingly technological, as one student struggled with the iPad camera, admitting, "I'm not good with iPads." Distressingly, however, the student sought no help in continuing with their task and overcoming a technological obstacle potentially meaningful to a future classroom. Another group professed to be stalled in completing their task because they could not connect to the classroom wireless router and had sat for 10 min, asking no one for help. When approached, they were ironically using their phone and cell data to check Facebook, an apparent schism to their understanding of mobile devices connecting the classroom to the outside world. In addition, most of the groups gave little attention to the final part of the activity where they were asked to create a digital artifact synthesizing online information and their observations hunting for moths. Rather than taking advantage of a rich communal opportunity to build understanding of both content and pedagogy, they deployed a familiar low-level literacy task by looking to find a paragraph from the webpage offered by the instructor and had one member paraphrase their "right" answer in a short audio recording.

The value of looking closely at this one event is to appreciate that the low-level engagement in this inquiry activity observed is less as a technological phenomenon, and more as a cultural one. It is attributable to naïve beliefs concerning science and the epistemological orientations of students toward the nature of the discipline itself. Many scholars have shown that teachers' beliefs about the discipline being factual, received, and authoritarian can negatively influence the way they approach academic tasks designed to promote more inquiry oriented representations (Hammer 1994; Lampert 1990; Pintrich et al. 1993; Roth and Roychoudhury 1994; Schoenfeld 1992; Stodolsky et al. 1991). While technology was intended to support a rich inquiry task, some of the PET's used cultural affordances of the tools as either an excuse to not engage, or as a means to sidestep the intended thinking of the task. Whatever the reason, PETs had used technological tools individually and collectively in brief and shallow spurts, which prevented them from gaining the full benefit of designed inquiry tasks. The technology activities were intended to facilitate collaborative participation and the use of investigative tools to approximate science as a discipline consistent with the NGSS. Instead, the tools were employed by PETs to reach the shallow goals of right answers, consensus, and authority driven decisions more quickly. Though we strongly support the use of mobility tools for teaching, it was clear that additional coaching and scaffolding will be required to foster appropriate usage of tools for learning. This scaffolding needs to address directly the anti-academic patterns of personal uses of the devices that function as silent disruptions of the class.

6.3.3 Enactments Online: Resistance to Using Social Technology to Inform Practice

As part of their preparation, PETs were asked to participate in a flipped-style classroom format. They were required to watch videos prior to class that documented technology-based science learning in the classroom from both teacher and student perspectives along with example science lessons and online resources for learning science content. To support their understanding, PETs were asked to interact with one another regarding how they made sense of children's science learning from examples provided. In the online setting, PETs were expected to contribute and comment on science education resources through social bookmarking tools like Diigo and an online discussion board as a way of building a co-constructed learning community. The social bookmarking tool would allow them to collect, annotate, organize and share their favorite web pages (Estellés et al. 2010). The use of these tools in education are founded in constructivist principles aimed at helping users go from an intrapersonal learning process to an interpersonal process by collecting, describing and organizing group resources (Estellés et al. 2010). Such tools are intended to strengthen collaborative education spaces online and mirror the participatory culture that exists in a growing amount of informal learning spaces (Jenkins 2006).

Unfortunately, similar to the in-class example, many cited technical obstacles to engaging these online tasks. Between 8 and 25% of students would respond to the weekly assessment with statement such as, "I didn't have time to watch the videos," or "My computer wouldn't connect," or "The link I tried wouldn't work." These students left this section of their weekly quiz blank, despite the fact that more than 75% of the students could have helped them to access the help and support provided online or send them the links they had used or guide them through the process of downloading and syncing course resources to their mobile device. This was not the only or even the primary obstacle, however. Many PETs approached their assignments meant for community building as mundane tasks. In class, they recounted completing online tasks to "get it over with," and did not employ tools in expert or thoughtful ways. When asked in class why they highlighted certain resources for their peers, a typical response was, "It was the first thing that came up, and no one had bookmarked it yet." The topics assigned to all PETs were intended to complement the example video lesson, readings, NGSS standard, and the weekly reflection where they were asked to write about their developing thoughts and concerns on how *they* might use technology to teach science. Evaluations, surveys, and focus groups revealed many PETs disliked the use of social bookmarking and discussion boards to share and discuss online teaching resources, calling it busywork, despite their reported propensity to do similar activities in their personal use of Facebook.

Contrary to using the venue to building a strong social network and sharing expertise, these Digital Natives struggled to find any purpose in posting their opinions, commenting on one another's reflections, or sharing relevant web resources.

Some PETs showed a strong disinterest in reading and resented having to comment on the posts of others:

> Lauren: "Honestly? ... I think that the discussion boards are just kind of one of those mundane tasks…especially responding to other people, I don't, find it that effective."

> Hope: I think the videos should be directly related to the readings. I need someone to pinpoint the important components of them – "Here's what to look for…"

What is being observed here is a posture of 'received knowing' (Clandinin, Davies et al. 1993) in that they expected to be shown directly anything they did not know. This is despite perceived need to be capable to find and use information as opposed to recalling it (Bransford et al. 2004, p. 4). Miller and McVee (2013) describe these two concepts as performance knowledge and propositional knowledge, respectively. Propositional knowledge refers to knowledge that already "exists" while performance knowledge consists of "knowing how to find, gather, use, communicate, and imagine new ways of envisioning assemblage of knowledge" (p. 3). They not only wanted to acquire their information 'really fast,' as Prensky (2001) suggested, some also wanted to minimize the intellectual effort necessary to make sense of it in the service of their coursework and ultimately, their teaching practice.

It should be noted that the in-class discussions and hands-on learning activities were well regarded and rated highly impactful in contrast to the virtual engagement of PETs in the flipped environment. Despite the fact that more than 90% of the students felt the collection of short classroom video excerpts of inquiry teaching were an intricate part of learning to teach science, at the end of the course 58% of the students reported that they did not show up prepared to class having watched the three video assignments. Prensky (2010) and others (Sams and Bergmann 2012) have argued this flipped environment is a design effective in leveraging class time for more meaningful discussions with deeper and greater learning. In contrast, 30% of the PETs admitted they didn't complete all the readings for class but twice that number reported they had failed to watch their brief video assignments (less than 10 min each). This was borne out not only by self-admission, but also the quality of journals and regular assessments. Weekly quizzes were intended to discriminate between thoughtful and low-level responses. As a cultural practice, the online learning component was not fully engaged by all PETs, suggesting there is still discomfort or disinterest with this particular use of the tools.

Though it is a commonly held belief that learning through online social media is a venue where younger learners excel and even prefer, this cohort of young learners support the opposite view. Face-to-face instruction was most highly valued despite the challenges highlighted earlier, and often because they overtly expressed they struggled or did not want to learn to learn independently through electronic means.

Table 6.1 Summary of most helpful practices for teacher development

What students found most helpful about technology	Tool	% Students
Actually working with children	–	95%
Teacher demonstrations	Vodcasts	95%
Videos of expert practice	Vodcasts	95%
Storytelling through video	iMovie	95%
Writing about my teaching	Discussion board	91
Hands on lab experiments	Probeware	86%
Creating media projects	iMovie	79%
Informal science outdoors/museums/etc.	–	67%
Writing about others' teaching	Discussion board	62%
Live data collection	Probeware	62%
Writing activities/journaling	Discussion board	58%
Computer simulations	iPads	49%
Browsing the internet	iPad	41%

6.3.4 Enactments That Engaged: Observing and Reflecting on Practice with Digital Video

Despite clear challenges, technology was leveraged in ways that was perceived to be beneficial by the PETs. Students were asked which tools and strategies were most helpful students and we reported the most helpful approaches. Of the reported uses of technology and constructivist strategies, those which were rated the highest included teacher demonstrations, digital storytelling, and a variety of inquiry based tools for scientific exploration. A summary of the reported most helpful usages are listed in Table 6.1.

Though tools were popular among most PETs, favoring technology was not universal nor were all uses of technology considered equal. It is very common to have PETs rate actual experiences teaching and engaging with children above most other components of teacher education design. Rated among these real life experiences structured into the methods course, however, were also some of the vital uses of technology, notable those which provided access to vicarious experiences and those which were used to make sense personal ones. In addition to rating the effectiveness of tools, the favored uses of technology, and which course components were valuable, PETs were also asked to rank their top five learning gains. Of the more than 30 potential options for ranking their top components for learning to teach, the most reported valuable outcomes for employing technology to teach science through inquiry were:

1. 'Understanding children's thinking'
2. 'Inspiring my teaching'
3. 'Understanding science content better'
4. 'Understanding the teaching is not uni-directional'
5. 'Learning to value the process of watching myself as a professional'

Looking at the many technological applications in the course, in hindsight it is easy to see why digital video was the best tool for supporting these goals, and why the social networking tools and flipped content methods were less effective. These insights are not afforded particularly well by the ways in which digital natives are accustomed to using those tools.

6.3.4.1 Watching Others Teach and Reflect with Digital Video

As part of the course, PETs were given a minimum of three weekly examples of vodcasts to facilitate vicarious experiences with other former PETs based upon collected artifacts during teaching. While there have been some reports of science educators using video case studies to vicariously immerse preservice teachers into issues surrounding elementary inquiry teaching (Abell et al. 1998; Beck et al. 2002; Lampert et al. 2016; de Mesquita et al. 2010; Sherin and van Es 2005), few accounts have engaged methods students in the process of telling their own evolution of thought using video vignettes. Even fewer accounts offer PETs the opportunity to engage with similar students in similar contexts to learn from and build upon the examples of PETs before them. PETs used the video editing tools to reflect upon children's thinking, and also examine, critique, and change their own actual teaching practices and the children's learning. The tools for digital editing gave them unprecedented access to evaluate their own teaching, which in many cases was their first attempt to teach science to children.

As preservice teachers began to use digital video for exploring children's beliefs and science learning, they revised their planned lessons and questioned their level knowledge of children and of science through class interactions, collaborative planning, and online reflections shared among peers. Vodcasts and videos directly challenged their preconceived notions of science teaching they learned in their past experience and their assignment to conduct, record, and edit a clinical interview surrounding a specific student misconception revealed many faulty assumptions among the PETs. For example:

> Tori: I definitely need to review my science content; I have forgotten many concepts about science. Videos helped document learning in real time and were good for reflecting and seeing what children learn. I learned to listen and let the students see why they are wrong or right. I learned there are other ways to get around your challenges, and there are tons of strategies that you can teach your students so that they will be successful, and make a difference.

6.3.4.2 Watching and Reflecting on Their Own Teaching

As a culminating task for the methods course, preservice teachers edited a final 10-min video depicting their learning process as an emerging science teacher. Teachers illustrated their understanding of children's thinking, defended their pedagogical choices, and provided evidence of their success as science teachers.

Rich discussions accompanied their shared videos depicting themselves employing a variety of strategies, exploring children's thinking through open-ended questions, providing evidence contrary to common-sense thinking, and soliciting commentary between children about competing ideas. Of all the technology supported activities, PETs overwhelmingly pointed to this combination of teaching of lessons in real classrooms and creating an iMovie reflection of the event as the most valuable part of the course. From an instructor's perspective, this was also the most valuable evidence of teacher learning and growth. Of the more than three hundred statements made regarding the learning activities they learned the most from, teaching and creating their reflection video was discussed the most (reported from more than 70% students more than once in open-ended feedback). Regardless of whether they felt they had expertly facilitated an inquiry lesson or not, or whether they felt they had enacted a lesson consistent with the NGSS, *all* students did attempt to write and facilitate inquiry lessons with children and reported that the activities corresponding to the reflection upon that event were the most beneficial. In fact, one student pointed out that there was value in seeing the limits of one's skills:

> Lori: The video reflections were a painful reminder how far I still need to progress if I hope to be a good teacher.

Such a perspective represents a growing maturity of perspective for elementary preservice teachers. Researchers argue that teachers' ability to reflectively position themselves within knowledge claims can lead to subsequent increased knowledge growth independent of discipline (Abell et al. 1998; Gustafson and Rowell 1995; Hofer and Pintrich 1997; Pintrich et al. 1993). The process of identifying one's weaknesses (pedagogical, content, or other) is an important stage of reflection and growth as a practitioner. Recognizing strong and weak lessons, failing and successful attempts are a sign of knowledge growth among PETs. For this reason, we interpret the recognition knowledge growth from examples of both failure and success as a positive form of knowledge reflection. During exit interviews and surveys, no fewer than eleven different preservice teachers cited failures alongside of example successful teaching examples as helpful prompts for learning to teach science.

> Tori: I have an example from which to start. The videos helped me learn to teach better by providing a variety of lessons to reflect on. When I first began this course, I had a lot of the misconceptions about science content that a lot of my students will have. The videos helped the most by showing how to get student misconceptions shared so that the teacher can address them. It provided an opportunity to see how others teach using inquiry-based learning. I had no previous experience to inquiry based learning before this semester. I think it is really important to learn from other teacher's or even experts mistakes, and understand that they too have failed and succeeded in multiple different ways just like your own.

Prospective teachers learned to identify more clearly the characteristics of inquiry-based lessons in their lesson critiques, article reviews, and peer-lesson evaluations. They began to write more articulate journal entries about teaching dilemmas and children's thinking. More teachers had become able to identify the real struggles surrounding the question of how to teach less content for greater understanding, and they express these revelations in journals that address misconceptions and difficult decisions about cutting certain content.

Considering the poor example teaching most PETs described in their autobiography, explaining the tales of woe that often led them away from a career in science, making the most of their perceived failure and missteps was an important lesson to these novice science teachers. They leveraged these bad experiences towards good ones. Sarah said it best revealing her next steps for becoming an excellent teacher,

> Videos showed me examples of teacher questioning, and allowed me to see the value in the student's own answers/questions. I was one of those students who I hated science in the past. I had awful teachers, and when it came down to it; did not learn anything. But I learned a lot of different strategies [from this approach] that changed my outlook on science and that I can hopefully use in my future classroom to change the outlook of a future student. I learn that I am not an idiot! I can teach science and I think I can be better than my teachers were for me.

It is hard to imagine how the positive learning outcomes in many of the PETs responses could have been achieved without the affordances of digital video recording, editing, and online delivery. These tools allowed a majority of these young PETs to engage what *they* thought was truly valuable: working with children to gain a deeper understanding of teaching and learning science, watching a variety teachers teach science in real situations, and critically, watching and reflecting on their own science teaching. This was especially important for elementary teachers who may have been uncomfortable teaching science, and had negative prior experiences in their own learning. The video not only showed them what could be improved, it showed them that it was not as difficult as they anticipated, and that standards-based science teaching was actually possibly in public school contexts, and was even interesting for both them and their students.

6.4 Understanding Enactments: Powerful Cultural Contexts at Work

It should be emphasized that enactments and attitudes observed do not suggest that there is something deficient about the PETs observed, or digital natives in general. Instead, it suggests a pervasive popular cultural use of technology, consistent with empirical evidence, that ICTs are predominantly designed to mediate personal communication and entertainment (Carr 2010; Wang et al. 2014). It was clear throughout our prolonged engagement with this cohort as well as our experiences teaching other younger generation teachers that PETs' technical agility across even familiar, unsophisticated collection tools we invoked in during their preservice programs did not demonstrate an immediate nor proficient disposition for learning to teach more effectively. Our data also suggests that the skills and knowledge culturally transferred by the use of technology for these purposes does not establish lasting knowledge, which transfers over to other knowledge domains—particularly those required for teaching and learning. Furthermore, it remains unchallenged that their common cultural uses and cognitive requirements could be incompatible with the intellectual demands of science teaching (Carr 2010). Given the response data we shared, and the support of over a decade of research finding limited uses of technology as a tool for learning (Cuban 1986; Wang et al. 2014), it seems reasonable to assume most PETs who participated in our approach were not exposed to the tools and methods of teaching and learning science in their 20 years of schooling. Indeed, from their own testimonials provided at the beginning the course, they recalled mostly traditional experiences with educational technology, and many reported struggles to be successful in science. Despite the obstacles however, when the technology aligned with their goals, preservice teachers exercised their technological knowledge and skills, employed the technology to develop new knowledge, and engaged technology to apply what they had learned toward developing, teaching, and reflecting upon lessons.

6.5 Recommendations for Moving Forward

Overall, skepticism is warranted regarding claims that preservice teachers are from an advanced, digital native generation who are savvier employing technology for teaching and learning (Bennet et al. 2008). Many current technology practices emphasize frequent social interaction, popular culture, and novel information, which are engaged uncritically compared to the intellectual traditions of developing rich understanding (Carr 2010). We can no longer proceed on the presumption that one kind of prowess in our technoculture translates to other contexts. We therefore recommend some specific changes in the preparation of elementary science teachers to develop the next generation of savvy science teachers.

Scaffold Public Collaboration with Technological Tools in methods classes. The use of specific tools for learning needs to be integrated and taught explicitly during teacher preparation courses and fieldwork. It needs to not only be employed, but also its purposes unpacked and specific pedagogical deployment tailored for explicit learning outcomes. For example, if blogs are used for the purpose of co-constructing knowledge, university instructors should be aware that blogs have a specific use and purpose outside the classroom for accomplishing different ends. University students need scaffolding in their writing to explore ideas, challenge one another's thinking, and thoughtfully weigh different opinions. Without such scaffolding, students will resort to their historical practices of documenting mundane self-centric observations (typified by interactions on Facebook and Twitter) and posting personal beliefs without careful examination. Likewise, in instances when novice teachers are using tools to gather evidence, they must be taught also the interpretive framework for that assists them in a more analytical approach to evaluating the world. Otherwise they will take graphs, images, and evidence as absolute representations without considering the implicit error and interpretative nature of scientific experimentation.

Contextualize Technological Tasks within the culture of today's upcoming teachers. The technological practices, learning tasks posed, scientific questions asked, and pedagogical approaches modeled in teacher education need to be situated in such a way to reveal new and deeper insights regarding children and content, and not just "ported" to online platforms. A deliberate, situated effort is needed if young teachers are to recognize the value of the tool and the specific learning intended as an outcome. For example, science concepts and data should not just be examined in abstraction in a software simulator. Rather, prospective teachers should use tools to test their own hypotheses, make predictions within their own local environment, and have the same rich meaningful learning experiences they would wish for their students. Such connections to real world events have been emphasized for decades in science education literature (Talbot-Smith et al. 2013). They are no less important preparing teachers to use technology. Furthermore, since elementary preservice teachers intend on teaching children, it becomes vital for them to see the tool in action in the lives of their children. There are authentic activities, such as documenting invasive species through mobile apps like *Leaf-Snap* which engages children in *citizen science* (Silvertown 2009). Preservice teachers should have extensive opportunities to view and facilitate children using these tools in afterschool programs, robotics clubs, in school labs, and other venues so that their perceptions of their use of tools as teachers is commensurate with their professional training.

Leverage Interests, Goals, and Research Based Approaches for teacher learning. Reflection upon teaching incidents needs to be a regular and central practice, especially given the ubiquitous nature of mobile tools and the interest of new teachers. Novice elementary teachers are predisposed toward children's learning and new generation of younger teachers use their camera functions more than any other phone apps capturing photos, selfies, and videos for social network purposes. At the same time, research has demonstrated that video reflection can be a key tool for transforming traditional teaching practice (Abell et al. 1998; Martin and Siry 2012; Roth 2009; Tomáš and Seidel 2013; Yerrick et al. 2005). This technocultural convergence of teacher interest and research driven practices should be capitalized upon. These generalizations are borne out from our data as well. Teachers we studied demonstrated that they enjoyed using their camera technology and looked forward to using it in their own classrooms. They also ranked the vicarious engagement through video excerpts working directly with children, teaching demonstrations, hands on activities, and activities related to narrating such experiences as the most impactful professional learning experiences. Teachers' interests and values need to be leveraged together with video self-reflection practices to inspire preservice teachers to become better teachers. Such a convergence of what educators want to know with what they need to learn can be leveraged to promote the values we espouse for future classrooms.

Provide Safe Contexts for Teacher Implementation. Finally, we need to heed the volumes of research that indicates the profound impact of mentors and models for teachers. If we are to move teachers from personal technology users to expert teachers seamlessly incorporating technology, we need to halt the duplicitous treatment of technology standards for teaching. If educational institutions adopt NETS and/or NGSS as *standards*, they should be treated as such—not simply as good ideas.

Yet we find veteran teachers not meeting standards continuing to use the same explanations they offered decades ago for not using technology (Czerniak et al. 1999; Gado et al. 2006; Pedersen and Yerrick 2001; Wang et al. 2014; Yerrick et al. 2011). Teachers who profess, "I'm the older generation," "We never did it that way," "I am technologically deficit," or even, "I don't have to," should not serve as mentors if our goal is to grow novice teachers to meet new standards. As teacher educators we need to be consciously and intentionally intolerant of efforts to demand new teachers use educational technology in unreceptive environments, unsupportive of our intended aims. It is an unfair expectation of novice teachers to act as change agents and placing novice teachers in precarious positions leads only to increased attrition among new teacher talent.

We believe it is possible to improve science teaching through expert use of technology and appropriate pedagogy. We also believe we will only succeed in this goal if we carefully examine the underlying assumptions regarding requisite teacher knowledge, tool-use, and learner characteristics. We believe that technological tools available to educators today should be leveraged far beyond implementation into existing practices. Literature is replete with evidence demonstrating minimal pedagogical impact for exorbitant technological investment (Cuban 1986, 2009; Oppenheimer 2007). Such a teacher education approach falls short of visions for transformative, societal-oriented changes representative of today's STEM reform visions. We agree with Carberry and Baker (2018) in Chap. 10 of this volume with regards to the importance culture plays in teacher training and practice, and embrace Waight and Abd-El-Khalik's (2018), in Chap. 7 of this volume) notions of considering participants' interactions, values, and beliefs around technology for contextualizing research as we consider the larger sociocultural context tools are employed. To this end, our study is but one small but illuminating example, demonstrating even the most basic and socially popular tools for communication and collaboration can be employed by pre-service teachers in higher education contexts to reveal dysfunctional, transmissional, and idiosyncratic cultural dispositions. These pre-service teacher dispositions and practice can often confound efforts to prepare elementary teachers to invoke inquiry in their future classrooms. We must therefore consider what sociocultural orientations teachers bring to bear on technological tools and not assume their social practices will directly lead them to expert teaching with those same tools. In this way, ours is a case study demonstrating the extended impact of accepting false assumptions surrounding 'next generation' teachers. It provides a cautionary tale of technology ineffectively employed for the transformation of elementary science education in a microcosm of obstructions, agendas, and challenges embedded in the motivations of tomorrow's teachers which must addressed by teacher education if we are to achieve new heights. Only then will we be able to leverage these tools to situate young teachers within novel and challenging forms of teaching and learning science, which foster a deep and lasting appreciation of how technology should play a role in reaching the next generation of students.

References

Abell, S. K., Bryan, L. A., & Anderson, M. A. (1998). Investigating preservice elementary science teacher reflective thinking using integrated media case-based instruction in elementary science teacher preparation. *Science Education, 82*(4), 491–509.

Adler, J.D. (2012). From campus to classroom: A study of elementary teacher candidates' pedagogical content knowledge.

Alan H. Schoenfeld, Ernst von Glasersfeld, (1992) Journal for Research in Mathematics Education 23 (3):290

Beck, R. J., King, A., & Marshall, S. K. (2002). Effects of videocase construction on preservice teachers' observations of teaching. *Journal of Experimental Education, 70*(4), 345–361.

Bell, D. (1979). The new class: A muddled concept. *Society, 16*(2), 15–23.

Bell, R. L., Matkins, J. J., & Gansneder, B. M. (2011). Impacts of contextual and explicit instruction on preservice elementary teachers' understandings of the nature of science. *Journal of Research in Science Teaching, 48*, 414–436. doi:10.1002/tea.20402.

Bennett, S., Maton, K., & Kervin, L. (2008). The 'digital natives' debate: A critical review of the evidence. *British Journal of Educational Technology, 39*(5), 775–786.

Boone, W. J., Abell, S. K., Volkmann, M. J., Arbaugh, F., & Lannin, J. K. (2011). Evaluating selected perceptions of science and mathematics teachers in an alternative certification program. *International Journal of Science and Mathematics Education, 9*(3), 551–569.

Bransford, J., Vye, N., Bateman, H., Brophy, S., & Roselli, B. (2004). Vanderbilt's AMIGO project: Knowledge of how people learn enters cyberspace. Learner-centered theory and practice in distance education: Cases from higher education, pp. 209-234. In T. M. Duffy, J. R. Kirkley (Eds) Learner-Centered Theory and Practice in Distance Education: Cases From Higher Education, Routledge.

Carberry, A. R., & Baker, D. R. (2018). The impact of culture on engineering and engineering education. In Y. J. Dori, Z. Mevarech, & D. Baker (Eds.), *Cognition, metacognition, and culture in STEM education* (pp. 217–239). Springer.

Carr, N. (2010). *The shallows: How the internet is changing the way we think, read and remember*. London: Atlantic Books Ltd..

Clandinin, D. J., Davies, A., Hogan, P., & Kennard, B. (1993). Finding new ways to work together: Living in the middle ground. Learning to teach, teaching to learn: Stories of collaboration in teacher education, 210-217.

Cohen, E. (1986). *Designing groupwork: Strategies for the heterogeneous classroom*. New York: Teachers College Press.

Cuban, L. (1986). *Teachers and machines: The classroom use of technology since 1920*. New York: Teachers College Press.

Cuban, L. (2009). *Oversold and underused: Computers in schools 1980–2000*. Cambridge, MA: Harvard University.

Czerniak, C. M., Lumpe, A. T., Haney, J. J., & Beck, J. (1999). Teachers' beliefs about using educational technology in the science classroom. *International Journal of Educational Technology, 1*(2), 1–18.

del Pozo, R. M., Porlán, R., & Rivero, A. (2011). The progression of prospective teachers' conceptions of school science content. *Journal of Science Teacher Education, 22*(4), 291–312.

de Mesquita, P. B., Dean, R. F., & Young, B. J. (2010). Making sure what you see is what you get: Digital video technology and the preparation of teachers of elementary science. *Contemporary Issues in Technology and Teacher Education, 10*(3), 275–293.

Hammer, D. (1994). Epistemological beliefs in introductory physics. *Cognition and Instruction, 12*(2), 151–183

Deniz, H., & Akerson, V. (2013). Examining the impact of a professional development program on elementary teachers' views of nature of science and nature of scientific inquiry, and science teaching efficacy beliefs. *Electronic Journal of Science Education, 17*(3), 1–19.

Estellés, E., Del Moral, E., & González, F. (2010). Social bookmarking tools as facilitators of learning and research collaborative processes: The Diigo case. *Interdisciplinary Journal of E-Learning and Learning Objects, 6*(1), 175–191.

Gado, I., Ferguson, R., & van't Hooft, M. (2006). Inquiry-based instruction through handheld-based science activities: Preservice teachers' attitude and self-efficacy. *Journal of Technology and Teacher Education, 14*(3), 501–529.

Grant, B. S., Cook, A. D., Clarke, C. A., & Owen, D. F. (1998). Geographic and temporal variation in the incidence of melanism in peppered moth populations in America and Britain. *Journal of Heredity, 89*(5), 465–471.

Gustafson, B. J., & Rowell, P. M. (1995). Elementary preservice teachers: Constructing conceptions about learning science, teaching science and the nature of science. *International Journal of Science Education, 17*(5), 589–605.

Harlow, D. B., Bianchini, J. A., Swanson, L. H., & Dwyer, H. A. (2013). Potential teachers' appropriate and inappropriate application of pedagogical resources in a model-based physics course: A "knowledge in pieces" perspective on teacher learning. *Journal of Research in Science Teaching, 50*, 1098–1126. doi:10.1002/tea.21108.

Hofer, B. K., & Pintrich, P. R. (1997). The development of epistemological theories: Beliefs about knowledge and knowing and their relation to learning. *Review of Educational Research, 67*(1), 88–140.

Hutchins, E. (1995). How a cockpit remembers its speeds. *Cognitive Science, 19*(3), 265–288.

Jenkins, H. (2006). *Convergence culture: Where old and new media collide*. New York : NYU press.

Lampert, M. (1990). When the problem is Not the question and the solution is not the answer: Mathematical knowing and teaching. *American Educational Research Journal, 27*(1), 29–63

Lampert, M., Heaton, R., & Loewenberg-Ball, D. (2016). Using technology to support a new pedagogy of mathematics teacher education. *Journal of Special Education Technology, 12*(3), 276–289.

Lave, J., & Wenger, E. (1991). *Situated learning: Legitimate peripheral participation*. Cambridge: Cambridge University Press.

Lei, J. (2009). Digital natives as preservice teachers: What technology preparation is needed? *Journal of Computing in Teacher Education, 25*(3), 87–97.

Martin, S.N., & Siry, C. (2012). Using video in science teacher education: An analysis of the utilization of video-based media by teacher educators and researchers. In *Second international handbook of science education* (pp. 417–433). Springer Netherlands.

McVee, M. B., Dunsmore, K., & Gavelek, J. R. (2005). Schema theory revisited. *Review of Educational Research, 75*(4), 531–566.

Miller, S. M., & McVee, M. B. (Eds.). (2013). *Multimodal composing in classrooms: Learning and teaching for the digital world*. New York: Routledge.

NGSS Lead States. (2013). *Next generation science standards: For states, by states*. Washington, DC: The National Academies Press.

O'loughlin, M. (1992). Rethinking science education: Beyond Piagetian constructivism toward a sociocultural model of teaching and learning. *Journal of Research in Science Teaching, 29*(8), 791–820.

Oppenheimer, T. (2007). *The flickering mind: Saving education from the false promise of technology*. New York: Random House.

Pedersen, J., & Yerrick, R. (2001). Technology in science teacher education: Survey of current uses and desired knowledge among science educators. *Journal of Science Teacher Education, 11*(2), 131–153

Pew Internet & American Life Project. (2015). *Teens, social media & technology overview 2015*. Available http://www.pewinternet.org/2015/04/09/teens-social-media-technology-2015/Last retrieved Sept 2015.

Pintrich, P. R., Marx, R. W., & Boyle, R. A. (1993). Beyond cold conceptual change: The role of motivational beliefs and classroom contextual factors in the process of conceptual change. *Review of Educational Research, 63*(2), 167–199.

Prensky, M. (2001). Digital natives, digital immigrants part 1. *On the horizon, 9*(5), 1–6.

Prensky, M. (2010). *Teaching digital natives: Partnering for real learning*. Thousand Oaks: Corwin Press.

Rakes, G. C., & Casey, H. B. (2002). An analysis of teacher concerns toward instructional technology. *International Journal of Educational Technology, 3*(1), 1.

Rideout, V. J., Foehr, U.G., Roberts, D. F. (2009). *Generation M: Media in the lives of 8-to 18-year-olds*. Henry J. Kaiser Family Foundation.

Roth, K. J. (2009). Using video studies to transform science teaching and learning: results from the STeLLA professional development program. In *The power of video studies in investigating teaching and learning in the classroom* (pp. 225–242).

Roth, W. M., & Roychoudhury, A. (1994). Physics students' epistemologies and views about knowing and learning. *Journal of Research in Science Teaching, 31*(1), 5–30.

Sams, A., & Bergmann, J. (2012). *Flip your classroom: Reach every student in every class every day*. Washington, DC: International Society for Technology in Education (ISTE).

Schraw, G., Crippen, K. J., & Hartley, K. (2006). Promoting self-regulation in science education: Metacognition as part of a broader perspective on learning. *Research in Science Education, 36*(1–2), 111–139.

Settlage, J., Odom, A. L., Pedersen, J. E. (2004). Uses of technology by science education professors: Comparisons with teachers' uses and the current versus desired technology knowledge gap. *Faculty Publications: Department of Teaching, Learning and Teacher Education, 81*.

Shaw, D. B. (2008). *Technoculture: The key concepts*. New York: Berg.

Sherin, M. G., & van Es, E. A. (2005). Using video to support teachers' ability to interpret classroom interactions. *Journal of Technology and Teacher Education, 13*(3), 475–491.

Shim, M. K., Young, B. J., Paolucci, J. (2011). Elementary teachers' views on the nature of scientific knowledge: A comparison of inservice and preservice teachers. *Electronic Journal of Science Education, 14*(2).

Shroyer, G., Riggs, I., Enochs, L. (2014). Measurement of science teachers' efficacy beliefs. In *The role of science teachers' beliefs in international classrooms* (pp. 103–118). Sense Publishers.

Silvertown, J. (2009). A new dawn for citizen science. *Trends in Ecology & Evolution, 24*(9), 467–471.

Stodolsky, S. S., Salk, S., & Glaessner, B. (1991). Student views about learning math and social studies. *American Educational Research Journal, 28*(1), 89–116.

Talbot-Smith, M., Abell, S. K., Appleton, K., & Hanuscin, D. L. (Eds.). (2013). *Handbook of research on science education*. New York: Routledge.

Tomáš, J., & Seidel, T. (Eds.). (2013). *The power of video studies in investigating teaching and learning in the classroom*. Bremen: BoD–Books on Demand.

Waight, N., & Abd-El-Khalik. (2018). Technology, culture, and values – Implications for enactment of technological tools in precollege science classrooms. In Y. J. Dori, Z. Mevarech, & D. Baker's (Eds.), *Cognition, metacognition, and culture in STEM education*.

Wang, S. K., Hsu, H. Y., Campbell, T., Coster, D. C., & Longhurst, M. (2014). An investigation of middle school science teachers and students use of technology inside and outside of classrooms: Considering whether digital natives are more technology savvy than their teachers. *Educational Technology Research and Development, 62*(6), 637–662.

Wertsch, J. V. (1998). Mind as action. New York: Oxford University Press.

Yerrick, R., & Hoving, T. (2003). One foot in the dock and one foot on the boat: Preservice science teachers' interpretations of field-based science methods in culturally diverse contexts. *Journal of Science Education, 87*, 390–418.

Yerrick, R., Parke, H., & Nugent, J. (1997). Struggling to promote deeply rooted change: The "filtering effect" of teachers' beliefs on understanding transformational views of teaching science. *Science Education, 81*(2), 137–159.

Yerrick, R., Ross, D., & Molebash, P. (2005). Too close for comfort: Real-time science teaching reflections via digital video editing. *Journal of Science Teacher Education, 6*(4), 351–375.

Yerrick, R., Ambrose, R., & Schiller, J. (2008). Ascribing legitimacy. *Electronic Journal of Science Education, 12*(2), 132–170.

Yerrick, R., McLaughlin, S., McDonald, S., Thompson, M. (2011). Collected from the cutting room floor: An examination of teacher education approaches to digital video editing as a tool for shifting classroom practices. *Journal of Contemporary Issues in Technology Education, 11*(1). Available: http://www.citejournal.org/vol11/iss1/currentpractice/article1.cfm.

Yerrick, R., Lund, C., & Lee, Y. (2013). Exploring simulator use in the preparation of chemical engineers. *Journal of Science Education and Technology, 22*(3), 362–378.

Chapter 7
Technology, Culture, and Values: Implications for Enactment of Technological Tools in Precollege Science Classrooms

Noemi Waight and Fouad Abd-El-Khalick

7.1 Introduction

Winner (2003) asked: "Where does one go to learn what one needs to know to write confidently about philosophy and technology" (p. 233)? In ensuing discussion, Winner explained that this learning involved a two-part process; it involved studying a conglomerate of factors that influence technology and developing understandings while immersed in the context. He explained how Mostert studied history, engineering, and economics and lived and traveled in a tanker in order to develop a better understanding. In this way, Mostert shared a keener awareness of how supertankers worked and, more importantly, was able to explain the related dynamics of supertankers. Instead of trying to understand supertankers at a distance or in isolation from context, Winner noted that the combined knowledge and resources, as well as rich instantiations of use in context, offered the most robust understandings of tankers (the technology in this case] in society. Using this approach, Winner suggested that one could learn more about technology in the context of "a school where computers are being introduced" (p. 224). Indeed, we advance that much can be learned about technology when we examine the nuances of context and give attention to the totality of factors that include interactions, resources, curricula, knowledge, and skills, as well as relevant organizational, economic, and political structures and social and cultural factors. We also advance that much can be learned about technologies in context when we give attention to the cultural milieu and related

N. Waight (✉)
Graduate School of Education, University at Buffalo, SUNY, Buffalo, NY, USA
e-mail: nwaight@buffalo.edu

F. Abd-El-Khalick
School of Education, The University of North Carolina at Chapel Hill, Chapel Hill, NC, USA

Y.J. Dori et al. (eds.), *Cognition, Metacognition, and Culture in STEM Education*, Innovations in Science Education and Technology 24,
https://doi.org/10.1007/978-3-319-66659-4_7

interactions that serve as the foundation of how technologies diffuse and are appropriated in learning environments.

Our own work so far (e.g., Waight et al. 2014; Waight and Abd-El-Khalick 2007, 2011) has involved extended observation and study of science teachers and science education teams and attempts to implement technologies—some of which originated in scientific research—in precollege science classrooms. These technologies included Internet-based databases, such as Biology Student Workbench, applications of computer and Internet platforms for research, modeling tools such as ChemViz and applications of Flash and NetLogo modeling tools in chemistry education, and use of microcomputer laboratory probes for data collection and analysis. We have identified a reoccurring theme in our studies: Technologies undergo many transformations and assume different roles and functions based on the user(s) and their "expertise," as well as context of deployment. In other words, the same technology may undergo different realizations in different contexts, even when the overarching goals for technology use and science teaching and learning are similar. How technologies are realized in context can be explained by the bounds of culture and cultural interactions within the environments where technologies are deployed.

A rather illustrative case of this theme stems from our examination of the implementation of Biology Student Workbench (BSW) (Waight and Abd-El-Khalick 2011). We found that this implementation featured several "best practices," which are consistently espoused in the science education literature (National Research Council 2011). First, there was substantial and sustained funding from organizations—including the US National Science Foundation, which employs rigorous criteria for evaluation of funded projects—that spanned more than a decade. Second, BSW's cycle from design to implementation involved a conglomerate of professionals with deep scientific, pedagogical, and technical expertise. These experts introduced multidisciplinary perspectives at all stages of the BSW process, which was led by scientists who developed the original tools on which BSW was modeled and who were prolific researchers in their respective domains of expertise. Third, the tool underwent various iterations of testing and development to ensure an appropriate platform for use by high school students. Fourth, curricular applications sought to align student learning with standards and social issues of the time (e.g., SARS, AIDS and Sickle Cell Anemia). Fifth, science educators and teachers were involved in all stages of development: They ensured that teachers' and students' voices were represented in the process. Sixth, consistent with one of the most common recommendations for technology implementation, science teachers received numerous opportunities for short- and long-term professional development (PD) opportunities. Finally, these PD opportunities were extended to include classroom visits, which involved technical, content, and pedagogical support. In some cases, teachers received 2 years of support via the GK-12 fellows program (see Waight and Abd-El-Khalick 2011).

Despite these best practices, our findings showed that the enactment of BSW in four science classrooms was, to say the least, less than optimal. The case of BSW illustrated a dynamic interplay between the context, culture, experiences, and expectations of social agents in several domains, including scientific research, the design environment of technologists, science education research, and the world of school science and practicing teachers. How the above factors interact is critical to our understanding of productive instances of adoption and/or instances that hinder the adoption and espoused implementation of technologies in science classrooms. This work expanded our repertoire of factors that require attention in decisions that promote design, development, and technology integration, in approaches that evaluate implementation, and in understandings of the complexity of technological practices. In this chapter, we seek to illustrate how culture and values underpin the dynamics of the social interactions of science teaching and learning in the context of technological tools. Toward that end, we draw on understandings of the nature of technology [NoT] (Waight and Abd-El-Khalick 2012), technology practice (Pacey 1983), notions of technology as prosthetics (Clark 2003), and the ecological nature of the learning environment (Zhao and Frank 2003). Collectively, these perspectives highlight the significance of culture and values as an important dimension in design, development, implementation, and enactment of technologies to support science teaching and learning.

7.2 Nature of Technology

Understandings of NoT provide a much-needed critical framework—and currently missing body of knowledge—to examine how technology interacts with individuals, society, culture, institutions, and economy (Waight and Abd-El-Khalick 2012). The NoT framework is significant for science education because it provides a lens to interrogate the artifact, its interactions, and the role of culture and context and engages the full cycle from design and development to enactment and discard. More specifically, NoT engages five core dimensions that help to explain realizations of technology in precollege science classrooms. These are the role of culture and values, notions of technological progression, technology as part of systems, technological diffusion, technology as a fix, and notions of expertise. In this chapter we elaborate on the significance of understanding the role of culture and values. We posit that the activity and interactions that manifest in the context of design, development, implementation, and enactment of technologies are social activities that are in constant flux. With this focus in mind, the chapter first develops a conceptualization of technology as process and activity. Second, we explore technology and culture as articulated by technology practice (Pacey 1983), notions of technology as prosthetics (Clark 2003), and understandings of technology and rationality (Volti 2010). Third, we discuss NoT with a specific focus on core dimensions related to science education. Fourth, we address the relationship of technology and science education. Specifically, we explore Zhao and Frank's (2003) ecological framework for technology implementation and outline an analysis of empirical studies to illustrate technology implementation in precollege science. Finally, we present a four-stage critical framework to expand opportunities for analytical and explanatory scrutiny of the role of culture and values in the process of design, development, adoption, implementation, and adoption of technologies in precollege science classrooms. We conclude with recommendations for researchers and educators.

7.2.1 Technology as Artifact, Process, and Activity

Philosophy of technology offers perspectives that expose the complexity of the instrument, technology as process and activity, associated interactions, and the role of knowledge and technique, as well as political, economic, and cultural implications. In this regard, philosophers of technology explained that technology is best understood in light of cultural structures and expectations (Ellul 1964; Illich 1973; Tenner 1996). Indeed, Rapp (1999) espoused a consistent theme on the impact of technological tools in society; he argued that any reasonable approach to technology should address natural and cultural aspects. Cultural aspects draw attention to interacting factors that emphasize structures, devices, and systems, as well as technique and skills needed to use technology (Tenner 1996). Likewise, Volti (2010) stressed the collective impact of technology: technology is "a combination of devices, skills and organizational structures" (p. 5). To complement these perspectives, Arthur (2009) organized three categories of meanings of technology and summarized technology as a "means to fulfill a purpose: a device, or method, or process. A technology does something. It executes a purpose" (p. 29). Note how these perspectives draw on an organized system of various interacting factors—technologies are viewed in conjunction with skills, knowledge, and cultural elements, which impact the purpose, execution, and outcomes of these technologies.

Other perspectives have emphasized the importance of technique, technical skill, and expertise. For example, Ellul (1964) focused on technology as technique: He noted technique as a "group of movements, of actions generally and mostly manual, organized and traditional, all of which unite to reach a known end for example, physical, chemical or organic" (p. 13). Tenner (1996) also stressed that everyday objects are both technology and technique. In this case, technology is described as consisting of the structures, devices, and systems we use. Defined from the perspective of power and dominance of technological expertise, McDermott (2009) stressed that the notion of dominance is embedded in a hierarchical system that privileges expertise and knowledge, both of which often are represented in only a small segment of the population. McDermott puts it this way; technology refers fundamentally to "systems of rationalized control over large groups of men, events, and machines by small groups of technically skilled men operating through organized hierarchy" (p. 87).

7.2.2 *Technology and Culture*

Pacey (1983) posed a fundamental question about technology: "Is technology culturally neutral?" (p. 3). In response, he claimed that one could argue that the machine's construction and its working principles could be viewed as culturally neutral. However, once a web of human interactions and activities are introduced as part of the machine's use, status, supply, organization, and its associated skills and knowledge, then arguments for neutrality dissipate. This position suggests that technology is deeply wrapped into ways of life and, thus, does not exist separate from or in a different compartment. If educational technology is to be of any use, Pacey argued, it "must fit into a pattern of activity which belongs to a particular lifestyle" (p. 3). In fact, Pacey argued that much of the confusion about technology lies in its faulty characterization, which often omits the associated activities and web of interactions. When technology is viewed as more than just an artifact or thing, as a process and activity, other aspects of the nature of technology become more salient.

7.2.2.1 Technology Practice: Technical, Cultural, and Organizational Facets

Pacey (1983) characterized process as technology practice: "technology-practice is the application of scientific and other knowledge to practical tasks by ordered systems that involve people and organizations, living things and machines" (p. 6). He argued that technology as a structure and practice have three dimensions: the cultural, organizational, and technical aspect. The cultural aspect addressed goals, values and ethical codes, beliefs in progress, as well as awareness and creativity. The organizational aspect addressed economic and industrial activity, professional activity, users and consumers, and trade unions. The technical aspect addressed knowledge, skill, and technique; tools, machines, and chemicals; and live ware, resources, products, and wastes. Altogether, understandings of technology practice offered an expanded perspective that engaged beliefs and values tied to the process of technology. These beliefs and values are often tied to the "users of equipment and their patterns of organization" (p. 8) and largely remain unexamined in discussion related to opportunities and challenges of technological implementation.

Pacey (1983) used three case studies to illustrate the synergy among cultural, organizational, and technical factors. The first case outlined the development of the snowmobile, which was used to clarify the misconceptions of value-neutral technologies. The first snowmobiles were designed with a rubber and steel crawler to transport seven passengers. Consequently, this design influenced more current models of the machine. In the United States and Canada, the snowmobile was used as a motorcycle for extra mobility during long winters and as recreation in holiday resorts. Meanwhile, in other countries, the snowmobile was an integral part of the livelihood of people. For instance, in the Swedish Lapland, the machines were used for reindeer herding, while the Eskimo trappers in the Canadian Arctic used them to harvest fox furs. Pacey suggested that whether a machine is used for recreation or to earn a living, it remained the same machine. However, to fully understand its function, one of the first steps was to understand how the machine was used, maintained, and organized in different communities. When viewed from this perspective, it was clear that the "machine designed in response to the values of one culture needed a good deal of effort to make it suit the purposes of another" (p. 8).

The second case focused on the breakdown of hand-operated water pumps in remote villages. Technology practice emphasized the technical aspect, which was evident with the design. Pump maintenance, on the other hand, reflected organizational structure, while the cultural aspect addressed interactions between the engineers who understood the technology and villagers who viewed this technology as indestructible. Collectively, technology practice exposed the nuance details of why certain pumps thrived while others rapidly deteriorated. In other words, the human aspect of technology—its cultural and organizational aspects—was more prominent in this case. With the third case, Pacey used the power-driven wheel and steam engine to illustrate how discussions of progression focused on the machine and its technical capabilities, all at the same time ignoring how human contributions influenced how these machines were used. Pacey stressed that it has always been a misunderstanding to believe that the steam engine shaped the expansion of the factory. Rather, the major contributions came from work organization. Merchants saw the benefit of creating a centralized workspace in order to control production, control against material embezzlement, achieve better quality, enforce longer work hours, and increase pace of work. The machine's introduction and impact was intertwined with a novel organization of work and workforce, which were as much an invention as their mechanical counterpart, that is, the steam engine.

The above cases highlighted how cultural factors and context of deployment and use provide alternative ways to think about technologies. That is, technologies were guided by specific goals and served specific populations, solved problems but simultaneously created new ones with different needs, and resulted in practices that were multidimensional and nonlinear. In fact, Pacey (1983) argued that multidimensionality and nonlinear technology practice result from human activity and associated cultural and organizational factors.

7.2.2.2 Technology as Prosthetics

Clark (2003) presented a view of cyborgs as human-technology symbiont, essentially, "thinking and reasoning systems whose minds and selves are spread across biological brain and nonbiological circuitry" (p. 3). He elaborated that humans share a natural tendency to create and exploit nonbiological devices. In other words, there is a natural tendency for humans to seek out devices to facilitate activities. Clark argued that when these devices become part of our everyday activities and needs, they become assimilated into our way of life. These devices, like prosthetics, function as extensions of our natural capacities. To illustrate the notion of prosthetic, Clark used the example of phones as "something you use and something that is part of you" (p. 9). These tools as extensions become even more important when information exchanged between the user and the tool is fluid and bidirectional. When this occurs, devices become transparent and in effect function as a "proper part of the user" (p. 103). Transparent technologies offer the best balance between need and use.

The notion of technology as prosthetics further reinforced how human interactions and beliefs create seamless boundaries between technology and user. What is more, transparency also highlights value structures tied to the cultural context of use. Here there is a need for caution since technologies may achieve different levels of transparency. So, while a technology may function as a prosthetic and extend human capacity, this same technology may remain obscure and external to the needs of another user. Prosthetics may thus reflect the cultural norms and expectations of a particular group or society.

7.2.3 Technology and Rationality

Technological development is based on a rational approach undergirded by a set of values and orientations that value progress as a measure of human success. These beliefs about progress reflect cultural values that view technology as synonymous with human progress. Rationality thus reflects our continuous appetite for technological solutions. Volti (2010) explained that "a society imbued with a rational ethos is dynamic and essentially optimistic" and may "alter existing ways of doing things to gain particular benefits" (p. 13). Technologies are created and function as solutions to various kinds of societal problems. Solutions may introduce new problems, sometimes more intractable than the original problem.

7.2.4 Nature of Technology (NoT) and the Relationship of Technology and Culture

Mitcham (1994) stressed the importance of philosophical thinking about technology: Such thinking interrogates the "making and using of artifacts" (p. 543), which occur in the context of various human interactions. The process of design, development, implementation, and adoption involve a web of interactions among numerous agents and stakeholders. As Mitcham urged, it is important to reflect how philosophy may aid our understandings of these processes. In Waight and Abd-El-Khalick (2011), we explained that philosophical understandings draw attention to the inherent nature of technology—essentially, understandings of NoT provide a framework to understand and evaluate transformations that technologies undergo as they move through and interact with different aspects of society (Heidegger 1977; Pacey 1983; Tenner 1996). Broadly, this framework addresses individuals, society, institutions, economy, politics, and culture. At the microscale, and most relevant to our science education context, the following core dimensions of NoT become particularly relevant: role of culture and values, notions of technological progression, technology as part of systems, technological diffusion, technology as a fix, and notions of expertise.

These core dimensions evolved from analyses that drew on empirical works, as well as works from philosophy and history (e.g., Illich 1973; Pacey 1983; Tenner 1996) and nature of technology (Arthur 2009). The analysis reflected empirical data garnered from examining the design process of Biology Student Workbench (BSW), which involved interviews with numerous stakeholders and classroom agents, classroom observations of implementation and enactment across four different classrooms, semi-structured interviews that focused on experiences of teachers and students with BSW, and semi-structured interviews that captured stakeholder reactions to the documented instantiations of BSW in science classrooms. Our findings revealed that despite the best practices involved in design and development, actual implementation and enactment were less than optimal and did not meet the vision, investment, and high expectations that key stakeholders had for BSW (see Waight and Abd-El-Khalick 2011). Below we summarize how the aforementioned core dimensions provided an alternative framework to interrogate these rather disappointing findings (for detailed discussion, see Waight and Abd-El-Khalick 2012).

With BSW, all of the activities associated with design, development, implementation, and enactment involved numerous interactions among stakeholders (technologists, computer scientists, scientists, educational researchers) and classroom agents (teachers and students). These interactions involved reciprocal exchange of knowledge, resources, and associated activities. In this chapter, we explain how culture and values function as the unifying dimension that undergirds the life cycle of technologies in science education precollege learning environments. Pacey (1983) reminds us that these interactions highlight the organizational, cultural, and technical aspects of technology. As he noted, only when we take into account these three axes of interactions, do we yield the best understanding of the successes and challenges of technology implementation and use. As with the case of hand-operated water pumps, challenges were often attributed to technical issues, while cultural elements were ignored. In contrast, instances of successful implementation—fewer breakdowns and better pump performance—were attributed to practices that accounted for organizational structure and cultural interactions. So, successful instantiations of pump maintenance involved individual and managerial responsibility and contributions of technical expertise.

Pacey's argument promoted the view that technology is a complex endeavor that requires attention to numerous factors, which are firmly rooted in understandings of the values, goals, beliefs in progress, ethical codes, and notions of creativity that guide cultural practices. When we take into account this complexity, it is clear that the often-espoused linear approaches to technology are misleading. The latter approaches tend to privilege outcomes that reflect technical expertise while ignoring human contributions and organizational factors. However, when cultural practices and values are taken into account, other factors such as different levels of expertise, contributions, and associated interactions emerge as important.

7.2.5 *The Core Dimensions Revisited*

When we revisit the NoT framework, dimensions that address technological progression, technological diffusion, technology as a fix, and notions of expertise are clearly observable in understandings of cultural practice. The impetus for new technologies is driven by human attributes that reflect an appetite for continual progress. Technological progression is driven by a need to make things better and faster, which is informed by our individual and collective cultural values. Pacey (1983) emphasized that the cultural aspects of technology are defined by the goals and values and beliefs in progress. In science education technological progression is fueled by the core idea that technologies provide opportunities for authentic experiences that replicate scientific and/or engineering practice. Scientific and engineering practice is a dynamic endeavor that keeps pace with cutting-edge technologies. For example, Bybee (2013) explained that STEM reform efforts are intended to preserve global STEM dominance and in particular develop STEM literacy, generate a workforce with twenty-first century skills and competencies and, match the pace of innovation in these respective domains. While these purposes are of tremendous value in the science education sphere, the associate rhetoric hardly captures the complexity of the process. What is more, these views, which tend to follow linear patterns, often omit that technological progression is uniquely a human endeavor, which is best understood in context and alongside the values that guide practice (Volti 2010).

Technological progression is closely associated with the rate of diffusion of technologies. Volti (2010) noted that speed of diffusion can be attributed to a conglomerate of factors: (1) advantages over existing technology, (2) compatibility with existing values of the transfer context (be it an organization, an institution, or a classroom), (3) ease or difficulty of understanding and applying the new technology, (4) ease in experimenting with the new technology or employing it on a trial basis (smooth learning curve), (5) extent to which positive results are immediately apparent, and (6) special kinds of people (e.g., with appropriate expertise, attitudes) who allow for the effective flow of technology. In this respect, diffusion involves transfer of technologies, knowledge, and ideas across contexts. The nature of transfer thus determines the success or failure of certain technologies. Volti further explained that successful transfer often is attributed to contexts that value and promote high rates of innovation. Thus, two sets of complimentary practices are at play during technological diffusions: valuing continuous progress, which is at the core of societies, and complimentary changes enacted by social agents to meet the needs of the new technology. Indeed, the hand pump case study illustrated that successful adoption and implementation occurred when complimentary changes associated with individual and managerial roles were enacted.

The need for constant improvement—making life better—is embedded in a belief that technology can "fix" societal issues. Technology is thus viewed and applied as a solution to various nontechnical problems. Without understanding the full scope of nontechnical social problems, technical "fixes" often take the form of superficial solutions that omit complexity. Traditionally, technical issues often take precedence over the social and cultural factors, which are at the core of technological fixes. For example, Sullivan (2008) and Verma et al. (2015) argued that exposure and experiences with robotics promote scientific literacy and, thus, counter students' disillusion and boredom with science. Similarly, Dori and Hameiri (2003) and Dori and Kaberman (2012) argued that computer-based models provide access to molecular phenomena that is inaccessible to novice learners. These calls illustrate how technologies are introduced into the learning environment with the intention of fixing issues associated with teaching and learning. Here, Volti (2010) cautioned that technological fixes might address particular issues while overlooking other major issues. Thus, in many cases where technological fixes are applied, markers of success and failure are difficult to define because of the numerous interacting factors associated with social and psychological causes. This analysis highlights how culture and values should be at the center of actions that forge adoption of new technologies, which propose to "fix" nontechnical issues.

The success of technological adoption and implementation is also a function of expertise, knowledge, and skills. Expertise can manifest itself as technical knowledge of the workings of and understandings of the black box, knowledge of applications, or knowledge associated with the discerning capabilities of a user. In all of these instances, the nature of expertise, knowledge, and skills determine the type of interactions, and realizations, that occur with a technology in the hands of experts and nonexperts (Waight and Abd-El-Khalick 2011). The various realizations can span the gamut of outcomes, from successful and effective to ineffective outcomes. Diffusion of the technology, knowledge, and skills may also determine realizations in context. When new technologies are introduced, various shifts may occur to accommodate the new tools. This process of change is complex and, in the case of hand pumps, may involve organizational and knowledge restructuring. While the above discussion, so far, has focused on technologies broadly, the next section addresses how these understandings may manifest in the context of technologies applied to improve (or fix issues!) with science teaching and learning.

7.2.6 Technology and Science Education

Research studies related to technology integration in science education have included intervention and naturalistic studies. Often, intervention studies seek to solve a problem and to better understand the problem in the context of science teaching and learning. While naturalistic studies have similar aims, they seek to understand the context in its original, unaltered state. This distinction is important because it may inform how researchers report on technology use in science classrooms. For intervention studies, generally, there is a preoccupation with reporting the effects of technology, and specifically, noting its role in successfully facilitating science teaching and/or promoting learning. This preoccupation often occurs at the cost of dismissing other factors that may inform best-case scenarios for successful implementation. In addition, while the outcomes are documented, the nuanced details of these outcomes—that is, the precise mechanisms that affect changes or behavior—are not always clear. So, the science education community is too often denied an understanding of the totality of impact of the technological process of design, development, adoption, implementation, and enactment.

It is here that understandings of NoT become pertinent. NoT is imperative to evaluate the totality of interactions associated with technology design, development, adoption, implementation, and enactment in precollege classrooms. Indeed, cultural factors, expectations, and values are implicated in all NoT core dimensions. Technology exists recursively, where it impacts and is impacted by many factors that determine how technologies are appropriated in context (Eglash 2004). Thus, success and sustainability of technologies in serving science teaching are dependent on the characteristics of users, context, and other key agents. This section first outlines the relevance of Zhao and Frank's (2003) ecological framework: Understandings of technology practice share clear lineage with understandings that emphasize ecological interactions of technology implementation and enactment in science education contexts. Next, we explore empirical studies that address implementation of inquiry-supported technologies, such as models and modeling, and computer-based tools, such as the Internet and software. We conclude with a summary that addresses the main themes of understanding technology practice and the ecological nature of technology implementation.

7.2.7 The Ecological Nature of Technology Implementation, Adoption, and Enactment

Zhao and Frank (2003) proposed an ecological framework that emphasized four "metaphorical equivalents to draw attention to the uses and interactions of computers: schools are ecosystems; computer uses are living species; teachers are members of a keystone species; and, external educational innovations are invasions of exotic species" (p. 811). Notions of schools as ecosystems draw attention to the complex nature of classroom interactions with biotic and abiotic components. Biotic components include relationships with teachers, students, parents, school administrators, and other stakeholders. Abiotic components focus on the supporting environment, and school and class resources, including physical setting, instructional resources, and curriculum, among others. The biotic and abiotic factors exist in reciprocal interactions that continuously modify the roles and nature of these relationships. The second aspect of the ecological framework addresses the notion of computers as living species. Computers are viewed as the invading species into the classroom milieu. This view is grounded in the notion that computers as living organisms also follow an evolutionary path (Basalla 1996). Zhao and Frank aptly described it this way: "Some of the technologies are judged to be more useful, or fit for the task, than others, and they survive while others perish that are judged to be less fit" (p. 812). Thus, understanding the uses of technology can provide valuable insight into those tools that have propensity to survive and replicate. In concert, it is also important to understand which factors facilitate survival and replication. These elements of the framework entail a set of crucial questions: Are technologies that survive ones that promote innovative approaches to learning, or alternatively, do the technologies that survive promote traditional modes of learning? Is survival most likely when there is balance among the technical, cultural, and organizational factors? In what ways do cultural and social factors influence survival?

The third aspect of the framework address teachers and their roles as key members in these interactions. The nature of the relationship among teachers themselves also informed how the invading species (i.e., computers) can survive or perish. In other words, teachers are generally focused on their individual classrooms, but may build social capital by assisting other teachers or seek their help, in this case, about uses of computers. The fourth aspect address innovations as exotic species, which when introduced into an environment (the classroom), intentionally or unintentionally affect the equilibrium of the environment. At this point, the environment may undergo various changes. Perhaps most relevant to the context of teaching and learning is that both species survive, but this may occur at a cost to the environment. For example, teachers and students may benefit from using computers, but this outcome may be dependent on characteristics related to teacher effort to learn and implement a technology effectively, and/or use of instructional time to participate in innovative learning. When this occurs, less time might be dedicated to activities that involve content learning, which may be more valued by the school culture. A second outcome might be the invader perishing and, thus, removed from the system. The third outcome involves both invader and existing species surviving, thus, undergoing mutual changes to accommodate benefits of both. The end result is that the environment and associated species would "go through a process of variation and selection and acquire new properties" (Zhao and Frank 2003, p. 813).

What this analysis reveals is that survival of a technology in the classroom is not just a matter of success or failure, but also of continuous changes among the biotic and abiotic factors—teachers, students, curriculum, and technology. The parallel with Pacey's (1983) technology practice is striking. Biotic and abiotic factors are observable in organization, cultural, and technical structures. As exemplified with the case of pumps, optimal use occurred when changes accommodated the benefits of human and material resources.

We follow with an analysis of the science education literature and, specifically, we explore a representative sample of studies that address implementation of inquiry-supported technologies, which include models and modeling, and computer-based technologies. We use these studies to illustrate how lessons that could be derived from NoT, technology practice, and ecological nature of technologies, are absent in the repertoire of findings, analysis, and discussion in these studies.

Such studies, we suggest, can benefit from understanding how the interactions of culture and values impact technology realizations in science classrooms.

7.2.8 Models and Modeling

A set of studies addressed student challenges related to visualization of macro and submicro scientific phenomena. These studies emphasize conceptual learning related to models and modeling, content understanding, student knowledge, and expertise. Models have been approached as tools that potentially can improve understanding of scientific phenomena, and mediate observations of unobservable phenomena via various levels of representations (Adadan et al. 2010; Barab et al. 2000; Liu and Hmelo-Silver 2009; Wu et al. 2001). Indeed, this research domain has traditionally focused on the cognitive processes involved in students' epistemological understandings of models and modeling (Sins et al. 2009; Stains and Talanquer 2008); how learning is facilitated through models in different disciplinary domains, such as chemistry (Dori and Hameiri 2003; Gobert et al. 2011); and how models improve understanding of abstract concepts (Ardac and Akaygun 2004; Crawford and Cullin 2004; Kozma and Russell 1997).

For example, Dori and Hameiri (2003) examined how students navigated the representations of chemistry phenomena. They documented how students were more successful with macroscopic representations, and experienced difficulties at the particulate level. To address these difficulties, a Multidimensional Analysis System (MAS) tool was implemented. The MAS tool was used to organize and scaffold the complexity of problems (e.g., macro ←→ symbol; micro ←→ symbol; process ←→ symbol). This approach allowed students to move from understandings of observable to particle level phenomena. Stains and Talanquer (2008) also addressed how expertise in chemistry facilitated students' classification of chemical reactions at the macro and submicro level. Similar to Dori and Hameiri, novice students engaged with surface processes, but struggled in their explanations of underlying phenomena. In contrast, experts with higher levels of knowledge and more experiences with modeling, moved easily through the process of problem solving.

In order to overcome some of the challenges cited by Dori and Hameiri (2003), Urhahne et al. (2009) suggested the use of 3D representations of chemical structures. From a conceptual perspective, imagery via models would relieve students from having to create these mental images. Wu et al. (2001) argued that it is necessary to guide students as they transfer ideas from 2D to 3D representations and seek to make linkages between macro and submicro explanations (Liu and Lesniak 2006). In fact, some of these studies exposed that the challenges at hand were associated with limited chemistry background knowledge and expertise with models and modeling, than only with ability to generate mental or other representations of chemical constructs (Kozma and Russell 1997; Zhang et al. 2006). Thus, it was suggested that immersion with background content could assist students' learning process.

All of the above studies were about students and their learning process. Teachers also experienced challenges with implementation and adoption of models in their science classroom. Like students, teachers revealed conceptual difficulties with modeling and transformations of phenomena at the submicro level (Crawford and Cullin 2004; Kikas 2004). In addition to content knowledge, teachers had to negotiate their pedagogical content knowledge (PCK) in order to meaningfully apply models in their instruction (e.g., De Jong et al. 2005; Koehler and Mishra 2008). Limitations with this knowledge resulted in reduced competency with models and modeling. In effect, teachers were more likely to forego use of these tools for teaching and learning. Factors related to teacher beliefs and attitudes were also critical in their decisions about using computer-based tools to model phenomena in the service of improving student understandings (Zacharia 2003).

7.2.9 Computer-Based Technology

A set of studies focused on modes of learning and instructional dissemination in the context of computer-based environments. For example, Hsu (2008) explored the impact of teacher- versus student-centered approaches in the context of a technology-enhanced learning (TEL) course, which addressed conceptual understanding of seasonal changes. The TEL theoretical framework incorporated cognition, integration of multimedia tools, and development of conceptual understanding. The TEL model focused on five phases: contextualization, sense making, exploration, modeling, and application. While there were overall gains in students' deep understandings of seasonal changes, student-centered approaches were more effective ($F = 28.05$, $p < 0.001$). Kim et al. (2007) also applied a pedagogical framework to evaluate teaching and learning in a technology-enhanced, inquiry-based science course. Their framework addressed the macro context, reform and standards, their teacher's community, the internal and external teacher connections for crosspollination of expertise, and the micro context of the classroom space where teaching and learning occurred.

Tao and Gunstone (1999) developed computer simulations to address students' conceptual development in mechanics. Specifically, they used computer simulations to present discrepant events and explore how these helped students confront alternative conceptions. The authors noted that the role of the simulations "was that of a tool, for providing discrepant events to effect conceptual change so that the change process can be investigated" (p. 862). While the study did not address the efficacy of the simulations, the findings revealed that students' conceptions were context dependent and unstable: student explanations vacillated between accepted scientific and alternative ideas. Finally, it was noted that only a few students provided consistent scientific explanations: they were able to identify the big ideas represented across different contexts of the phenomena. What is important in this study is that the authors dismissed the tool—computer simulations that facilitate conceptual change—in light of uneven findings. Instead, they highlighted how students' conceptions may be context dependent and, thus, resistant to change.

Jaakkola et al. (2011) compared learning about electricity in two different environments: A simulation versus an environment that combined simulation and real circuits. The study also investigated how learning outcomes were mediated via implicit (only procedural guidance) and explicit (structure and procedural guidance) instruction. The findings revealed that explicit instruction during the discovery process in the context of simulation only supported student understandings of electricity. In contrast, explicit instruction used in the combined environments of simulation and real circuits did not improve students' understandings. Explicit procedures seem to have sabotaged the role of inquiry. Instead, the authors noted that the inquiry process was best facilitated via implicit instruction. Importantly, there were higher gains for student understanding in the combined over the simulation-only environment.

Finally, Marbach-Ad et al. (2008) explored how computer animations and illustrations influenced learning of molecular genetics. They compared three groups: the control group received traditional lecture format, while the experimental groups received instruction with a computer simulation or illustrations. The authors found no significant differences among the three groups. However, responses to an open-ended questionnaire revealed that students in the computer animation group demonstrated more understanding of genetics concepts when compared with the other two groups.

7.2.10 Summary

The above studies addressed conceptual and pedagogical challenges for K-12 science teaching and learning. Modeling and other computer-based interventions were offered as a potential "fix" for these challenges. Researchers focused on charting teacher and student conceptual understanding of science content, models, and modeling, as well as teachers' pedagogical approaches. The studies found that difficulties persisted and increased with the complexity of abstract, scientific phenomena. However, the persisting challenges were directly linked to teacher and student expertise and knowledge, or lack of modeling skills. Indeed, an examination of how findings were presented reveals a consistent theme: implementation and enactment of models and modeling was often addressed in isolation of the broader context of teaching and learning. In other words, these studies did not address how technical, cultural, and organizational factors were implicated in the process of conceptual understanding. Also, there was apt silence on the disruption of the classroom ecology and how the invading tool (e.g., computer modeling) altered existing classroom environment. Notions of expertise and knowledge remained a reoccurring concern across all these studies. Thus, this begged the question, which biotic and abiotic factors were relevant to conceptual change in the context of models and modeling?

Given our understandings of technical knowledge, which knowledge and skills were scaffolded in this learning context? Pacey (1983) and Zhao and Frank (2003) stressed that both technology and users have to benefit in order for the technology to survive and become sustainable in the learning environment. Empirically, our task is to highlight how these benefits are related to the axes of technology practice and how technology practice is represented in interactions within the ecological framework of technology integration in science teaching and learning. Teaching and learning in the context of technological tools is a social action mediated via a multiplicity of interactions that are informed by culture and values.

7.3 Technology Practice: Applications for the Science Classroom

The above analysis illustrates how culture and values is a constant in design, development, adoption, implementation, and enactment of technological tools in precollege science classrooms. Notably, understandings of culture and values emerge from concurrent understandings of technology practice (Pacey 1983), NoT in science education (Waight and Abd-El-Khalick 2012), how technologies function as prosthetics (Clark 2003), and the ecological nature of technology implementation (Zhao and Frank 2003). This suggests the need for more critical evaluation of the nuances of practice in the context of technological tools to provide a more informed and holistic account of the impact of these technologies. As analysis of the literature exposed, it is not enough to just focus on conceptual learning, or instructional approaches, or the affordances of particular technologies, or notion of technologies as a fix, or that technologies are important because they represent progression and innovation. Instead, it is important to stress how conceptual learning materializes in the context of the social and cultural interactions fostered by teacher- and/or student-centered approaches, how values inform teaching and learning orientations, how technologies extend our capabilities and function as prosthetics, and how the ecology of classrooms reorganizes to accommodate the presence of new learning tools. We contend that this kind of analysis is poised to present the most complete representation of the culture of technology in context.

In this section, we propose a much-needed critical framework that expands opportunities for analytical and explanatory examination of the role of culture and values in the process of design, development, adoption, implementation, and enactment of technologies in precollege science classrooms. The framework comprises four stages: (1) preplanning, (2) intent, (3) adoption and enactment, and (4) outcomes. We delineate the major subcomponents at each stage, which explicate how aspects of culture are manifested in classroom practice. Importantly, two major themes were reoccurring across all four stages. The first theme highlights the purpose of the technology, while the second focuses on the outcomes of each stage of the framework.

This framework is particularly relevant for preservice science teacher educators and for practicing teachers, who need guidance in their approach to and evaluation of the impact of implementation and enactment of technology. The nuanced planning framework and guiding questions may also be instructive for science education researchers.

7.3.1 *Preplanning: The Purpose of the Technology*

The process starts by identifying the purpose of the technology: Why is this particular technology being considered for implementation in a particular context? Context can involve grade level, science content, associated state standards (e.g., Next Generation Science Standards, NGSS Lead States 2013) and, location of the school and classroom milieu. Next, preplanning identifies the organizational structure that supports the technology at the school and classroom levels. Organizational structures draw attention to infrastructural supports, such as network capabilities, bandwidth, and access at the school and classroom levels. These structures also include personnel support and their distribution across school and individual classrooms. How does the expertise of the support group align with the needs of the science education context? How are responsibilities negotiated among the different stakeholders at this stage? Once this information is identified, it is important to revisit how the initial purpose for implementation and adoption aligns with the organizational structure of the school and classroom (see Fig. 7.1).

Information on the organizational structure will also illuminate the culture of technology at the school and classroom level. The technical infrastructure and level of support can shape instructional capacities and expectations. In this case, one might pose the following questions: In what ways is the purpose of the technology aligned with instructional expectations and capacities? In what ways will the purpose of the technology promote or constrain interactions in the context of teaching and learning? These questions allow us to anticipate the range of interactions that may result from knowledge of the organizational structure and purpose of the technology.

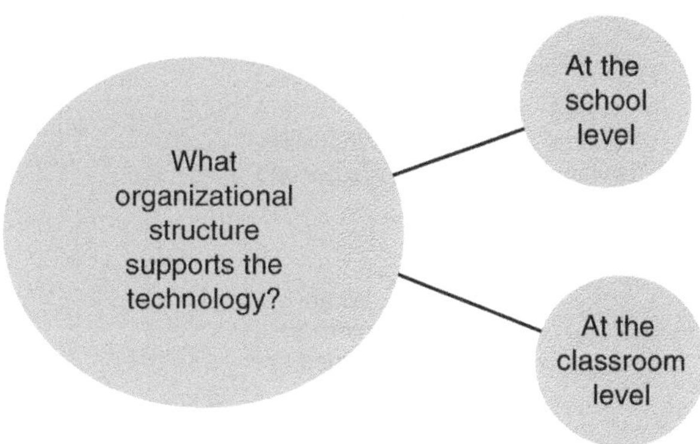

Fig. 7.1 Preplanning and framing the purpose of technology

7.3.2 *Intent: Planning*

Once the organizational structure is defined and the purpose revisited, then the process can address intent (see Fig. 7.2). This stage of planning is guided by two major themes. The first is concerned with resources available to support the technology. It is crucial to start with identifying resources to address the specific needs of teachers and students. Identifying these resources draws attention to the nature of technical knowledge needed to understand the purpose of the technology. Technical knowledge defines the background knowledge required to implement, use, and apply a particular technology. While technical knowledge may be distinct for teachers and students, teachers may have the additional responsibility of fully understanding how students will realize the technology. Since teachers are required to blend technical and pedagogical knowledge, planning at this stage might suggest different types of classroom realizations.

Technical knowledge also is informed by the repertoire of skills to support the use of technologies. These include skills to manipulate the inner workings of the machine and/or knowledge that render an application (e.g., software—a modeling tool) or hardware (e.g., temperature probe) usable. For example, these skills may include troubleshooting ability, understanding different platforms, and/or translating patterns generated by graphs. It should also be noted that specific platforms and technological applications might introduce the need for specific skills—skills that are for specific technologies. Here it should be noted that technical skills are not fixed and, instead, constantly evolve.

In science education, content knowledge is at the heart of teaching and learning with technologies. Indeed, the reoccurring argument for implementation is based on

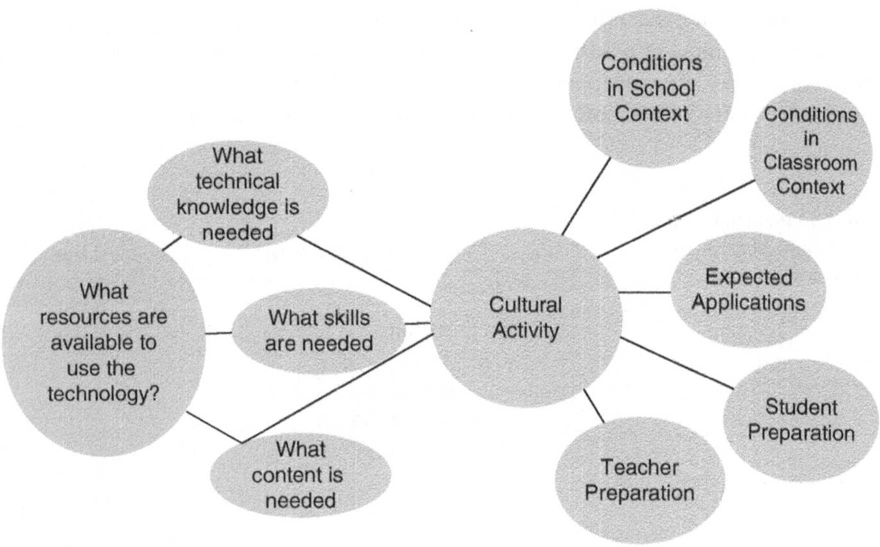

Fig. 7.2 Intent: planning

the capabilities of technologies to promote the use of scientific practices and in turn advance scientific understandings of phenomena. This view leverages content, skills, and habits of mind, which prepare students to conduct scientific investigations via inquiry-based approaches to teaching and learning. Thus, the intent stage helps anticipate the mutually reinforcing technical knowledge, content knowledge, and skills necessary to enact and use the technology. When these aspects are explicitly delineated, stakeholders can evaluate how the purpose of the technology and organizational structure align with the needed knowledge and skills. Of significance, technical knowledge can also inform the interactions that may coexist between biotic and abiotic factors. With this information, one can evaluate the conditions that may promote or limit the intended outcomes for the technology.

Knowledge resources also are linked to the cultural activity associated with implementation and enactment. Cultural activity brings to bare five related dimensions that illustrate the dynamic, evolving nature of technology use in science classrooms: Under what conditions is technology adopted at the school level and at the classroom level, the applications of technology, and how teachers and students are prepared to use the technology. According to Pacey (1983), cultural activity includes the goals, values, ethical codes, and beliefs in progress, as well as awareness and creativity (see Fig. 7.1). For example, in defining the conditions under which technologies are adopted at the school and classroom level, it is important to understand how the purpose for the technology is mediated via the goals and values of the school and classroom culture, and the beliefs of the school and classroom agents who operate within these structures.

The prompts that address conditions of adoption at the school and classroom level reinforce the message that access and resources determine proclivity for success. These conditions also address how the mission and philosophy of the school promote innovation and progressive approaches to teaching and learning. For example, a school that espouses a progressive mission with learner-centered approaches will be better poised to promote technology-supported inquiry approaches. Conversely, a school that serves a traditional mission, one that promotes a scripted curriculum, may negate approaches to learning that best foster implementation and enactment of technologies. Moving to the classroom, conditions in the classroom can be impacted by teachers' and students' values and belief structures, which can function to advance or negate the school's philosophy. Classroom conditions, furthermore, are influenced by the curriculum and pedagogical support available to teachers.

Classroom conditions ultimately determine how technologies manifest in learning spaces. When we understand what is possible within the scope of a progressive and learner-centered approach, we can plan for a repertoire of potential realizations. In understanding how a technology might be used in the classroom, we can assist teachers to anticipate and prepare for challenges. The result is a structure that identifies conditions and aligns these conditions with possible uses. What this means is putting in place teacher and student supports that create the best opportunities for optimal outcomes of technology enactment.

Teacher preparation is one of the most common recommendations for successful technology use in classrooms. So, the intent stage requires deep understanding of how teachers are prepared to use technologies for science teaching. This entails documenting the PD approaches, modes of dissemination, emphasis on knowledge and skills, emphasis on procedural use of technology versus open-ended inquiry approaches, and the extent to which the PD is tailored to target standards and pedagogical needs of teachers. In particular, will the PD afford opportunities for teachers to enact the technology in a safe environment and assist them with a specific repertoire of uses and implementation? While this list is not exhaustive, it addresses fundamental aspects of teacher preparation.

Students, like teachers, also need to understand the why and how of technology implementation and enactment. We argue that students should be prepared in parallel with teachers to understand the broad capabilities of the technology at hand. Proponents of technology implementation should include students in the same PD opportunities afforded to teachers. The advantages for teachers can be numerous. For one, immersing students with the purpose and benefits of a technology can expand the scope of potential applications. Second, such engagement can relieve some of the expectations and pressures placed on teachers. Students can contribute additional resources in the teaching and learning process. Third, understanding the benefits and challenges—from the perspectives of students at this stage—can serve as yet another lens to gauge possible issues and the potential of implementation. For example, if the goal is to enhance student understanding of chemical phenomena via computer-based modeling, PD can provide information about how students access, decode, and translate visual representations in the context of models. Rather than uncovering these challenges after implementation, this approach would expose these challenges early in the process so that modifications can be addressed before instruction. We remain cognizant that PD will not expose or address all of the challenges; however, we argue that opportunities in the early stages can help mitigate some of the most critical needs. Likewise, early intervention can also inform opportunities that can be used to motivate and foster student interest.

7.3.3 Adoption and Enactment

Enactment draws attention to outcomes of implementation. In other words, this stage focuses on how technology impacts teaching and learning and how the latter impact the technology. Specifically, this stage helps understand how a technology is used and integrated in instruction, how students are expected to use it, how the technology is actually used, how the technology is aligned with assessment, and the specific roles of the student and teacher agents. Here, some components addressed in both the preplanning and intent stages are revisited. Conditions of enactment are revisited to identify actual features and characteristics of the environment where the technology is used. In light of the conditions, the actual classroom applications also

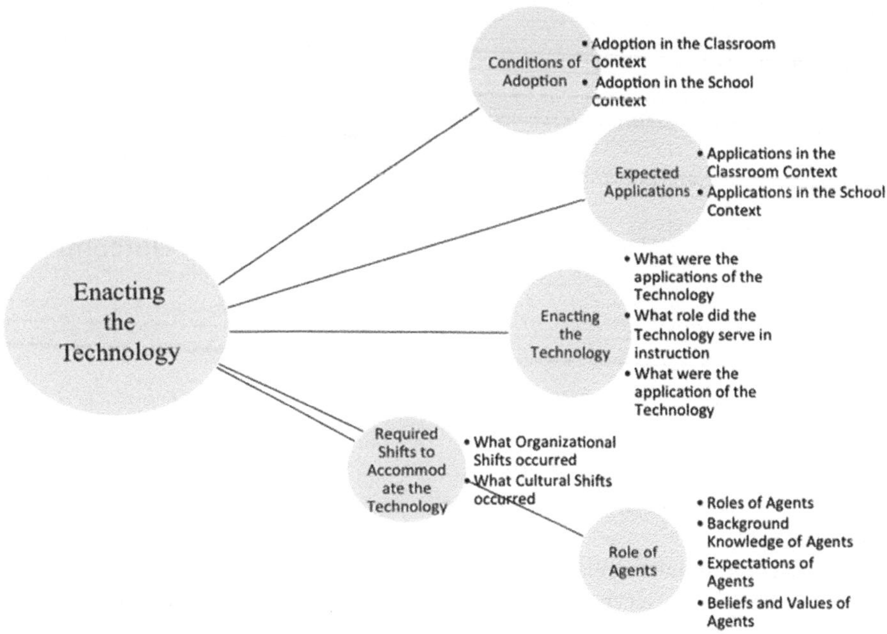

Fig. 7.3 Adoption and enactment

are assessed. For example, if classroom conditions promoted innovative approaches with a learner-centered focus, then the expectation is that realizations would foster inquiry-based approaches. In sum, how technology is enacted would focus on the role of the technology, how it was used, and thus how is the technology realized. This stage provides yet another opportunity to revisit how the purpose of the technology is realized in actual practice (see Fig. 7.3).

7.3.4 *Outcomes*

Enactment and outcomes are directly related to the required organizational and cultural shifts needed to accommodate technologies in the science classroom (see Fig. 7.4). These shifts are examined within the context of the interactions of biotic and abiotic factors. Specifically, there is need to examine what shifts occur with the agents' role, background knowledge (content, technical, and skills), as well as expectations, beliefs, and values. Essentially, all aspects of enactment can be examined from a collective organizational and cultural lens. When viewed collectively, outcomes of technology implementation and enactment follow a cyclical pathway that is consistently informed and reinforced by the conditions of preplanning, intent, and enactment. In sum, facets of technology practice—technical, cultural, and organizational structures, the nature of technology, and the ecological nature of

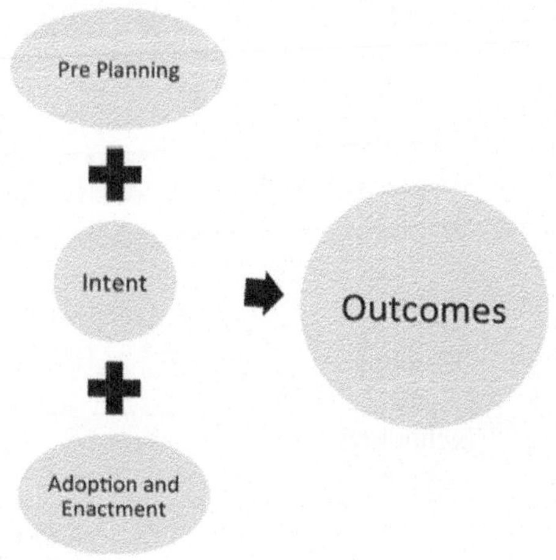

Fig. 7.4 Outcomes of implementation, adoption, and enactment

technology—reemerge at every stage in the process of adoption, implementation, and enactment.

Our analysis simultaneously supports and challenges Yerrick et al.'s (2018) recommendations for preservice elementary teachers' (PET) preparation for classroom teaching and learning, outlined in Chap. 6 of this book. Yerrick and colleagues examined PETs practices and reflections in a technology-rich science methods course. The findings were particularly striking because of the expectations attached to digital natives—a savvy generation exposed and immersed in the rapid progression of information technologies—expected to transform the technology culture of science teaching and learning. Instead, it was documented that PETs revealed apathy toward engagement and learning with technological tools. The authors explained that tendencies toward traditional modes of learning versus innovative learner-centered approaches were cultural as opposed to technological. In other words, teachers' expectations for classroom learning were not consistent with modes of learning (i.e., use of blogs, podcasts, online discussions) experienced via the methods course. Teachers' use of information and communication technologies in their personal and social sphere was not compatible with the culture of teaching and learning.

The significance of our holistic approach was further reinforced by our examination of Yerrick et al. (2018, in Chap. 6 of this volume) findings and recommendations, especially in light of our understanding of the nature of technology and the role of culture and values. We argue that an understanding of the cultural, organizational, and technical aspects of technology (Pacey 1983) must be part and parcel of any analysis of design, development, implementation, and enactment in precollege science classrooms. Each aspect of this framework must feature prominently in the processes that interrogate practice and research. For example, there was an emphasis on those domains that teachers need: knowledge, skills, scaffolding, mentoring, and safe spaces. In contrast, we argue that understandings of technologies must extend beyond the bounds of the context and agents of teaching to include other variables and forces—cultural and organizational—that may have significant impact. Consistent with our discussion above, the analysis might benefit from an examination of the incompatibilities of the cultural and organizational structures in the social and entertainment sphere versus a teaching and learning context. In terms of culture, we might ask students to identify and collate their last 10–15 communicative posts that serve as a representation of their interactions on social media or blogging sites. This data could serve as a snapshot of the kinds of interactions that manifest in these spaces. Next, they can examine these posts for evidence of conceptual, critical, and/or thought-provoking commentary. Some pertinent questions might include the following: In what kinds of ways are these posts compatible or incompatible with tasks and expectations that address teaching and learning? In what ways are the organizational structures supportive of these various practices? Inherently, the same technologies could be "transparent" for students in one cultural setting and simultaneously become "opaque" in a different cultural setting (Clark 2003). Note how this observation departs from Carberry and Baker's (2018) claims—outlined in Chap. 10 of this book—that culture is least important when technology transfer occurs between industrialized nations. Evidence from our own work (e.g., Waight and Abd-El-Khalick 2011), as well as Yerrick et al. (2018), indicates a constant interplay with aspects of culture and organization. We believe that identifying this gray matter of incompatibilities will be more useful in "fixing" and transforming how we teach with technologies. Our framework above is nuanced and provides guidelines for a more holistic approach to science teaching and learning.

7.4 Recommendation for Researchers and Educators

Like with the case of any technology, a scientific technology (e.g., one that allows three-dimensional visualization and manipulation of molecular arrangements of proteins) would have been developed to serve a particular purpose (e.g., test the compatibility of certain drugs with specific cell surface receptors), shaped by the knowledge and technical expertise of those who developed the technology (funding agency officers, programmers, scientists, etc.), and manifested its utility in a variety of ways and contexts (e.g., basic research, pharmaceutical research, and beyond originally envisioned possibilities). More importantly for our purposes, the very development of scientific technologies and their successful implementations also reflect the beliefs and values of broader societal agents and agencies (voting public, scholars, media personnel, politicians, bureaucrats, etc.) in the beneficial returns of their investment and trust in these technologies (e.g., drugs). The core message of this chapter is rather straightforward: When technologies—many of which proved incredibly useful in the practice of professional science and scientists—are introduced into classrooms (whether in original or modified renditions) for purposes of improving science teaching and learning, due consideration should be given to the "baggage" these technologies bring along. This baggage includes the specific purpose and context, knowledge and expertise of the specialized agents, and structures and modes of organization underlying the institutions—financial, professional, and social—that ushered and nurtured the use and adoption of these technologies, complete with their underlying values and beliefs, their cultures.

A foremost implication for science educators and science education researchers who are considering the adoption of such technologies in precollege science classrooms is to avoid thinking of a technology as artifact—even if this thinking extends to understanding the inner workings and assumptions underlying the technical functioning of the technology. When thought of as inert or artifact, we have seen from some of the literature explored in this chapter that less-than-desired outcomes of implementation are blamed on the agents involved, including teachers, school technical support staff, and students, and/or less-than-adequate technological infrastructure available to a school, classroom, or teacher. There is no doubt that all these factors matter. Examination of the four-stage model presented above shows that it speaks to questions of technological infrastructure and support, teacher and student knowledge, and facility with using the technology, among other things. The much-needed shift is to think of technology as both *artifact* and *process*. In this latter sense, technology takes a life of its own, which compelled us to favor the biological and ecological analog used by Zhao and Frank (2003) to best illustrate what introduction of technologies into a classroom ecology might entail. This shift would entail a shift in the kinds of research and practical questions we ask of less-than-optimal implementations and outcomes of technologies that often seem to have much promise to realize active, authentic science learning contexts in precollege classrooms. The fault might not be with the technology or classroom agents: This fault might lie with the interaction or incompatibility of those agents' values and beliefs and associated structures prevailing in their learning environment with the values, beliefs, and structures of the people and organizations who created and seem to have thrived in realizing the same technologies. Questions for research and practice at this nexus surely are more challenging but likely to advance the current research agenda in this domain, as well as enhance chances of successful implementation.

Another implication stems from answers to the abovementioned questions. Specifically, instead of asking why a certain technology (be it Biology Workbench or 3D simulations) failed to transform teaching and learning in a science classroom, the more viable and prudent question would be the following: How should the technology be transformed in order to promote the sort of desired learning? What technological modifications and associated pedagogical transformations are needed to ensure such success? As a case in point, the second author recalls a discussion with the scientist who was the thrust behind the funding and development of BSW (which outlaid a student-friendly skin on the original Biology Workbench, which was being extensively and successfully used by biological researchers around the globe). The scientist wanted to expose students to the larger world of scientific research and associated possibilities by providing meaningful access to the Biology Workbench database. We argued that compared to the structure and values of science classrooms at the time (using hardcopy static textbooks, rewarding student abilities to respond to factual questions, teachers' preference for structured learning environments), even a carefully selected slice of the Workbench would constitute a tremendous and larger world for high school students to explore. Eventually, BSW provided students with access to the whole scientific database. We found that more than a decade after substantial and sustained efforts to realize the potential of BSW that the few remaining instantiations were highly structured, point-and-click fortes for students are the peripheries of Workbench.

Finally, it is crucial to stress that we too share the view that technologies have incredible potential in elevating science teaching and learning in precollege classrooms. Nonetheless, we advocate for making the technology itself and its interactions with students' and teachers' values and beliefs (in addition to knowledge and technical skills) an object for investigation in research, and for consideration in the professional education of teachers, and the planning for technology design, adoption, and implementation. The four-stage framework presented above provides a roadmap to initiate holistic examinations of, and planning for, adopting and realizing the promise of technologies in school science education.

References

Adadan, E., Trundle, K. C., & Irving, K. E. (2010). Exploring grade 11 students' conceptual pathways of the particulate nature of matter in the context of multi representational instruction. *Journal of Research in Science Teaching, 47*, 1004–1035.

Ardac, D., & Akaygun, S. (2004). Effectiveness of multimedia-based instruction that emphasizes molecular representations on students' understanding of chemical change. *Journal of Research in Science Teaching, 41*, 317–337.

Arthur, W. B. (2009). *The nature of technology*. New York: Free Press.

Barab, S. A., Hay, K. E., Barnett, M., & Keating, T. (2000). Virtual solar system project: Building understanding through model building. *Journal of Research in Science Teaching, 37*, 719–756.

Basalla, G. (1996). *The evolution of technology*. New York: Cambridge University Press.

Bybee, R. W. (2013). *The case for STEM education: Challenges and opportunities*. Arlington: National Science Teachers Association Press.

Carberry, A. R., & Baker, D. R. (2018). The impact of culture on engineering and engineering education. In Y. J. Dori, Z. Mevarech, & D. Baker (Eds.), *Cognition, metacognition, and culture in STEM education* (pp. 217–239). Springer.

Clark, A. (2003). *Natural born cyborgs: Minds, technologies, and the future of human intelligence*. New York: Oxford University Press.

Crawford, B. A., & Cullin, M. J. (2004). Supporting prospective teachers' conceptions of modeling in science. *International Journal of Science Education, 26*, 1379–1401.

De Jong, O., Van Driel, J. H., & Verloop, N. (2005). Preservice teachers' pedagogical content knowledge of using particle models in teaching chemistry. *Journal of Research in Science Teaching, 42*, 947–964.

Dori, Y. J., & Hameiri, M. (2003). Multidimensional analysis system for quantitative chemistry problems: Symbol, macro, micro and process aspects. *Journal of Research in Science Teaching, 40*, 278–302.

Dori, Y. J., & Kaberman, Z. (2012). Assessing high school chemistry students' modeling sub-skills in a computerized molecular modeling learning environment. *Instructional Science, 40*, 69–91.

Eglash, R. (2004). Appropriating technology: An introduction. In R. Eglash, J. L. Croissant, G. Di Chiro, & R. Fouche (Eds.), *Appropriating technology: Vernacular science and social power* (pp. 1–28). Minneapolis: University of Minnesota Press.

Ellul, J. (1964). *The technological society*. Toronto: Alfred A. Knopf.

Gobert, J. D., O'Dwyer, L., Horwitz, P., Buckley, B. C., Tal Levy, S., & Wilenksy, U. (2011). Examining the relationship between students' understanding of the nature of models and conceptual learning in biology, physics, and chemistry. *International Journal of Science Education, 33*, 653–684.

Heidegger, M. (1977). *The question concerning technology and other essays*. New York: Harper & Row, Publishers.

Hsu, Y. (2008). Learning about seasons in a technologically enhanced environment: The impact of teacher-guided and student-centered instructional approaches on the process of students' conceptual change. *Science Education, 92*, 320–344.

Illich, I. (1973). *Tools for conviviality*. New York: Harper & Row.

Jaakkola, T., Nurmi, S., & Veermans, K. (2011). A comparison of students' conceptual understanding of electric circuits in simulation only and simulations-laboratory contexts. *Journal of Research in Science Teaching, 48*, 71–93.

Kikas, E. (2004). Teachers' conceptions and misconceptions concerning three natural phenomena. *Journal of Research in Science Teaching, 41*, 432–448.

Kim, M., Hannafin, M. J., & Bryan, L. (2007). Technology-enhanced inquiry tools in science education: An emerging pedagogical framework for classroom practice. *Science Education, 96*, 1010–1030.

Koehler, M. J., & Mishra, P. (2008). Introducing TPCK. In AACTE (Ed.), *Handbook of technological pedagogical content knowledge (TPCK) for educators* (pp. 3–29). New York: Routledge.

Kozma, R. B., & Russell, J. (1997). Multimedia and understanding: Expert and novice responses to different representations of chemical phenomena. *Journal of Research in Science Teaching, 34*, 949–968.

Lead States, N. G. S. S. (2013). *Next generation science standards: For states, by states*. Washington, DC: The National Academies Press.

Liu, L., & Hmelo-Silver, C. E. (2009). Promoting complex systems learning through the use of conceptual representations in hypermedia. *Journal of Research in Science Teaching, 46*, 1023–1040.

Liu, X., & Lesniak, K. (2006). Progression in children's understanding of the matter concept from elementary to high school. *Journal of Research in Science Teaching, 43*, 320–347.

Marbach-Ad, G., Rotbain, Y., & Stavy, R. (2008). Using computer animation and illustration activities to improve high school students' achievement in molecular genetics. *Journal of Research in Science Teaching, 45*, 273–292.

McDermott, J. (2009). Technology: The opiate of the intellectuals. In A. H. Teich (Ed.), *Technology and the future* (pp. 87–97). Boston: Wadsworth Cengage Learning.

Mitcham, C. (1994). *Thinking through technology: The path between engineering and philosophy*. Chicago: The University of Chicago Press.

National Research Council. (2011). *A framework for K-12 science education: Practices, crosscutting concepts, and core ideas*. Washington, DC: National Academy Press.

Pacey, A. (1983). *The culture of technology*. Cambridge, MA: MIT Press.

Rapp, F. (1999). The material and cultural aspects of technology. *Society for Philosophy & Technology, 4*, 45–57.

Sins, P. H. M., Savelsbergh, E. R., van Joolingen, W. R., & van Hout Wolters, B. H. A. M. (2009). The relation between students' epistemological understanding of computer models and their cognitive processing on a modeling task. *International Journal of Science Education, 31*, 1205–1229.

Stains, M., & Talanquer, V. (2008). Classification of chemical reactions: Stages of expertise. *Journal of Research in Science Teaching, 45*, 771–793.

Sullivan, F. R. (2008). Robotics and science literacy: Thinking skills, science process skills and systems understanding. *Journal of Research in Science Teaching, 45*, 373–394.

Tao, P., & Gunstone, R. F. (1999). The process of conceptual change in force and motion during computer-supported physics instruction. *Journal of Research in Science Teaching, 36*, 859–882.

Tenner, E. (1996). *Why things bite back: Technology and the revenge of unintended # consequences*. New York: First Vintage Books Edition.

Urhahne, D., Nick, S., & Schanze, S. (2009). The effect of three-dimensional simulations on the understanding of chemical structures and their properties. *Research in Science Education, 39*, 495–513.

Verma, G., Puvirajah, A., & Webb, H. (2015). Enacting acts of authentication in a robotics competition: An interpretivist study. *Journal of Research in Science Teaching, 52*, 28–295.

Volti, R. (2010). *Society and technological change*. New York: Worth Publishers.

Waight, N., & Abd-El-Khalick, F. (2007). The Impact of technology on the enactment of inquiry in a technology enthusiast's sixth grade science classroom. *Journal of Research in Science Teaching, 44*, 154–182.

Waight, N., & Abd-El-Khalick, F. (2011). From scientific practice to high school science: Transfer of scientific technologies and realizations of authentic inquiry. *Journal of Research in Science Teaching, 48*, 37–70.

Waight, N., & Abd-El-Khalick, F. (2012). Nature of technology: Implications for design, development, and enactment of technological tools in school science classrooms. *International Journal of Science Education, 34*, 2875–2905.

Waight, N., Liu, X., Gregorius, R. M., Smith, E., & Park, M. (2014). Teacher conceptions and approaches associated with an immersive instructional implementation of computer-based models and assessment in secondary chemistry classrooms. *International Journal of Science Education, 36*, 467–505.

Winner, L. (2003). Social constructivism: Opening the black box and finding it empty. In R. C. Scharff & V. Dusek (Eds.), *Philosophy of technology: The technological condition* (pp. 232–243). Malden: Blackwell.

Wu, H., Krajcik, J. S., & Soloway, E. (2001). Promoting understanding of chemical representations: Students' use of a visualization tool in the classroom. *Journal of Research in Science Teaching, 38*, 821–842.

Yerrick, R., Radosta, M., & Greene, K. (2018). Technology, culture and young science teachers—A promise unfulfilled and proposals for change. In Y. J. Dori, Z. Mevarech, & D. Baker (Eds.), *Cognition, metacognition, and culture in STEM education* (pp. 117–138). Springer.

Zacharia, Z. (2003). Beliefs, attitudes, and intentions of science teachers regarding the educational use of computer simulations and inquiry-based experiments in physics. *Journal of Research in Science Teaching, 40*, 792–823.

Zhang, B., Liu, X., & Krajcik, J. S. (2006). Expert models and modeling processes associated with a computer-modeling tool. *Science Education, 90*, 579–604.

Zhao, Y., & Frank, K. A. (2003). Factors affecting technology uses in schools: An ecological perspective. *American Educational Research Journal, 40*, 807–840.

Chapter 8
Engineering Cognition: A Process of Knowledge Acquisition and Application

Senay Purzer, Tamara J. Moore, and Emily Dringenberg

8.1 Introduction

The year 2020 is an important crossroad in engineering as part of a vision of engineering education for the new century articulated by the National Academy of Engineering (2004). In fact, as we write this chapter for this book, the undergraduate institutions throughout the world have already admitted their graduates of 2020. This vision for 2020 includes the education of engineers who have strong analytical skills, practical ingenuity, creativity, communication skills, business acumen skills, leadership with high ethical standards, and abilities for lifelong learning.

In establishing the context for recent reform efforts in engineering education, we find that it is useful to address the rich and interesting history of the development of engineering education, which has consistently been propelled by the need to meet the demands of society. Here we focus on the historical developments in the United States while acknowledging the rich history in other regions of the world especially in primary and secondary education such as in Australia, Europe, New Zealand, and Canada (for more details, see Lachapelle and Cunningham 2014; Corlu et al. 2017, in Chap. 10).

Historically, the twentieth century has seen a number of reports (e.g., Grinter 1955; Mann 1918; Wickenden 1930; Hammond 1940), each evaluating and criticizing the focus of undergraduate engineering education on practical skills and demanding the need for promoting basic sciences at the core of engineering

S. Purzer (✉) • T.J. Moore
School of Engineering Education, Purdue University, West Lafayette, IN, USA
e-mail: senay@purdue.edu

E. Dringenberg
School of Engineering Education, Purdue University, West Lafayette, IN, USA

Department of Engineering Education, Ohio State University, Columbus, OH, USA

Y.J. Dori et al. (eds.), *Cognition, Metacognition, and Culture in STEM Education*, Innovations in Science Education and Technology 24,
https://doi.org/10.1007/978-3-319-66659-4_8

education (for more detailed information on these reports, see ASEE 2015). The emphasis for engineers to be trained in theoretical coursework was also promoted by the continued industrialization of the United States as well as the military needs of both the Cold War and World War II. The most recent report, published in 2013, on Transforming Undergraduate Education in Engineering (TUEE), continues to emphasize the importance of learning the fundamentals of engineering science, an opinion shared by a diverse range of stakeholders such as those from academia, industry, as well as students and parents (NAE 2013).

In precollege or K-12 education, among the four subject areas covered under the umbrella of STEM (science, technology, engineering, and mathematics) education, much attention has historically focused on science and mathematics (Cajas 2001; Sneider and Purzer 2014; Lachapelle and Cunningham 2014). It is with the publication of *Engineering in K-12 Education* by the National Academy of Engineering that researchers and policy makers started to talk more explicitly about engineering in precollege education (NRC 2012; Purzer et al. 2014b). Especially, engineering has become a topic of heightened interest through the publication of *A Framework for K-12 Science Education* (NRC 2012) followed by the *Next Generation Science Standards* (Achieve 2013). In these reports, engineering provides new opportunities for improving science education and the development of twenty-first century skills (Purzer et al. 2014a). In fact, engineering is not only the sole goal as argued in previous policy discourse (NAE 2009) but also the integrator of knowledge among other STEM disciplines (NAE and NRC 2014).

It is also important to discuss the evolution of engineering education research as part of the historical developments in engineering education. In the last 15 years, engineering education research has experienced significant developments with the publication of a specific report in the *Journal of Engineering Education* (Streveler and Smith 2006; Adams et al. 2006). Among one of the key disciplinary-based research communities, engineering education researchers focus on five broad research areas: (1) engineering epistemologies, (2) engineering learning mechanisms, (3) engineering learning systems, (4) policy including diversity and inclusiveness in engineering, and (5) engineering assessment. Engineering education researchers explore research questions impacting a broad range of learners from early childhood and K-12 education to undergraduate and graduate education to practices in the workplace.

These relatively recent developments in policy, education, and research provide an opportunity for a deeper understanding of engineering cognition and new research directions, which this chapter attempts to address.

8.2 What Is Engineering?

A popular definition of engineering is often described as a means to distinguish engineering from basic sciences stating that engineering is the application of natural sciences and mathematics to practical solutions. Expanding this definition slightly, the Accreditation Board for Engineering and Technology defines engineering as "the profession in which a knowledge of the mathematical and natural sciences, gained by study, experience, and practice, is applied with judgment to develop ways to utilize, economically, the materials and forces of nature for the benefit of mankind" (ABET 1998). In alignment with these definitions, the modern engineering education system in the United States reflects the idea that engineering is founded on theoretical concepts from engineering science and mathematics. Yet, these definitions are too simple and incomplete. In addition, the emphasis on theoretical foundations is often criticized, pointing out the need for students to prepare for engineering practices they will engage in after graduation (Sheppard et al. 2008).

Antonio Dias de Figueiredo presents a more nuanced description of engineering knowledge with four dimensions representing various roles of the engineer: engineer as scientist (basic sciences), engineer as sociologist (social sciences), engineer as designer (design), and engineer as doer (practical realization) (Figueiredo 2008). In alignment with Figueiredo's description of engineering, Radcliffe (2015) states that engineering problems take root in complex sociopolitical contexts and hence require solutions that depend upon a nuanced, deep understanding of social sciences and humanities. Radcliffe argues for a perspective embracing two STEMs, the *traditionally defined STEM* (**s**cience, **t**echnology, **e**ngineering, and **m**athematics), and coins the term the *contemporary STEM* (social **s**ciences, the ar**t**s, **e**ducation, and the hu**m**anities). Knowledge of the contemporary STEM, Radcliffe argues, is as critical as the traditional STEM (technical knowledge) if students are to learn to proficiently define and formulate, let alone solve, wicked real-world problems.

Interestingly, in the 1860s, the emphasis of engineering education was on practical skill development rather than engineering sciences. In the United States, the number of engineering schools has increased drastically since 1862 (Sheppard et al. 2008). This surge for engineering schools was motivated by the passing of the Morrill Land Grant Act, which allowed states to donate public lands to establish universities with a focus on agriculture and mechanic arts. These pioneering schools of engineering emphasized practical skills such as operating machinery. Today we see the premises of the historical practices as reflections of two tensions. The first tension is between valuing knowledge acquisition over knowledge application, assuming that acquired knowledge is easy to apply to real-world problems. The second tension is between emphasizing knowledge application and knowledge acquisition with the assumption that deducing new knowledge from practical experiences is a straightforward process. Solving engineering problems, however, deals with both and requires going back and forth between acquiring knowledge and applying knowledge. Our model of engineering cognition reflects this cohesive perspective with a combination and integration of the teaching of the theoretical underpinnings and engineering sciences with the teaching of practical experiences. This suggests a curricular shift from well-defined problems with single correct solutions to solving problems with real-world implications. Even the TUEE report includes a recommendation to "review and update the curriculum with current and emerging industrial practice in mind, and … to teach and demonstrate the fundamentals in the context of engineering design and real-world examples" (NAE 2013, p. 13). In fact, real-world problems are wicked and ill defined, as Loui (2016) states, "Engineering is the art of designing imperfect solutions to meet incomplete specifications by applying unrealistic theories to produce incomprehensible calculations that use inaccurate data based on imprecise measurements."

8.2.1 Clarification of Terms Common to Engineering

Educators and researchers in engineering education commonly use terms such as design, design process, design inquiry, and design practices. Similarly, in science education, the terms inquiry, scientific process, scientific inquiry, and science practices are used commonly. However, the term "design" should not be thought as limited to engineering, neither the term "inquiry" as limited to science.

Design is a ubiquitous problem-solving activity performed by individuals in various fields and across diverse disciplines such as industrial design, software design, architecture, dance choreography, clothing design, and engineering (Adams et al. 2016; Daly et al. 2012). Designs in these diverse fields have common aspects but also distinctly different goals that reflect the designer's disciplinary focus (Purzer and Fila 2014). Hence, researchers should recognize that not all design is engineering design and yet much can be learned from the broader body of research on design cognition that STEM education researchers can build on.

Similarly, in science education, "inquiry" is a commonly used term. "Inquiry" in the science education community is often implied to refer to scientific inquiry; however, there is a broader meaning of the term inquiry, which applies to many diverse fields beyond natural sciences. In a recent article, Purzer et al. (2015) included a disclaimer that the two terms, design and inquiry, have specific meanings in K-12 STEM education. They argue that design, with a set of strategies, is itself an inquiry. Yet, the terms inquiry and design have specialized meanings in STEM education. While design researchers in engineering education agree that design is an aspect of engineering cognition that distinguishes it from science and mathematics (de Figueiredo 2008; Atman et al. 2007; Sheppard et al. 2008; Daly et al. 2012), we invite researchers to carefully use these terms in their specific meanings (*engineering design* and *scientific inquiry*) or in their general meanings (*design* and *inquiry*), to avoid confusions.

8.2.2 Three Perspectives on Engineering Design

Engineering design can be described in diverse ways: as a problem type, through the strategies used by designers, and as a type of cognitive activity. As shown in Fig. 8.1, these diverse ways of describing design are not mutually exclusive, but rather, represent the complexity of design cognition. In this chapter, we discuss multiple

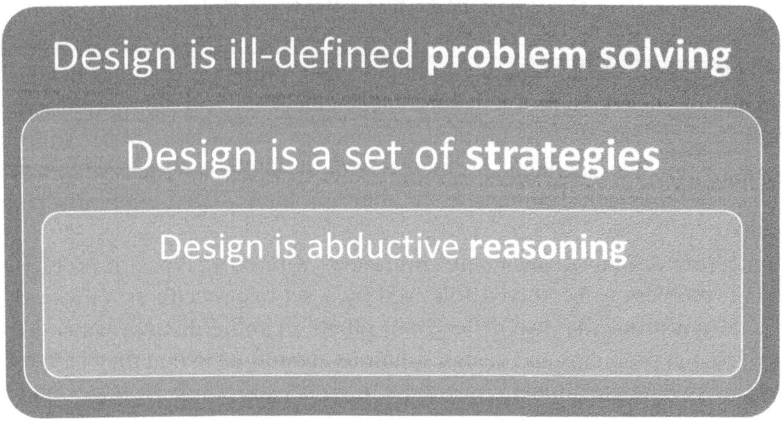

Fig. 8.1 Three perspectives on engineering design

perspectives of engineering design rather than a single definition to emphasize the diverse, multifaceted views of design similar to the argument Erduran and Dagher (2014) make about the nature of science.

In a broad sense, engineering design is a type of ill-defined and ill-structured problem solving (Jonassen 2000). Another way of thinking about design is as a set of strategies used to solve problems. Through a meta-literature review, Crismond and Adams (2012) identified nine strategies critical to teaching design. Several of these strategies include delaying decisions to deeply understand the challenge, conducting experiments and gathering information to build knowledge about the system and existing solutions, using multiple representations to explore and investigate design ideas, and iterating in a managed way via feedback with the goal of improving solutions. While there are numerous different design process models that typically represent a cycle (scope problem, generate ideas, test ideas, communicate ideas, etc.), these models lean on a set of common design strategies. Design process models guide designers about "how" with specific actions that are linear, cyclic, or iterative (e.g., testing design ideas after building prototypes). Design strategies, however, address higher-level behaviors while tackling the question of "why" (e.g., conducting experiments and gathering information to build knowledge about the system and existing solutions).

When considered as a cognitive activity, more contemporary studies argue design as a form of abductive reasoning, which is a form of reasoning where parsimonious explanations are formed from observations (Peirce 1932; Dong et al. 2014). Abductive reasoning allows designer to approach a problem despite limited information and resources (Roozenburg 1993). To understand engineering design thinking, we can look at these three aspects broadly, separately, or as a whole.

8.2.2.1 Engineering Design as Ill-Defined Problem Solving

The broadest form of thinking about engineering design is defining design as ill-defined and ill-structured problem solving. Engineers are often required to develop a solution to problems with incomplete information (Gainsburg 2006). Additionally, engineering practice includes problems with multiple subgoals that are often implicit to be considered alongside the main problem stated explicitly (Jonassen et al. 2006).

Engineering design carries many similarities with other types of problems with an explicit problem to be solved followed by a set of procedures. However, engineering design problems also differ from others in some distinct ways. First, the nature of design problems and design solutions are unique in that they lack information and involve nontechnical constraints (e.g., social, political) (Goel and Pirolli 1992). Hence, engineering design problems require a deep understanding of the problem context and efforts to scope and clarify the underlying issues that often are not evident in a given problem statement. Hence informed designers spend a great deal of time trying to understand the problem before producing new solutions (Atman et al. 2005; Crismond and Adams 2012).

Second, there is rarely a single best solution to an engineering problem because engineering problems lie within social contexts and inherently have competing variables, such as constraints of money and time or criteria that are subjective (Frezza et al. 2013; Hatchuel and Weil 2003; Zannier et al. 2007). According to Frezza et al. (2013), solutions should be timely and sufficiently complete – that is, the solutions should be "good enough" to address the multiple needs of the client and end users. Further, a solution that works in one context may not be viable or feasible in another. Frezza and colleagues further state, it "is not the one single, best right answer, that comprises good engineering. Rather the best *engineering* answer(s) are judged pragmatically, and routinely involve social context (e.g., a company or a customer)" (Frezza et al. 2013, p. 6).

8.2.2.2 Engineering Design as a Set of Strategies

There is a set of design strategies agreed upon by design researchers to differentiate informed designers, novices, and experts (Crismond and Adams 2012). This set of common strategies focuses design as an interdisciplinary but coherent activity beyond ill-defined problem solving, which is broadly defined (Lawson 2006). A critical component of expertise development is an understanding of design language and the informed design strategies as well as the utilization of these strategies (Cross 1982, 2001).

Within the engineering design literature, prior studies have focused on four areas of research: (1) comparison of expert and novice behaviors involving short problem-solving tasks (Atman et al. 2005), (2) longitudinal studies that examine progressions of student learning and engagement during design tasks (Purzer et al. 2011b), (3) classroom studies that examine science learning through design (Apedoe et al.

2008; Hmelo et al. 2000; Kolodner 2002; Kolodner et al. 2003; Mehalik et al. 2008; Schnittka and Bell 2011), and (4) case studies that follow few students over a long period of time (Kittleson and Southerland 2004; Purzer 2011; Tonso 2007). There are also an emerging number of studies that look at the effects of learning environments, instruction, and project type on the quality of student design outcomes. The literature related to the former two areas agrees that students face challenges in certain areas of design such as problem definition and information gathering. The next two areas of literature highlight aspects of curriculum, teaching practices, or student team interactions that support or hinder student learning.

One way we can distinguish learners and educators with deep understanding of informed design is by identifying if they explain or think about design as a set of discrete steps followed step by step to solve a problem or in more nuanced, iterative ways. Thinking about design merely as a series of steps is faulty and misleading. In order to inform the teaching and learning process, it is important to gain an understanding of the underlying cognitive processes when solving engineering design problems. For example, we can differentiate informed designers from those who are new to design by observing how they interpret different design process diagrams or models (Purzer et al. 2011a).

Dubberly's (2004) compendium of design models presents 26 different design models. To a novice designer or educator, the sheer abundance of models would be overwhelming. However, an expert would recognize the underlying and common structure of these design process models to be coherent while acknowledging different conceptualizations of design to meet different fields of design. An expert would recognize that all 26 different design models that Dubberly presents carry common attributes with a set of strategies that start with a goal or a need, which is guided by input from stakeholders, scoped iteratively, and ends with an outcome (a product, a process, or a system).

In Fig. 8.2, we represent common features of engineering design process models with a focus on key design strategies. This model starts with a goal and ends with an outcome. The path to the desired outcome includes the use of design strategies such as gathering various types of information and data (through experimentation or studying the users), forming explanations (about relationships between design features and criteria), and macro- (revisiting problem statement during idea generation, revisiting idea generation during detailed design) and micro-iterations (testing retesting, re-generating additional ideas). This design model aligns with Lu's (2014) four drivers of design: problem driven, information driven, solution driven, and knowledge driven.

Such a representation of engineering design must be accompanied with a model such as Crismond and Adams' (2012) informed design teaching and learning matrix that articulates key behaviors: engaging in problem framing, doing background research, balancing benefits and trade-offs, conducting valid tests and experiments, and managed and iterative designing. These examples are just a select few from the rich literature on design cognition. Finally, the iterative nature of design is necessary to allow feedback and input to inform design process and decisions, an aspect of the abductive reasoning process that we discuss in detail in the following section.

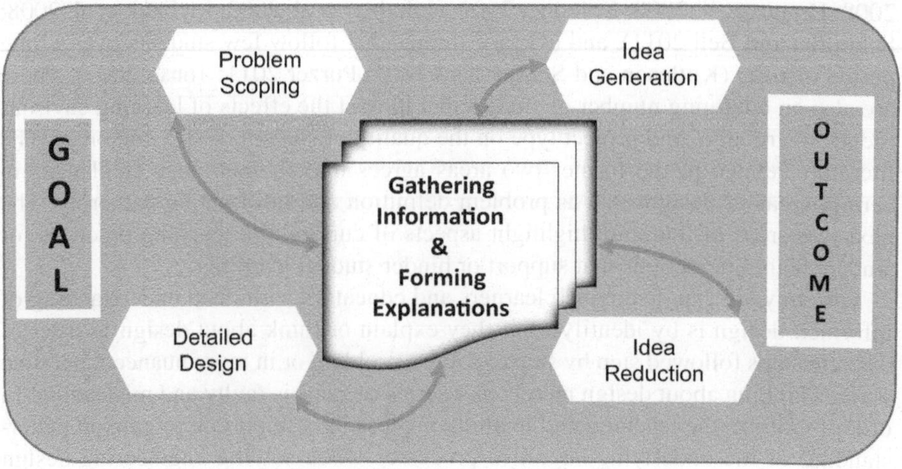

Fig. 8.2 An engineering design process model focused on design strategies

8.2.2.3 Engineering Design as Abductive Reasoning

Earlier we discussed the role of problem scoping, information gathering, and experimenting as critical attributes of engineering design. Such generative form of reasoning distinguishes design thinking from other forms of cognition (Dong et al. 2014). Dorst (2011) argues that design cognition relies on abductive reasoning, in addition to the deductive and inductive reasoning that are often used in scientific discovery. Abductive reasoning is a form of discovery through logical reasoning (Dong et al. 2014). This type of reasoning is needed as the designer develops a solution while also figuring out the working principles that explain the relationship between design forms or features and the intended design criteria or functions (Kroll and Koskela 2014).

Abductive reasoning is needed when the design problem has a clear value to be reached (determined by the user or client), but the solution to be generated as well as the working principle to guide the designer to the desired value are unknown (Dorst 2011). In this sort of situation, which is closely associated with conceptual design or idea generation, designers must "frame" the problem or identify themes of the desired value in order to decide "how" their design can provide a solution that is of sufficient value (Dorst 2011). However, the use of a "frame" to clarify the designer's way of looking at a design problem is not sufficient to then determine a solution. Rather, creative design requires the coevolution of the problem and the solution (Dorst and Cross 2001; Dorst 2003).

With these three different but related definitions of engineering design, we highlight the diverse, multifaceted views of engineering and emphasize the need to show its complexity as a cognitive activity. In the following section, we present a model

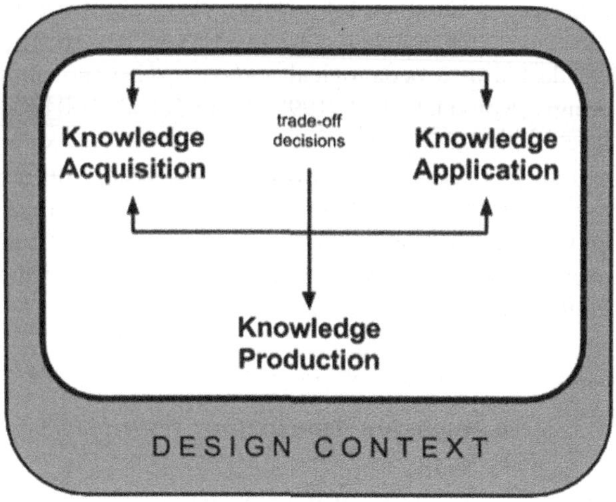

Fig. 8.3 Engineering cognition

for engineering cognition that carries this complexity with an ongoing iteration between acquiring knowledge and applying knowledge to lead to producing new knowledge.

8.3 Our Definition of Engineering Cognition

Engineering is often a reaction to a novel problem, a novel context, a novel set of users, or combination of these factors. Engineering cognition is the thinking that is required to generate, synthesize, and utilize knowledge necessary to arrive at a viable solution. Engineering cognition emerges through the interaction between knowledge acquisition and knowledge application resulting in the production of a set of problem-specific knowledge vital to shaping a design solution. However, these ways of thinking are not necessarily distinct. Figure 8.3 is a graphical representation of our general model for engineering design cognition. This cognition takes place within a specific context to ensure the solution meets the needs and wants of users.

8.3.1 Knowledge Acquisition Within Engineering Cognition

Engineering design problems are often nebulous, necessitating the designer to acquire new knowledge in order to generate a reasonable solution to the problem at hand. By focusing on the need to acquire new knowledge, solving engineering design problems requires and promotes information literacy (Fosmire and Radcliffe 2013). A notable number of studies associated with design education focus on learning environments that support learning new knowledge through engineering design experiences. The focus of new knowledge, in these studies, is often the learning of science or mathematical concepts through design. These studies that use engineering design as a pedagogical approach have shown evidence of learning through design (Kolodner et al. 1998; Fortus et al. 2004; Svarovsky and Shaffer 2007; Apedoe et al. 2008; Wendell and Lee 2010). An example of knowledge acquisition is a student recognizing ways sound travels in different mediums by designing a musical instrument. However, simply engaging in design is not sufficient because knowledge acquisition through engineering design does not occur if there is no explicit discussion of content that students are expected to acquire (Crismond 2001; Walkington et al. 2014).

8.3.2 Knowledge Application Within Design Cognition

The assumed purpose of the knowledge that students gather during the problem-scoping portion of their engineering design process is to be applied to the problem in an effort to generate solutions. Transfer of learning becomes evident when learning in one context affects a related performance in another context (Perkins and Salomon 1992). This is one of the most fundamental ideas of formal education, as we intend for the learning that occurs in the classroom to be used later in a meaningful way to solve ill-defined and ill-structured problems. Engineering education at the university level often expects students to transfer (apply) their knowledge of mathematics and science to solve engineering problems (Froyd and Ohland 2005). An example of knowledge application is a student selecting a specific type of material for sound dampening based on his or her prior knowledge on acoustics and ways sound travels in different mediums.

Knowledge application within design can refer to a range of applications of prior knowledge. On one hand, application may simply be using a previous concept, tool, or procedure in a similar way in a new context. On the other hand, transfer can refer to the learners' ability to see a design problem as one that is structurally similar to previous experiences and bringing to bear relevant knowledge and experiences in developing a solution to the problem. Calls for this higher level of transfer are prevalent throughout the literature. Application of knowledge, especially, becomes apparent during idea generation as students generate ideas and make associations based on their prior knowledge.

Our model connects knowledge acquisition with knowledge application with two double-headed arrows (see Fig. 8.3). This is in an attempt to acknowledge that the boundaries between acquiring new knowledge and applying that knowledge in a productive manner are not easily distinguished. In addition, this process is iterative and ongoing throughout the engineering design process facilitated by the need to balance design benefits and trade-offs (Purzer et al. 2015). It is the repeated interaction of these two pieces that eventually leads to the production of new and synthesized knowledge as well as an adequate design solution.

8.3.3 *Knowledge Production*

Knowledge production occurs as a result of the ongoing iteration between knowledge acquisition and application which results in a deeper understanding of the problem context and the performance of a variety of design features. Design, at its core, necessitates the process of knowledge production essential in managing ambiguity, novelty, and complexity of engineering problems. Knowledge production can be reinforced and inhibited by aspects of design problems students are presented with. Effective design problems or challenges require (1) the exploration of the problem context and the user of the solutions to be produced, (2) data gathering that enables the exploration of design features in relationship to design constraints and criteria, and (3) balancing benefits and trade-offs.

8.3.4 *Design or Problem Context*

Finally, the outer frame encompassing the pieces of our model for engineering cognition is the design context, which the designer explores broadly first as part of problem scoping. We would argue that a design problem does not really require engineering design cognition if there is not a specific design context that frames the problem and informs the designer of how their design solution will meet the needs of the eventual user. We acknowledge that the separation is artificial – it is impossible to delineate exact moments where engineering designers switch from applying some prior knowledge to acquiring relevant knowledge needed to produce a design solution, but we have done so to emphasize the importance of both parts of engineering cognition.

As a concrete example to illustrate the meaning of this model, when students are asked to design an energy-efficient building, we would expect them to start their design process by asking a set of questions about the problem context (*Where will this building be located? What/who will this building be used for? What are the budget constraints?*) and about the underlying scientific or design principles (*How does the size of a roof impact energy efficiency? Where would the best locations be for windows?*). All of these questions represent the start of knowledge acquisition. At the same time, these questions can be influenced by the application of students' prior knowledge or experience with a related problem. The answers to these initial questions can then turn into knowledge application as the student is reviewing the performance of a solution and making revisions by modifying the roof or changing the location of a window. In design settings, however, these two (knowledge acquisition and knowledge application) often occur simultaneously due to the need to balance design benefits and trade-off that reflect attempts to meet multiple criteria and constraints at once. The student may know or learn that the U-values (a measure of thermal transmittance) of windows are higher than walls and hence might design an energy-efficient and low-cost building with very small or no windows. By doing so, however, the design will lose its curb appeal, a potential user need. In such design situations, students must make key decisions about the relative importance of design criteria (knowingly or unknowingly) when to seek out new knowledge and when to apply knowledge.

8.4 Toward a Cohesive View of Engineering Cognition in Teaching and Learning

Having proposed our model of engineering design cognition, we would like to provide some of the implications in terms of teaching and learning to inform both the precollege and undergraduate levels of engineering education. We provide a discussion of the ways that existing curriculum models may not measure up to the cognition required for engineering practice, provide some recommendations for designing tasks to promote engineering design cognition, and say a word on what this means in terms of classroom assessment. One may argue our cognition model is too complex for undergraduate students, let alone K-12 students. Yet, we challenge educators toward a cohesive view of engineering education.

We encourage educators to start by evaluating the learning experiences that they create for their students based on how and where these learning experiences account for specific aspects of our model. This suggests asking questions such as: *Are students required to understand a design context? Are there opportunities for students to interact with end users, understand their needs? Is there sufficient time for students to iteratively acquire and apply knowledge to create a unique design solution?*

While we argue that engineering cognition occurs through ongoing connections between knowledge acquisition and application, often teaching practices fall short when the components of engineering cognition, design to apply knowledge or acquire knowledge, are taught or experienced in a distinct manner. These practices, common in the classroom, limit effective formation of engineering thinking. Traditionally, undergraduate engineering programs tend to treat this relationship between acquiring knowledge and applying knowledge as a linear transition. The opportunities for working on ill-defined and ill-structured problems are most frequently offered to students in the latter half of their studies often through capstone design (Stevens et al. 2008). Students move from well-defined problem solving and analogous transfer on problems that demonstrate theoretical scientific concepts to design projects that require an extensive consideration of the design context. However, this approach rests on the expectation that students will be able to determine and apply relevant knowledge from their prior education and also assumes students retained said knowledge from prior instruction. In addition, this transition can become a source of discomfort and frustration for students (Stevens et al. 2008), who have limited opportunities to both acquire and apply their knowledge toward solving engineering design problems. Learners' approaches to solving well-defined, puzzle-like problems (Simon 1978) differ from their approaches to solving complex problems without a single solution (Jonassen et al. 2006). Consideration of the type of problem being solved is relevant as evidence suggests that different cognitive processes and skills are needed when dealing with different kinds of problems (Schraw et al. 1995; Shin et al. 2003).

A similar pattern separating the two components of engineering cognition is also common in K-12 classrooms. Design for *knowledge acquisition* is an approach common to integrated STEM and design-based science. While these approaches target specific science and mathematics concepts, the contextual factors that are essential for engineering can be left out due to lack of time or for the sake of promoting science learning. This model may also not account for prior knowledge that inevitably bias or inform students' design decisions (Goldstein et al. 2016).

Similar to how experts solve problems within their domain with extensive tacit knowledge and prior experiences (Chi et al. 1982), student experiences can be designed around their prior knowledge and experiences. These experiences can be formed with an understanding of Jonassen's (2000) categorization of problem types depending on factors such as the inputs, success criteria, context, amount of structure or constraints, and abstractness of the problem. The exploration of the problem context is especially pertinent to design problems as there are no universal best solutions and the quality of a solution depends on how it meets the contextual factors.

We caution about teaching engineering through two discrete approaches to learning; that is, teach all of the background knowledge that may be needed to address the engineering challenge and then present the engineering challenge as an end of unit activity or a capstone project. We call this the content transfer approach. Another approach is the process transfer, where students engage in multiple design challenges with opportunities to practice design in multiple contexts. We urge educators to check the assumptions associated with these approaches such as that students can transfer content learned in one context to another and can transfer process skills or informed design practices across different problems. Table 8.1 presents the pros and cons of each approach when engineering design is used to address one part of the framework.

8.4.1 Normative and Nonnormative Views About Engineering Design

Engineering design can be represented in ways that are aligned or misaligned with the engineering cognition we argue for. Often educators explain or think about design process as a set of steps followed to solve a problem. Thinking of design as a series of steps is misleading and would lead to misconceptions. Engineering design challenges used in the classroom, especially at K-12, often miss the point of

Table 8.1 Comparison of fragmented approaches to engineering education

Perspective	Pros	Cons
Design for knowledge application	Students engage in the application of their prior understanding of science and mathematics concepts while working on a design project	This rests on the expectation that students will be able to determine and apply relevant knowledge from their prior education and also assumes that students retained the said knowledge from prior instruction
Design for knowledge acquisition	Students acquire science and mathematics concepts while solving structured engineering challenges	The challenges need to be structured to elicit target learning objectives, and hence the contextual factors that are essential for engineering tend to get left out for lack of time and the sake of promoting science learning. This model may also not account for prior knowledge that inevitably influence students' design decisions

ill-nature of design by front-loading all necessary information including materials, constraints, and background information. Design problems require the designer to "discover" the problem and the continuous evolution of problem definition along with solution development.

Jonassen (2000) provides a broad definition of design problems as ill structured and situated in a context, which is in alignment with the literature in engineering design cognition. However, his examples such as making a paper airplane emphasize the need to produce an artifact. It is important to note that while engineering design problems can include artifacts, many engineering problems require the design of a process or a system. In fact, system and process design are prevalent in chemical engineering, civil engineering, and industrial engineering. The focus on artifact design also reinforces the emphasis on the functionality of the final prototype. Whether a solution is a process or a product, prototypes are necessary for communicating or testing solutions, but not as a means to design. The emphasis should be on explaining and justifying how design features represented in a prototype are linked to design performance. Educators in any opportunity as they can should highlight that:

1. Engineering design can result in solutions that describe a process or a system (e.g., wastewater treatment process, city transportation system), not just artifacts or products.
2. Design problems are ill defined requiring the designer to "discover" the problem while generating solutions.
3. Prototypes are necessary for communicating or testing solutions, not as a means to design.
4. Engineering requires justifications with evidence, how design features are linked to design criteria including the trade-offs made by the designers.

Finally, students need multiple opportunities to experience the diverse facts of engineering design and develop abilities to transfer competencies across these proj-

Table 8.2 Normative and nonnormative views of design and approaches to design education

Nonnormative views and approaches	Normative views and approaches
Presentation of design process as a linear or cyclic process (single-sided arrows, a numbered process, etc.)	Presentation of design as an iterative process illustrated with double-sided arrows, arrows that revisit previous stages, etc.
Front-loading all necessary information including materials, constraints, and background information	Design problems are ill defined requiring the designer to "discover" the problem
Imagining the only design outcome to be a product	Engineering design can result in solutions that describe a process or a system (e.g., wastewater treatment process, city transportation system)
Putting too much emphasis on the functionality of the final prototype	Prototypes are necessary for communicating or testing solutions, not as a means to design. The emphasis should be on justifications, how design features are linked to design criteria including the trade-offs made by the designers

ects. Students who are learning to design will consider design in terms of their specific experiences (Brown et al. 1989). If their first introduction to design is within a project that engages them in designing a cardboard table, they will tend to think about design within that context and as a process that ends in a physical object. During a second design project, when they are asked to design a process for sorting recyclables, they would be exploring design of a process or a system. As students continue to participate in other engineering design problems that have a wide variety of end goals, they will come to see the nature of engineering design rather than just the processes that were required in each of their separate experiences. Table 8.2 presents various views – both normative and nonnormative – of design and approaches to design education.

Table 8.3 presented a classic design-build challenge, the egg drop used in K-12 classrooms (Purzer et al. 2013), to illustrate what we would consider weak and strong elements of the task with respect to engaging engineering design cognition. The tasks reported out in Table 8.3 are meant to be very short explanations and illustrations of tasks related to an egg-drop challenge, not implementable curriculum. We discuss variations on a classic design-build challenge by comparing the context in which the design challenge is set in place, the extent to which explanations from data are encouraged, and whether the practices of design trade-offs are expected.

We recommend that educators introduce the problem context upfront. Through the problem-scoping activities, students should help in the identification of the kinds of background knowledge they need in order to address the problem. Students should engage in learning activities that provide that background, yet keep the engineering design challenge in mind as the context for their learning. When moving into the iterative stages of solution generation, students should use the background learning as well as results from their tests of their prototypes as evidence for their

Table 8.3 Rethinking the traditional egg-drop challenge

	Weaker task	Stronger task	Reasoning
Context	Design a solution to protect an egg dropped from roof of building, ~10 ft high	Design a solution to safely transport eggs from a farm to a local, natural grocer. (Appropriate tests and simulations of the effects of this transportation to be determined by the students and teacher collaboratively)	The task should require students to consider a real-world context and the needs of the users. The task should also introduce the possibility of competing criteria
Building explanations from data	Build and test up to three iterations – show your design changes	Conduct research and experiment or analyze existing data in order to explain the relationships between specific design features and performances of prototypes	Students should make their design cognition explicit by writing explanations (their knowledge production)
Trade-off decisions	Determine your best design	Create a systematic way of determining the optimal solution (e.g., a solution that is effective in protecting the eggs and low cost)	Students should synthesize the produced knowledge to make trade-off decisions. Students should also be asked to articulate any ethical or unintended consequences a new solution might create

design decisions. The integration of the engineering with the background knowledge is imperative for the learning gains desired through implementing engineering in the classroom.

8.4.2 *Assessment and Engineering Design*

Embracing our model for engineering cognition has implications for assessment. The National Academy of Engineering published a report, *Tech Tally*, that included reviews of 28 assessment instruments including multiple-choice tests and performance-based activities intended to measure aspects of design and problem-solving ability (NAE and NRC 2006). Modes of assessment traditional within undergraduate engineering and K-12 education, such as multiple-choice tests, are unable to assess the aspect of design cognition that requires knowledge acquisition and application, which are a crucial part of developing engineering design cognition. As Bransford and Schwartz (1999) argue that assessment that is solely for sequestered problem solving and cannot do justice to students' ability to apply their knowledge.

Tech Tally highlights performance assessments used in the United Kingdom where design has been a mandatory part of the precollege curriculum since the 1990s. The United Kingdom's *Assessment of Performance in Design and Technology* project includes a 90-min paper-and-pencil design task students are asked to solve. The student work is then evaluated, with a rubric that focuses on design capabilities, communication skills, and conceptual understanding of underlying concepts. Such a focus on performance puts considerable weight on how students *use* their knowledge, whether they recognize when they are missing key information and how skillfully they gather new data. In the long term, computer-based simulation can be useful in the assessment of such capability with examples of novel assessment methods in large-scale or embedded forms. The Technology and Engineering Literacy (TEL) assessment of the National Assessment of Educational Progress (NCES 2016) is a computer-based assessment administered to a national representative sample in the United States in 2014. Another body of literature focuses on computer-supported embedded assessment systems using learning analytics built on computer-aided design software with abilities to track student design actions in the background as they design (Xie et al. 2014; Vieira et al. 2016).

While we have not discussed the social aspects of design in detail, we do not wish to diminish the fact that design is a social endeavor. Assessment also takes a different form when evaluating design work that is done in student teams rather than individually. In such situations, educators struggle in identifying individual learning outcomes and may rely on approaches such as peer assessment (Dutson et al. 1997). Yet, peer assessments typically measure the social aspects of the learning process rather than individual learning outcomes. Individual learning should be assessed based on students' ability to conduct experiments and explain meaningful relationships between design features and the performance of prototypes that guide their design solution.

Finally, effective assessment prepares for future learning and guides students toward competencies that cut across multiple projects such as the informed design behaviors such as seeking and evaluating new, relevant information (Purzer and Douglas in press). Hence, student assessment should reflect how well their performance aligns with the elements of informed design practices. The final design solutions should not be evaluated simply based on how they function but based on students' ability to justify the design criteria and constraints it was designed to meet and the trade-offs students attempted to address.

8.5 Recommendations for Educators

While we define engineering cognition as the interaction and iteration between acquiring knowledge and applying knowledge, engaging students in these cognitive behaviors focusing on knowledge production is not a straightforward task. It is essential that the design challenges students asked to formulate and solve are crafted in a way to allow opportunities to engage such cognition through problem scoping, ideation, and explicit decision-making.

Real-world engineering problems involve a novel problem, a novel context, a novel set of users, or a combination of these that necessitate knowledge production at the heart of design. Such authentic experiences often refer to design projects in engineering (Dym et al. 2005). Hence, effective design tasks should include at least one of these aspects so that the solution generated is not conventional. Experts, for example, spend a substantial amount of time in knowledge building activities, so an engineering design task should provide opportunities to do so. In the classroom, this type of problem scoping may look like presenting the student teams with a brief problem statement or memo from a client that has the basic needs outlined but only some of the criteria and constraints. Then throughout the design challenge, the students ask questions to the client multiple times to continue to understand the needs and wants of the client and end users. Students need to develop abilities to scope and formulate problems. It is important that they have opportunities to grapple with defining the needs, constraints, and criteria rather than being provided with all of them at the beginning of the challenge in a definitive manner.

Educators must keep in mind that the value of an engineer's work lies in a timely, sufficiently complete solution (Frezza et al. 2013). This focus on sufficiency as opposed to a "global best solution" shifts the emphasis away from single-solution thinking and toward evidence-based decisions. The social and ethical implications of new solutions should also be discussed. Hence, classroom assessment practices should be planned carefully so that judgments of student learning are not inaccurately based communication skills or whether a prototype fails. What is most critical is to assess students' ability to provide appropriate technical and contextual evidence in support of their design solutions. Furthermore, uses of effective pedagogies, that clearly promote argumentation and evidence-based reasoning during the justification of design decisions, help students make this knowledge production explicit.

Given these recommendations, we believe this chapter provides critical insights for undergraduate and K-12 education in engineering, curriculum and assessment development, and in-service and preservice teacher professional development.

8.6 Recommendations for Researchers

STEM education reform has long been informed by prior research on learning and cognition in STEM fields which has traditionally focused on science and mathematics. As an integrative field, engineering education research has a unique position in shaping the future of STEM education research. Future research should focus on engineering design behaviors specifically focusing on the interaction between knowledge acquisition and knowledge application in the context of engineering design problems (e.g., Purzer et al. 2015). Such research must be informed by the rich body of literature on design cognition that are often not read or built on among precollege STEM education researchers.

Future research should also examine educators' teaching practices, pedagogical content knowledge, and abilities to scaffold integrative learning. Research should focus on the role of different types of design problems and its context in impacting student learning. At the heart of understanding teacher practices as well as student learning is understanding the impacts and uses of assessment in and outside the classroom. Future research should examine teachers' challenges when assessing design learning (Hynes et al. 2014) and evaluate professional development strategies necessary for supporting effective assessment in the classroom.

8.7 Concluding Remarks

The road ahead in STEM education research is exciting. Similar to how engineering creates a perfect environment for learning, engineering education opens doors to a new era of research in STEM education. Our definition of engineering with multiple dimensions representing abilities beyond sole application of science and mathematics aligns with Carberry and Baker's (2018) call for broadening the definition of engineering and engineering education in Chap. 10. We also present a history of the US engineering education system complementing the history of the European perspective presented by Corlu and colleagues in Chap. 11. Our chapter agrees with Crippen and Antonenko's (2018) claim that problem solving is a learning activity in Chap. 5; yet, we argue that knowledge building in itself is not sufficient for solving engineering problems. Similar to Crawford and Capps in Chap. 2 of this book (2018), we clarify common terms used in engineering to caution STEM educators and researchers with various uses of terms in this interdisciplinary space. Furthermore, we spell out common normative and nonnormative engineering practices in the classroom and argue that the assessment of integrated learning with engineering requires a performance-based and competency-based approach.

As we define engineering cognition as a process of managing ambiguity and complexity through recurring knowledge production, we also raise many questions to be answered through research. *Can the interplay between knowledge acquisition and knowledge application be promoted with specific interventions such as by introducing a set of trade-offs? How do students handle ambiguity and complexity? How can teachers be prepared to help students synthesize and use information gathered during problem scoping? How do teachers approach classroom assessment when teaching engineering?* Many more questions need to be explored to help the recent STEM reform efforts be effective and useful in the classroom. Research in engineering cannot be based on silos of disciplinary research but rather through a synthesis of historical and contemporary research in science, technology, engineering, and mathematics education. This book is one example of efforts where researchers in all parts of STEM education are engaged.

Acknowledgments This work presented in this manuscript is based upon work supported by the National Science Foundation (NSF) under Grant DUE #1348547 and EEC #1150874. Any opinions, findings, and conclusions or recommendations expressed in this paper, however, are those of the authors and do not necessarily reflect the views of the NSF.

References

Accreditation Board for Engineering and Technology. (1998). *ABET accreditation yearbook*. Baltimore: ABET.

Achieve. (2013). *Next generation science standards*. http://www.nextgenscience.org/next-generation-science-standards. Accessed 12 Nov 2014.

Adams, R., Aldridge, D., Atman, C., Barker, L., Besterfield-Sacre, M., Bjorklund, S., & Young, M. (2006). The research agenda for the new discipline of engineering education. *Journal of Engineering Education, 95*, 259–261.

Adams, R. S., Cardella, M., & Purzer, Ş. (2016). Analyzing design review conversations: Connecting design knowing, being and coaching. *Design Studies, 45*(Part A), 1–8. https://doi.org/10.1016/j.destud.2016.03.001.

American Society for Engineering Education. (2015). ASEE reports. http://www.asee.org/member-resources/reports. Accessed 1 Dec 2014.

Apedoe, X. S., Reynolds, B., Ellefson, M. R., & Schunn, C. D. (2008). Bringing engineering design into high school science classrooms: The heating/cooling unit. *Journal of Science Education and Technology, 17*(5), 454–465.

Atman, C. J., Cardella, M. E., Turns, J., & Adams, R. (2005). Comparing freshman and senior engineering design processes: An in-depth follow-up study. *Design Studies, 26*(4), 325–357.

Atman, C. J., Adams, R. S., Cardella, M. E., Turns, J., Mosborg, S., & Saleem, J. (2007). Engineering design processes: A comparison of students and expert practitioners. *Journal of Engineering Education, 96*(4), 359–379.

Bransford, J. D., & Schwartz, D. L. (1999). Rethinking transfer: A simple proposal with multiple implications. *Review of Research in Education, 24*, 61–100.

Brown, J. S., Collins, A., & Duguid, P. (1989). Situated cognition and the culture of learning. *Educational Researcher, 18*(1), 32–42.

Cajas, F. (2001). The science/technology interaction: Implications for science literacy. *Journal of Research in Science Teaching, 38*(7), 715–729.

Capps, D. K., & Crawford, B. A. (2018). Teacher cognition of engaging children in scientific practices. In Y. J. Dori, Z. Mevarech, & D. Baker (Eds.), *Cognition, metacognition, and culture in STEM education* (pp. 9–32). Springer.

Carberry, A. R., & Baker, D. R. (2018). The impact of culture on engineering and engineering education. In Y. J. Dori, Z. Mevarech, & D. Baker (Eds.), *Cognition, metacognition, and culture in stem education* (pp. 217–239). Cham: Springer.

Chi, M. T. H., Glaser, R., & Rees, E. (1982). Expertise in problem solving. In R. Sternberg (Ed.), *Advances in the psychology of human intelligence* (Vol. 1, pp. 7–76). Hillsdale: Erlbaum.

Corlu, M. S., Svidt, K., Gnaur, D., Lavi, R., Borat, O., & Corlu, M. A. (2018). Engineering education in higher education in Europe. In Y. J. Dori, Z. Mevarech, & D. Baker (Eds.), *Cognition, metacognition, and culture in stem education* (pp. 241–259). Cham: Springer.

Crawford, B. A., & Capps, D. K. (2017). Teacher cognition of engaging children in scientific practices. In Y. J. Dori, Z. Mevarech, & D. Baker (Eds.), *Cognition, metacognition, and culture in stem education* (pp. 9–32). Cham: Springer.

Crippen, K., & Antonenko, P. D. (2018c). Designing for collaborative problem solving in STEM. In Y. J. Dori, Z. Mevarech, & D. Baker (Eds.), *Cognition, metacognition, and culture in stem education* (pp. 89–116). Cham: Springer.

Crismond, D. (2001). Learning and using science ideas when doing investigate-and-redesign tasks: A study of naïve, novice, and expert designers doing constrained and scaffolded design work. *Journal of Research in Science Teaching, 38*(7), 791–820.

Crismond, D. P., & Adams, R. S. (2012). The informed design teaching and learning matrix. *Journal of Engineering Education, 101*(4), 738–797.

Cross, N. (1982). Designerly ways of knowing. *Design Studies, 3*(4), 221–227.

Cross, N. (2001). Design cognition: Results from protocol and other empirical studies of design activity. In C. Eastman, W. Newstatter, & M. McCracken (Eds.), *Design knowing and learning: Cognition in design education* (pp. 79–103). Elsevier: Oxford.

Daly, S. R., Adams, R. S., & Bodner, G. M. (2012). What does it mean to design? A qualitative investigation of design professionals' experiences. *Journal of Engineering Education, 101*(2), 187–219.

Dong, A., Garbuio, M., & Lovallo, D. (2014). *Robust design review conversations*. Proceedings of the 10th Design Thinking Research Symposium. West Lafayette: Purdue University Press. Retrieved from http://docs.lib.purdue.edu/dtrs/2014/Impact/3/

Dorst, K. (2003). The problem of design problems. *Expertise in Design, 6*, 135–147. Design Thinking Research Symposium. Creativity and Cognition Studios Press: Sydney, Australia.

Dorst, K. (2011). The core of 'design thinking' and its application. *Design Studies, 32*(6), 521–532.

Dorst, K., & Cross, N. (2001). Creativity in the design process: Co-evolution of problem–solution. *Design Studies, 22*(5), 425–437.

Dubberly, H. (2004). How do you design? Retrieved from http://www.dubberly.com/articles/how-do-you-design.html/

Dutson, A. J., Todd, R. H., Magleby, S. P., & Sorensen, C. D. (1997). A review of literature on teaching engineering design through project oriented capstone courses. *Journal of Engineering Education, 86*(1), 17–28.

Dym, C. L., Agogino, A. M., Eris, O., Frey, D. D., & Leifer, L. J. (2005). Engineering design thinking, teaching, and learning. *Journal of Engineering Education, 94*(1), 103–120.

Erduran, S., & Dagher, Z. R. (2014). *Reconceptualizing nature of science for science education*. Netherlands: Springer. doi:10.1007/978-94-017-9057-4.

Figueiredo, A. D. (2008). *Toward an epistemology of engineering*. Workshop on Philosophy and Engineering, The Royal Academy of Engineering, London, November 10–12, 2008 . Available at SSRN: https://ssrn.com/abstract=1314224

Fortus, D., Dershimer, C., Krajcik, J., Marx, R., & Mamlok-Naaman, R. (2004). Design-based science and student learning. *Journal of Research in Science Teaching, 41*(10), 1081–1110.

Fosmire, M., & Radcliffe, D. F. (2013). *Integrating information into the engineering design process*. West Lafayette: Purdue University Press. http://docs.lib.purdue.edu/purduepress_ebooks/31/.

Frezza, S., Nordquest, D., & Moodey, R. (2013). Knowledge-generation epistemology and the foundations of engineering. In *Proceedings of the IEEE Frontiers in Education Conference*. Oklahoma City: IEEE.

Froyd, J. E., & Ohland, M. W. (2005). Integrated engineering curricula. *Journal of Engineering Education, 94*(1), 147–164.

Gainsburg, J. (2006). The mathematical modeling of structural engineers. *Mathematical Thinking and Learning, 8*(1), 3–36.

Goel, V., & Pirolli, P. (1992). The structure of design problem spaces. *Cognitive Science, 16*(3), 395–429.

Goldstein, M. H., Purzer, Ş., & Adams, R. S. (2016, August). *More than a vehicle to learning science: The role of design in P-12 Education*. Paper presented at the 3rd P-12 Engineering and Design Education Summit. Chicago.

Grinter, L. E. (1955). Report on evaluation of engineering education. *Journal of Engineering Education, 46*(1), 25–63.

Hammond, H. P. (1940). *Report of committee on aims and scope of engineering curricula*. Pittsburgh: Society for the Promotion of Engineering Education.

Hatchuel, A., & Weil, B. (2003). A new approach of innovative design: An introduction to CK theory. In *DS 31: Proceedings of ICED 03, the 14th International Conference on Engineering Design*. Stockholm, Sweden.

Hmelo, C. E., Holton, D. L., & Kolodner, J. L. (2000). Designing to learn about complex systems. *The Journal of the Learning Sciences, 9*(3), 247–298.

Hynes, M., Purzer, Ş., & Moore, T. J. (2014). *Teachers' attempts assessing middle school engineering design work*. Proceedings of the 121st American Society for Engineering Education Annual Conference and Exposition. Indianapolis: ASEE. Retrieved from https://peer.asee.org/teachers-attempts-assessing-middle-school-engineering-design-work

Jonassen, D. H. (2000). Toward a design theory of problem solving. *Educational Technology Research and Development, 48*(4), 63–85.

Jonassen, D. H., Strobel, J., & Lee, C. B. (2006). Everyday problem solving in engineering: Lessons for engineering educators. *Journal of Engineering Education, 95*(2), 139–151.

Kittleson, J. M., & Southerland, S. A. (2004). The role of discourse in group knowledge construction: A case study of engineering students. *Journal of Research in Science Teaching, 41*(3), 267–293.

Kolodner, J. L., Crismond, D., Gray, J., Holbrook, J., & Puntambekar, S. (1998). Learning by design from theory to practice. In *Proceedings of the International Conference of the Learning Sciences*. Retrieved from https://www.cc.gatech.edu/projects/lbd/htmlpubs/lbdtheorytoprac.html

Kolodner, J. L. (2002). Facilitating the learning of design practices: Lessons learned from an inquiry into science education. *Journal of Industrial Teacher Education, 39*(3), 9–40. Retrieved from http://scholar.lib.vt.edu/ejournals/JITE/v39n3/kolodner.html/.

Kolodner, J. L., Crismond, D., Gray, J., Holbrook, J., & Puntambekar, S. (1998). Learning by design from theory to practice. In *Proceedings of the International Conference of the Learning Sciences*. Retrieved from https://www.cc.gatech.edu/projects/lbd/htmlpubs/lbdtheorytoprac.html

Kolodner, J. L., Crismond, D., Fasse, B. B., Gray, J. T., Holbrook, J., Ryan, M., et al. (2003). Problem-based learning meets case-based reasoning in the middle-school science classroom: Putting a learning-by-design curriculum into practice. *The Journal of the Learning Sciences, 12*(4), 495–548.

Kroll, E., & Koskela, L. (2014). On abduction in design. In J. S. Gero & S. Hanna (Eds.), *Design computing and cognition DCC'14* (pp. 357–376). Dordrecht: Springer.

Lachapelle, C. P., & Cunningham, C. M. (2014). Engineering in elementary schools. In Ş. Purzer, J. Strobel, & M. Cardella (Eds.), *Engineering in pre-college settings: Synthesizing research, policy, and practices*. West Lafayette: Purdue University Press.

Lawson, B. (2006). *How designers think: The design process demystified*. London: Routledge.

Loui, M. (2016). Quotations. Retrieved from http://publish.illinois.edu/loui/quotations.

Lu, C.-C. (2014). The relationship between student design cognition types and creative design outcomes. *Design Studies, 36*, 59–76.

Mann, C. R. (1918). *A study of engineering education: Prepared for the joint committee on engineering education of the national engineering societies. (No. 11)*. Boston: Merrymount Press.

Mehalik, M. M., Doppelt, Y., & Schunn, C. D. (2008). Middle-school science through design-based learning versus scripted inquiry: Better overall science concept learning and equity gap reduction. *Journal of Engineering Education, 97*(1), 71–85.

National Academy of Engineering. (2004). *The engineer of 2020: Visions of engineering in the new century*. Washington, DC: National Academy of Engineering.

National Academy of Engineering. (2009). *Engineering in K-12 education: Understanding the status and improving the prospects*. Washington, DC: National Academies Press.

National Academy of Engineering. (2013). *Transforming undergraduate education in engineering: Phase I: Synthesizing and integrating industry perspectives*. Arlington: American Society for Engineering Education. Retrieved from http://www.asee.org/TUEE_PhaseI_WorkshopReport.pdf.

National Academy of Engineering & National Research Council. (2006). In E. Gamire & G. Pearson (Eds.), *Tech tally: Approaches to assessing technological literacy*. Washington, DC: National Academies Press.

National Academy of Engineering & National Research Council. (2014). *STEM integration in K-12 education*. Washington, DC: National Academies Press.

National Center for Education Statistics. (2016). Technology and engineering literacy assessment. Retrieved from https://nces.ed.gov/nationsreportcard/tel/

National Research Council. (2012). *A framework for K-12 science education: Practices, crosscutting concepts, and core ideas*. Washington, DC: National Academies Press.

Peirce, C. S. (1932). *Elements of logic*. Cambridge, MA: Harvard University Press.

Perkins, D. N., & Salomon, G. (1992). *Transfer of learning. International encyclopedia of education* (2nd ed.). Oxford: Pergamon Press.

Purzer, Ş. (2011). The relationship between team discourse, self-efficacy, and individual achievement: A sequential mixed-methods study. *Journal of Engineering Education, 100*(4), 655–679.

Purzer, Ş. & Douglas, K. (in press). Assessing early engineering thinking and design competencies in the classroom. In L. English & T. Moore (Eds.), *Early engineering leaning*. Springer.

Purzer, Ş., Fila, N. D., & Dick, E. C. (2014). *A cross-case analysis of disciplinary identities communicated through design reviews*. Proceedings of the 10th Design Thinking Research Symposium. West Lafayette: Purdue University Press. Retrieved from http://docs.lib.purdue.edu/dtrs/2014/Identity/5/

Purzer, S., Cardella, M. E., & Hsu, M.-C. (2011a). Elementary teachers' views about teaching design, engineering, and technology. *Journal of Pre-College Engineering Education, 1*(2), 31–39. Retrieved from. https://doi.org/10.5703/1288284314639.

Purzer, S., Hilpert, J. C., & Wertz, R. E. (2011b). Cognitive dissonance during engineering design. In *Proceedings of the IEEE Frontiers in Education Conference (FIE)*. Rapid City: IEEE.

Purzer, Ş., Duncan-Wiles, D., & Strobel, J. (2013). Cost or quality? Teaching fourth and fifth graders about engineering optimization and trade-offs. *Science and Children, 50*, 34–39.

Purzer, Ş., Moore, T., & Baker, D., Berland, L. (2014a). *Engineering in the Next Generation Science Standards (NGSS): A position statement on the implementation of engineering as covered in the NGSS*. Position Paper Invited by the National Association for Research in Science Teaching, A Worldwide Organization for Improving Science Teaching and Learning Through Research. https://portal.narst.org/Login.aspx?LoginRedirect=https://www.narst.org/members/NGSSblog/index.cfm. Accessed 29 Nov 2014.

Purzer, Ş., Strobel, J., & Cardella, M. (Eds.). (2014b). *Engineering in pre-college settings: Synthesizing research, policy, and practices*. West Lafayette: Purdue University Press.

Purzer, Ş., Goldstein, M. H., Adams, R. S., Xie, C., & Nourian, S. (2015). An exploratory study of informed engineering design behaviors associated with scientific explanations. *International Journal of STEM Education, 2*(9), 1–12. https://doi.org/10.1186/s40594-015-0019-7.

Radcliffe, D. (2015). A tale of two STEMs. Prism. Retrieved from http://www.asee-prism.org/last-word-dec-3/

Roozenburg, N. F. (1993). On the pattern of reasoning in innovative design. *Design Studies, 14*, 4–18.

Schnittka, C., & Bell, R. (2011). Engineering design and conceptual change in science: Addressing thermal energy and heat transfer in eighth grade. *International Journal of Science Education, 33*(13), 1861–1887.

Schraw, G., Dunkle, M. E., & Bendixen, L. D. (1995). Cognitive processes in well defined and ill defined problem solving. *Applied Cognitive Psychology, 9*(6), 523–538.

Sheppard, S., Macatangay, K., & Colby, A. (2008). *Educating engineers: Designing for the future of the field* (Vol. 3). San Francisco: Jossey-Bass Inc Pub.

Shin, N., Jonassen, D. H., & McGee, S. (2003). Predictors of well structured and ill structured problem solving in an astronomy simulation. *Journal of Research in Science Teaching, 40*(1), 6–33.

Simon, H. A. (1978). Information-processing theory of human problem solving. *Handbook of Learning and Cognitive Processes, 5*, 271–295.

Sneider, C., & Purzer, Ş. (2014). The rising profile of STEM literacy through national standards and assessments. In Ş. Purzer, J. Strobel, & M. Cardella (Eds.), *Engineering in pre-college settings: Synthesizing research, policy, and practices*. West Lafayette: Purdue University Press.

Stevens, R., O'Connor, K., Garrison, L., Jocuns, A., & Amos, D. M. (2008). Becoming an engineer: Toward a three dimensional view of engineering learning. *Journal of Engineering Education, 97*(3), 355–368.

Streveler, R. A., & Smith, K. A. (2006). Conducting rigorous research in engineering education. *Journal of Engineering Education, 95*(2), 103–105.

Svarovsky, G. N., & Shaffer, D. W. (2007). Soda constructing knowledge through Exploratoids. *Journal of Research in Science Teaching, 44*(1), 133–153.

Tonso, K. L. (2007). *On the outskirts of engineering: Learning identity, gender, and power via engineering practice*. Rotterdam: Sense Publishers.

Vieira, C., Goldstein, M. H., Purzer, Ş., & Magana, A. (2016). Using learning analytics to characterize student experimentation strategies in the context of engineering design. *Journal of Learning Analytics, 3*(3), 291–317. doi: http://dx.doi.org/10.18608/jla.2016.33.14.

Walkington, C. A., Nathan, M. J., Wolfgram, M., Alibali, M. W., & Srisurichan, R. (2014). Bridged and barriers to constructing conceptual cohesion across modalities and temporalities: Challenges of STEM integration in the precollege engineering classroom. In J. Strobel, Ş. Purzer, & M. Cardella (Eds.), *Engineering in pre-college settings: Research into practice*. West Lafayette: Purdue University Press.

Wendell, K. B., & Lee, H. S. (2010). Elementary students' learning of materials science practices through instruction based on engineering design tasks. *Journal of Science Education and Technology, 19*(6), 580–601.

Wickenden, W. E. (1930). *Report of the investigation of engineering education, 1923–1929* (Vol. I and II). Pittsburgh: Society for the Promotion of Engineering Education.

Xie, C., Zhang, H., Nourian, S., Pallant, A., & Hazzard, E. (2014). A time series analysis method for assessing engineering design processes using a CAD tool. *International Journal of Engineering Education, 30*(1), 218–230.

Zannier, C., Chiasson, M., & Maurer, F. (2007). A model of design decision making based on empirical results of interviews with software designers. *Information and Software Technology, 49*(6), 637–653.

Chapter 9
Metacognition and Meta-assessment in Engineering Education

Niva Wengrowicz, Yehudit Judy Dori, and Dov Dori

9.1 Metacognition, Meta-assessment, and Engineering Education

Modern engineering education programs aim to enrich students with the necessary knowledge, skills, and attitudes for becoming successful young engineers (Crawley et al. 2011; National Science Foundation [NSF] 1998; Rugarcia et al. 2000). Johri and Olds (2011) noted three distinguishing characteristics of engineering education: use of representations or models, alignment with professional practices, and emphasis on design. De Graaff and Christensen (2004) claimed that active learning and engineering education constitute a natural pair, so an engineer should be educated to design and trained to construct solutions to problems in the real world. NSF determined that engineering education should promote, among other things, teamwork, project-based learning (PBL), and close interaction with industry. The goal of engineering education programs of such nature is to train students to be able to conceive, design, implement, and operate complex value-added engineering products, processes, and systems in modern, team-based environments. These insights formed the basis for innovative educational frameworks aimed at producing

N. Wengrowicz (✉)
Faculty of Industrial Engineering and Management, Technion, Haifa, Israel

Faculty of Education in Science and Engineering, Technion, Haifa, Israel
e-mail: nivawen@technion.ac.il

Y.J. Dori
Faculty of Education in Science and Technology, Technion, Haifa, Israel

The Samuel Neaman Institute for National Policy Research, Haifa, Israel

D. Dori
Faculty of Industrial Engineering and Management, Technion, Haifa, Israel

The Samuel Neaman Institute for National Policy Research, Haifa, Israel

Y.J. Dori et al. (eds.), *Cognition, Metacognition, and Culture in STEM Education*, Innovations in Science Education and Technology 24,
https://doi.org/10.1007/978-3-319-66659-4_9

the next generation of engineers, such as CDIO – conceive-design-implement-operate (Crawley et al. 2008), and for formulating the higher engineering education standard by the Accreditation Board for Engineering and Technology (ABET 2014).

This modern approach to engineering education emphasizes the importance of active and hands-on learning, which exposes students to experiences that engineers are likely to encounter during their professional lives. To enable these kinds of experiences, a typical engineering-oriented syllabus contains significant elements of project-based learning (Dym et al. 2005; Lewis et al. 1998) and an outcome-based assessment process (Olds et al. 2005).

9.1.1 Cognition, Metacognition, and Engineering Education

Cognition, metacognition, and motivational skills are key components of self-regulated learning (SRL), a concept which refers to the ability of students to understand and control their learning environments. Learners improve their learning by governing the interaction between cognitive, metacognitive, and motivational components (Herscovitz et al. 2012; Schunk and Zimmerman 2003).

Cognition is concerned with what students know and what they can do with this knowledge. It includes skills necessary to encode, memorize, and recall information (Schraw et al. 2006) or to apply and transfer knowledge from one context to another (Dori and Sasson 2013). According to the taxonomy of Bloom and his followers (Anderson et al. 2001), the classification of students' cognitive performance and thinking skills includes six hierarchical cognitive levels: remembering, understanding, applying, analyzing, evaluating, and synthesizing or creating new knowledge. The last three levels are considered higher-order thinking skills, as they involve problem solving, critical thinking, and presenting new ideas, and they usually have more than one correct answer for a given problem (Resnick 1987; Zohar and Dori 2003). Developing students' higher-order thinking skills requires teaching and learning methods that are significantly different than those needed for remembering and understanding facts and concepts.

Metacognition includes skills that enable learners to understand and monitor their cognitive processes. As Fig. 9.1 shows, metacognition is concerned with knowledge of cognition – what students know about their knowledge – and regulation of cognition – what students can do with this knowledge to better control their learning (Herscovitz et al. 2012; Vos and De Graaff 2004; Vrugt and Oort 2008).

Flavell and Wellman (1977), and following them also Cross and Paris (1988), classified the metacognitive knowledge into three types: (1) declarative (or personal) knowledge, which focuses on knowing one's own learning skills, strategies, and factors that influence cognition performance; (2) procedural (or task) knowledge, which refers to knowing how certain skills and strategies work and how they should be applied to improve learning; and (3) conditional (or strategic) knowledge, which is knowing why and when to apply the various skills and strategies to that end.[1]

[1] The terms personal, task, and strategy knowledge of cognition were coined by Flavell and Wellman (1977). Cross and Paris (1988) changed those terms to declarative, procedural, and conditional.

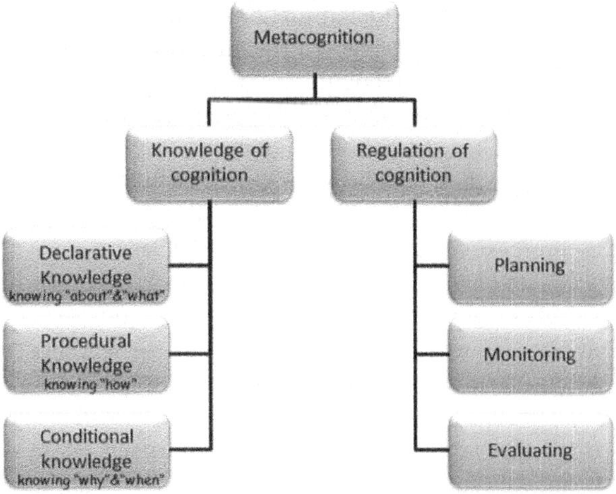

Fig. 9.1 Metacognition components (adapted from Schraw et al. 2012; Herscovitz et al. 2012)

Regulation of cognition refers to a set of skills, such as planning, monitoring, and evaluation, which help learners control their learning. Planning involves the appropriate selection of strategies and allocation of resources. It includes goal setting, activating relevant background knowledge, and budgeting time. Monitoring includes self-testing skills that increase the awareness and comprehension of the task. Through monitoring, learning can be controlled, as the learner considers how to complete a task, and whether a selected strategy is working. Evaluation refers to appraising the learning product and the efficiency of the learning process. When students evaluate their learning, they may ask themselves what peers would think about their work. If they were to carry out a similar learning activity, they might consider planning differently and review their strategies in order to improve their performance (Schraw et al. 2006; Vrugt and Oort 2008). Pintrich (1999) differentiated between regulation of cognition activities that help the students in planning their use of cognitive strategies on the one hand and resource management activities that assist students in managing and controlling their learning environment on the other hand. Resource management encompasses regulating students' time, effort, study setting, and the use of help-seeking strategies to better achieve their needs and goals.

The application of metacognition enables one to monitor and regulate the cognitive processes – to evaluate the current learning state and to plan and allocate limited learning resources with optimal efficiency (Carr and Strobel 2012). Extending the conceptual discussion on metacognition, Ford and Yore (2012) claimed that critical thinking, metacognition, and reflection originated from and grew out of three different education disciplines: philosophy, psychology, and progressive education, respectively. Since each one of these three disciplines contains elements of the other two, the authors proposed that convergence of metacognitive thinking with reflective and critical thinking can improve the overall level of one's thinking.

De Graaff and Christensen (2004) clarified that the goal of engineering education is to provide a well-adapted learning environment, designed to encourage students to develop their metacognitive skills, and that this goal is more important than trying to compress large amounts of content knowledge into students' heads. Knowledge of cognition would allow students to improve how they learn and approach new problems, while regulation of cognition may enable them to acquire the abilities to plan and monitor their assignments. Both these qualities facilitate the attainment of key professional engineering competencies.

Vos and De Graaff (2004) claimed that active learning tasks, such as working on projects in engineering courses, not just require metacognitive knowledge and skills but also encourage the development of the learners' metacognition. Brodeur et al. (2002) clarified that working on projects provides students with opportunities to monitor their own learning and assess their own progress. Newell et al. (2004) suggested that taking a metacognitive approach while working on projects in teams encourages students to become conscious of their team skills. Thus, metacognition may be valuable for improving an individual's relationship not only to her or his own learning processes but also to the learning processes of others and to the collaborative learning process. Others found that metacognition plays important roles in enhancing engineering teams' performance, supporting students' engagement in the learning process, and helping students solve problems effectively (Brodeur et al. 2002; Carr and Strobel 2012; De Graaff and Christensen 2004; Lawanto 2009; Mills and Treagust 2003; Newell et al. 2004; Schraw et al. 2006; Vos and De Graaff 2004).

Based on several decades of research literature, Lin (2001) concluded that there are two basic approaches to developing students' metacognitive skills: training in strategy and designing supportive learning environments. There are also two kinds of content that can be used while focusing on developing metacognitive skills: domain-specific content and self-as-learner content.

Veenman (2012) pointed out three principles for the successful instruction of metacognitive skills: (1) Metacognitive instruction should be embedded in the context of the task; (2) learners should be informed about the benefit of applying metacognitive skills; and (3) instruction and training should be repeated over time rather than being a one time directive.

The first and third skills were integrated in the courses we describe in the sequel of this chapter.

Vos and De Graaff (2004) claimed that creating innovative and active learning curricula for engineering courses that are designed to develop students' metacognitive skills differs from designing curricula for achieving only cognitive objectives. They noted that there is no one good answer to the question of how to design a task meant for developing one's metacognition. Kuhn (2000) focused on the link between metacognition and the development of the three higher order thinking skills – analysis, synthesis, and evaluation. She explained that higher order thinking involves reflecting on what is known and how that knowledge can be verified, and these are metacognitive processes. Vos and De Graaff (2004) suggested adopting some metacognitive principles as part of designing higher order thinking tasks.

These principles might include self-examination to check understanding, finding needed information independently, modeling reality, designing new products and processes, developing alternative ways to solve a given problem, comparing alternatives, and choosing the optimal one. Several of these ideas can or should be applied in systems engineering education and more specifically in model-based systems engineering.

Newell and his colleagues (2004) added metacognitive principles related to teaming skills, such as discussing with students the strengths and weaknesses of each team member, possible sources of conflict, consideration of how different people process information, and ways to bridge differences in learning preferences.

The innovative curricula of the information systems engineering course and model-based systems engineering course that we have investigated and present in this chapter have adopted metacognitive principles in order to develop our students' cognitive skills and content knowledge along with their metacognitive skills.

9.1.2 Project-Based Learning and Engineering Education

The term *project* is universally used in engineering practice as a *unit of work*, usually defined on the basis of the client (Mills and Treagust 2003). Project-based learning (PBL) is a teaching and learning method focusing on developing a project in order to engage students in sustained, cooperative investigation (Bransford and Stein 1993). PBL combines academic knowledge with real-world applications and is used in a number of domains, including engineering. The PBL approach is part of progressive education and has been around for many years, tracing its origins back to the work of Dewey (1934). PBL is related to the prevailing trend of the constructivist *learning by doing* approach, according to which learning occurs through participation in meaningful activities that are part of a community of practice (Johri and Olds 2011).

The project serves as a means to make learning more meaningful than studying the theory alone, and it is readily applicable to real-world problems. When carrying out PBL, students can learn both content and thinking skills. Projects can be divided into two main kinds: problem-oriented projects and design-oriented projects. In problem-oriented projects, which focus on "know-why," students solve theoretical problems. In design-oriented projects, which focus on "know-how," students design and/or construct products. In both project kinds, students need to exercise higher order thinking skills, such as analysis, synthesis, and evaluation of knowledge (Dym et al. 2005). Using higher order thinking skills in order to carry out a project successfully requires also the application of metacognitive skills, such as reflecting on what is known and how that knowledge can be verified (Kuhn 2000).

The benefits of learning through working on projects include enhanced student participation in active and self-learning processes, improved communication skills, and promotion of critical thinking by enabling students to form original opinions and express individual standpoints (Hadim and Esche 2002).

Both the information systems engineering and the model-based systems engineering courses included students' projects and the assessment of those projects. Students were asked to construct conceptual models using two leading conceptual modeling languages that were taught in the two courses: Unified Modeling Language (UML) and Object-Process Methodology (OPM).

9.1.3 Assessment and Peer Assessment in Two Engineering Courses

Assessment is defined as the process of collecting and gathering data or evidence about the impact of education (Olds et al. 2005; Orem 2012; Ory 1992). Assessment of students' outcomes related to a learning unit can be divided into two major types: formative and summative. Formative assessment is conducted during the learning process in order to improve learning and expand success (Popham 2004; Topping 1998). Such assessment is intended to help students develop their knowledge, higher order thinking skills, and other personal and professional skills (Boud 1990; Brown and Knight 1994; Dori 2003; Topping 1998). Summative assessment is conducted at the end of the learning unit in order to make evaluation – judgments or decisions about the level of attained achievements and often determination of individual students' success or failure in demonstrating mastery of knowledge of the studied content matter (Orem 2012; Topping 1998). At higher levels, students' assessment is commonly applied to measure and evaluate their cognitive and metacognitive quality of performance and thinking skills by examining their learning outcomes.

Undergraduate engineering PBL courses in general, and mandatory courses with a large amount of participants in particular, are challenging. The challenge arises due to the fact that unlike large lecture-hall classes with hundreds of passive listening students, PBL courses require engaging a large body of students in performing different projects in small teams, advising each team, monitoring teams' progress in a formative assessment mode, and measuring their performance in a summative assessment mode.

Effective and efficient assessment and evaluation of students' content knowledge and their level of thinking skills, as reflected in projects they carry out in such courses, call for creative approaches to cope with the need to devote significant amounts of time and attention to keeping track of, guiding, monitoring, and finally evaluating a large number of different projects at both the team and the individual levels. In our large-scale PBL course, each project required the students to conceptually model a real complex Web-based system in two different modeling languages, each producing a set of some dozen diagrams. The challenge is even bigger when one considers the need to combine the performance of the project team as a whole on the one hand and the individual's contributions to the project and her/his personal learning outcomes on the other hand. Key to overcoming this challenge is the incorporation of *peer assessment for team performance* and *meta-assessment for individual performance* as integral parts of the course.

Peer assessment has been defined as an arrangement in which individuals consider the amount, level, value, worth, quality, or success of the products or outcomes of learning of peers of similar status (Topping 1998). This is a formative assessment method, in which students evaluate each other's work. Peer assessment is a reflective learning activity, which increases the students' time on task and can help them consolidate, reinforce, and deepen their understanding by letting them experience reviewing, summarizing, clarifying, giving feedback, diagnosing misconceived knowledge, and identifying missing knowledge. Peer assessment might help students develop higher-order thinking and improve their metacognitive skills (Bedford and Legg 2007; Kollar and Fischer 2010; Topping 1998).

The peer assessment categories and their related criteria are defined in advance and should conform to the task requirements. Using peer assessment for evaluating peer projects presents some obstacles, such as imposing on students to assess their peers' projects. Indeed, researchers found that students in large courses in Hong Kong avoided grading each other in assessment activities (Liu and Carless 2006). To address the challenge of students' reluctance to rate each other, we added to peer assessment the *student-oriented* meta-assessment element – the element described below, in which the course team grades the quality of the students' peer assessment.

9.1.4 Meta-assessment in the Two Engineering Courses

Meta-assessment, i.e., assessment of assessment, is a technique for systematic evaluation of the assessment process itself (Orem 2012). Although several educational institutes use meta-assessment, it has not received sufficient attention in the literature (Fulcher et al. 2012). Examining the current literature in which this term is discussed, we found that meta-assessment has two different interpretations or meanings. The first is *going beyond assessment* – a formative assessment approach for examining the elements of assessment, the necessary and sufficient conditions for effective assessment, and the needs and implications of the assessment on stakeholders (McDonald 2010). We call this meta-assessment type *assessment-oriented* meta-assessment. The second meaning relates to formative assessment of organizational assessment practices, which can help institutions explain and improve the quality of assessment of a school, a university, a department, or a curriculum of some program (Fulcher and Good 2013). We call this meta-assessment type *organization-oriented* meta-assessment. In this research, we define and apply a third type of meta-assessment, which relates to assessment of students' assessment by their teachers. We call this meta-assessment type *student-oriented* meta-assessment. The focal point of *student-oriented* meta-assessment, employed in this study, is the assessment of how students assess their peers' conceptual models presented in their team projects.

Assessment in its various forms in our course had three objectives: (1) to endow the students with higher order thinking and metacognitive skills by assessing others' work after submitting their teams' work, (2) to enable assessment of a large amount of projects without the need of the course team to spend a lot of time on reading all the projects thoroughly, and (3) to evaluate and grade the quality of each student's arguments in the peer assessment, as reflected in her/his grading explanation and models comprehension. The last objective was achieved by the *student-oriented* meta-assessment – assessment carried out by the course teaching team, in which we assessed the quality of students' individual peer assessment.

We classify the peer assessment in the course as formative assessment, since it has provided us with three significant benefits that make our educational approach to teaching large-scale PBL courses viable, feasible, and valuable. First, students learn to assess and evaluate works of others. These are important cognitive and metacognitive skills they will need to apply frequently as engineering professionals in general and as information and systems engineers in particular. Second, having students assess their peers' projects frees precious course teaching team time, which is better devoted to mentoring the teams throughout the semester. Third, since the peer assessment is individual, the meta-assessment can be individual too. This solves the part of the puzzle that called for combining the team performance assessment with that of the individual. The meta-assessment goes a step beyond peer assessment. It makes the overall individual assessment more robust by using the peer team assessment as the basis for the individual meta-assessment.

9.2 Conceptual Modeling

Model-based systems engineering (MBSE), used in the courses described in this chapter, is an emerging approach to coping with the complexity of current and future systems. Conceptual models represent visually and/or textually human thoughts, ideas, designs, and purposes. MBSE is a necessary tool for coherent thinking, sharing ideas, providing common ground for communication, designing projects, and solving problems jointly. Conceptual modeling helps understand a complex problem and its potential solutions through abstraction and is therefore an important component in systems engineering. MBSE facilitates the construction and communication of complex systems (Thomas 2004), as it provides means for coordination and caters to common understanding among colleagues and customers.

Evaluating the quality of a conceptual model is a major issue that professionals in the field of systems engineering tackle (Akoka et al. 2008). Beside the evaluation of syntax and structure correctness, which is generally used to evaluate students' outcomes, there are semantics-oriented criteria, which can and should serve to assess undergraduates' conceptual models (Akoka et al. 2008; Lindland et al. 1994; Mohagheghi and Aagedal 2007). Our evaluation instruction to the students included four criteria: model correctness, model completeness, documentation, and model clarity and understandability. Completeness of a conceptual model means that the model contains all the requirements included in the scope (Lindland et al. 1994). The documentation focuses on the contribution of the documentation to the understanding of the considerations that guided the construction of the model and on the documentation appropriateness (Mohagheghi and Aagedal 2007). Model clarity and understandability (MCU) are key quality characteristics of conceptual models (Akoka et al. 2008; Selic 2003). Understandability, i.e., a model's ability to be easily understood, is a model property that has been investigated quite intensely (Cruz-Lemus et al. 2009).

Our peer and meta-assessment techniques focused on evaluating and comparing the MCU, correctness, completeness, and documentation of a given conceptual system model expressed in two different modeling languages: Unified Modeling Language (UML) and Object-Process Methodology (OPM). The models were constructed in a large-scale undergraduate course by teams of students, based on reverse engineering a complex Web-based system and authoring an appropriate scope and requirements document for that system.

9.2.1 *Unified Modeling Language*

UML (the Unified Modeling Language), developed by Object-Management Group (Covert 2012; OMG UML 2015), is the current de facto software modeling language. Developed by Rumbaugh, Booch, and Jacobson in 1996 as a nonproprietary modeling language (Covert 2012; Dori 2002b), UML consists of 14 diagram types – sevenstructural and seven behavioral. Researches (Cruz-Lemus et al. 2010; Zugal et al. 2012), who analyzed the understandability of UML, have identified many related factors, including the size of the model, control flow complexity, and the impact of hierarchy and modularity on model understandability. Meta-analysis of UML understandability (Cruz-Lemus et al. 2009) concluded that UML understandability results are mainly affected by subjects' previous experience and the size and complexity of the UML diagrams modeled.

9.2.2 *Object-Process Methodology*

Object-Process Methodology, OPM (Dori 2002a) is a holistic formal graphical and textual paradigm for the representation, development, and life cycle support of complex systems. OPM is an ISO standard 19450 (ISO 2015) for automation systems and integration. OPM enables representing systems using a highly compact set of concepts in a single diagram type and equivalent natural language. The graphical OPM model is translated on the fly to a subset of natural English, complementing the visual representation with a textual one. OPM has proven to be better in visual specification and comprehension quality when used for representing complex systems compared to OMT (Object Modeling Technique), a UML predecessor (Peleg and Dori 2000). OPM's formal yet intuitive graphics and text combination make it ideal for communicating and collaborating knowledge and ideas, even between inexperienced and novice users and domain experts who are not systems engineers. By using a single holistic hierarchical model for representing structure and behavior in the same diagram type, clutter and incompatibilities can be significantly reduced even in highly complex systems, thereby enhancing their understandability (Reinhartz-Berger and Dori 2005).

9.3 Two Cases in Point: The Undergraduate and Graduate Engineering Courses

The research described in this chapter was designed to achieve two goals: (1) teach PBL engineering courses with emphasis on developing students' higher-order and metacognitive thinking skills and (2) plan, implement, and test an innovative pedagogical approach for teaching such courses by developing and validating two specially designed Web-based assessment tools – a peer assessment tool and a meta-assessment tool.

The research was conducted in two courses with complementary settings: (1) a mandatory, large-scale undergraduate fifth semester course, titled Specification and Analysis of Information Systems at the Faculty of Industrial Engineering and Management at the Technion, Israel Institute of Technology, during the Fall 2012–2013 Semester, and (2) a small-scale graduate course, titled Model-Based Systems Engineering at the Systems Design and Management professional Master's program at Massachusetts Institute of Technology (MIT SDM 2015) during the Spring 2014 Semester.

To achieve the research goals, in both courses, we employed a PBL approach followed by peer assessment and meta-assessment. Using the peer assessment tool, students are expected to individually develop their higher order thinking and metacognitive skills while assessing the conceptual models developed by other student groups. The meta-assessment tool is designed to enable the teaching team to evaluate individual students' cognitive and metacognitive skills, as reflected by their performance in the peer assessment.

The two courses differ from each other in three aspects: (1) size: large vs. small number of students in class; (2) level: basic vs. advanced course; and (3) culture: Israeli vs. American students.

9.3.1 *The Undergraduate Course: Specification and Analysis of Information Systems*

Industrial engineering and information systems engineering undergraduate students at the Technion are required to study the Specification and Analysis of Information Systems course. The objectives of this large-scale course are to familiarize the students with analysis, conceptual modeling, design, and assessment of systems in general and of information systems in particular. The course met once a week for a three-hour lecture session and a two-hour recitation session. UML and OPM were taught every other week alternately.

In this PBL course, we tasked students with reverse engineering a widely used Web-based system, such as Gmail, Dropbox, or eBay, and derive a requirements document for that system as if it does not exist and needs to be developed. Twenty three groups of six students (each divided into two teams of three) were required to model the system by using the two different conceptual modeling languages, UML and OPM, described above.

After defining the requirements document for their reverse-engineered system and getting feedback from the course team, the two teams of three students each within each group modeled the same system in a crossover method: The first team started to model the system using OPM, while the second team started to model the same system using UML. Then, around the middle of the semester, the teams swapped, and each team continued elaborating and refining the model that the other team in the same group had started. Since both models were constructed in part by all the six team members, each student in each team had the opportunity to practice modeling in both OPM and UML.

After submitting the final project, each student was required to individually perform peer assessment. The peer assessment consisted of two phases. The first phase, which is not the focus of this chapter, called for each student to look into and collect findings about each one of the two peer projects assigned randomly to her or him. The second phase, discussed here, is a task that relied on the findingsfrom the first phase. This assessment focused on detailed structured comparison of the models in the two projects using a rubric comprised of a list of categories and criteria with examples. Adequately performing this task required each student to demonstrate good command and deep understanding of both languages in order to be able to provide valuable and insightful comments on the four peer project models (as there were two projects, each containing both a UML model and an OPM model) and rank them accordingly.

To collect the large amounts of the rubric-based data efficiently, we developed this peer assessment as a dedicated Web-based tool. Figure 9.2 presents a screenshot of the student's interface of the comparison form of our tool. Using this tool, students had to individually compare and assess each one of the four models in the two projects based on four categories: (1) model clarity and understandability (MCU), (2) model completeness, (3) model correctness, and (4) documentation. The students had been exposed to the categories and their related criteria beforehand during

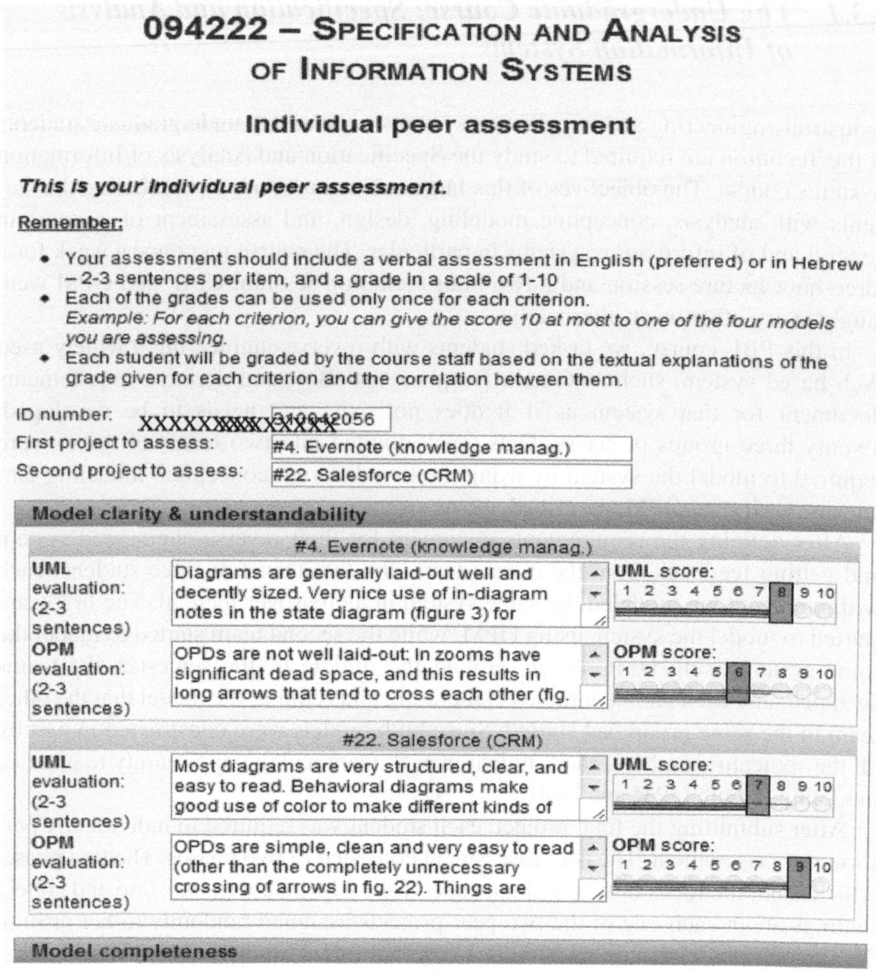

Fig. 9.2 A screenshot of the Web-based peer assessment system

a dedicated course session, and the categories were also discussed during other sessions in the course in preparation for the peer assessment.

Each model in each project was assessed individually by 12 students. To increase grading homogeneity and make it difficult for students to collaborate on this online individual assessment assignment, the peer assessment tool automatically tasked each student with assessing a unique pair of projects. A 24-hour time window was allotted for this individual peer assessment home assignment, and students were allowed to consult any learning materials.

9.3.2 *The Graduate Course: Model-Based Systems Engineering*

The Model-based Systems Engineering course at MIT's SDM professional program is a graduate elective course aimed at familiarizing systems engineers with 5–10 years of experience with principles and working knowledge of model-based systems engineering.

Similar to the undergraduate course described above, the students in this small-scale graduate PBL course were asked to specify a requirements document for a system. Unlike the other course, though, the students were asked to specify a new system related to their professional field. The three topics students elected to model for their three projects were a crop health monitoring system, a satellite managing system, and a pedestrian potential injury reducing system. The student teams were required to model these new systems in the same two conceptual modeling languages studied in the undergraduate course: Systems Modeling Language, SysML (which is a *profile* or extension of UML for systems engineering rather than for software engineering; OMG SysML 2015), and OPM. The course met twice a week for a one and a half-hour session. SysML and OPM were taught every other session alternately. Small groups of students constructed models of the systems using both languages.

Eight students attended this course. They were divided into two groups of three and one group of two students. After specifying the requirements document for their new system and getting feedback, the groups modeled their system of choice using SysML and OPM. After submitting the final project, each student was asked to individually perform the same two phases of peer assessment: collecting evidence on the models and comparing and ranking them. Each student was taskedwith assessing the two projects of the other two groups. Thus, each model in each project was assessed individually by five or six students. A 72-hour time window was allotted for this individual peer assessment home assignment, and, as in the other course, students were allowed to consult any learning materials.

9.4 Research Tools

We developed three Web-based research tools to measure the effectiveness of our approach: a peer assessment tool for the students, a meta-assessment tool for the course staff to assess students individually based on their peer assessment, and a reflection tool, also for the students.

The Peer Assessment Tool The main research tool was the individual peer assessment tool, which has two parts: (1) an online form for each assessed project (this part is not described in this chapter) and (2) an online comparison form, in which the students were asked to record their verbal evaluation and judgment arguments based on evidence and reasoning and give a corresponding grade on a scale of 1–10 (see Fig. 9.2).

Each student had to assess two models for each project, one expressed in OPM and the other in UML (in the first course) or SysML (in the second course). The tool prevents the assessing student from giving the same grade to two or more projects, so the student cannot give a high grade to all the models and must compare and rank the models. Thus, the highest possible set of grades can be 10, 9, 8, and 7. For example, in Fig. 9.2, the assessing student gave 8 in the MCU category to the Evernote UML model and 6 to the OPM model. The Salesforce UML model received a 7, and the OPM model received a 9. Each student had to also provide a short written assessment that clarifies and justifies the given grade, based on the criteria of each category and findings from the work.

In order to perform the task, each student had to apply metacognitive and high-order cognitive processes. Planning the execution of such a task and allocating the optimal amount of time for each subtask are metacognitive processes, as each student must take in account her or his familiarity with their own abilities, experience, and the level of content knowledge. Furthermore, students are required to monitor and evaluate strategies as they perform this complex peer assessment task using a mix of metacognition, reflection, and critical thinking. The peer assessment requires high order cognitive skills. Evaluating the quality of each model and comparing it to other models of two different systems in two different modeling languages require critical thinking that is based on reflection on what they had learned and practiced.

Another challenge is involved when this complex comparison raises conflicts between different findings. Consider the following two projects: one is an elaborate model that describes the details of the behavior of a system in a structured and methodical way, but has some syntax errors. The other is a simplistic model of a different system that is impeccable in terms of its syntax. The student is required to create a virtual mental common denominator for enabling a sensible comparison and deciding which model should get a higher score for correctness and explain why that score was given. Such a task does not have a clear single correct answer, so the student must exercise metacognitive thinking in order to make this decision. The decision of which model should receive a higher score requires both reflection on what was studied in the course and critical thinking about the relative importance of each finding in each model in the context of the other projects being compared.

After comparing the two projects, we asked the students to compare the two languages taught based on their experience with their project and the two projects assessed (see Fig. 9.3). This task involved reflection on what they know, generalization of what they had learned, and critical thinking that allows verifying conclusions based on three different sources.

The Meta-assessment Tool The students' written explanation was read, reread, and analyzed from a descriptive-interpretive perspective. We created a grading scale, presented in Table 9.1, which is based on themes that had emerged from students' explanation, combined with cognitive and metacognitive skills that were identified in the text. Table 9.1 presents the rubric with examples of several of the possible grades with definitions and explanations. Three researchers were involved in data analysis and definition of the grading scale of the meta-assessment in order to estab-

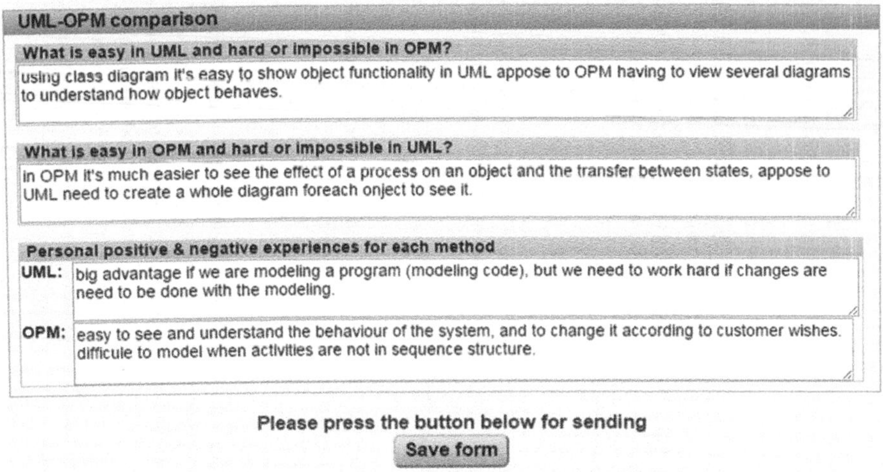

Fig. 9.3 A screenshot of the Web-based peer assessment system – student's interface of language comparison

lish its content validity. Two of the researchers were experts in science and engineering education and one in model-based systems engineering.

Ranking of the peer assessment quality was conducted by five raters, who were members of the course teaching team. They evaluated the same 10% of students' answers, following the rubric and the grading scale. This process established the inter-rater reliability of the meta-assessment tool. Correlations between raters ranged from 0.78 to 0.93, and all were significant ($p < 0.01$). Based on this grading scale, we developed the Web-based meta-assessment system, which enables the course team to assess students' peer assessment (see Fig. 9.4).

The Course Reflection Tool In the online reflection tool, the students were asked to comment on their course experience. Among other things, students were requested to reflect on their teamwork and their peer assessment task. While writing the reflection report, students were making sense of their experience and used a variety of metacognitive skills, including planning, monitoring, and evaluating. Tasks of this nature can help develop students' regulation of cognition. Each statement was classified into a specific metacognitive component [in square brackets] based on Schraw et al. (2006) and Herscovitz et al. (2012), or resource management based on Pintrich (1999).

The teamwork Students' reflections on teamwork in the large-scale undergraduate course emphasized teaming experience:

- "We coordinated our expectations." [planning to avoid future conflicts]
- "Each member of the group knew the material." [declarative knowledge]
- "We have created a good communication that helped the coordination." [planning to streamline activities]

Table 9.1 Meta-assessment grading scale

Grade	Evaluation definition	Examples	Evaluation Explanation
Very Low = 20	Poor understanding:	Lines cross each other.	Repeating one of the items of the criteria list.
	No or very poor explanations		
	No summary of		Neither frequency details nor examples.
	No examples		Missing the "where?" and "what does it mean?" answers.
Satisfactory = 60	Basic understanding:	Overall the work is nice, but some arrows and links cross each other (SD0, SD1) and make it hard to read.	Repeating two basic criteria with three examples and simple explanation.
	Few criteria considered		
	Few examples with little or no explanation	The SD1.1 diagram included more than 5 processes and it is overloaded.	
	Low text-grade correspondence		
Good = 80	Good understanding:	You need to improve the CU of this model. Most of the criteria are followed except- missing the top-level diagram (very important for simplifying), the SD1 diagram is overloaded and too small (a thing that could be avoided if it weren't overloaded).	Three different aspects with examples and explanations that reinforce the main claim.
	Enough criteria considered		
	Sufficient examples with good explanation		
	Good text-grade correspondence		
Excellent = 100	Excellent understanding:	There is unfolding mixed with in zooming (Fig 2.3, 2.4), For better understanding- it should be separate. Maybe I would use more tagged structural links for better explanation (Fig. 6).	Finding generalization with examples and recommendations.
	High analysis ability		
	Good model examples		
	Recommendations for improvement		
	Ability to synthesize and conclude	A few mistakes that don't seem to indicate a big lack of understanding (one object is not connected to a process, XOR link missing-Fig.6), overall- the model seems to be correct.	Finding generalization with examples and conclusions.
	Different types of diagrams examples		
	Drawing conclusions based on criteria		Demonstration of system thinking skills: student can understand the meaning of his finding as part of the whole picture.
	Ability to evaluate and compare		
	Comparing to the other assessed models for grading explanation	(Fig 2.5, 2.7). Overall- more mistakes than the other OPM team, BUT their diagrams are more detailed and informative, they use correct states of objects and tagged structural links which makes it better in my opinion.	Critical thinking, multiple examples, comparison with other models.

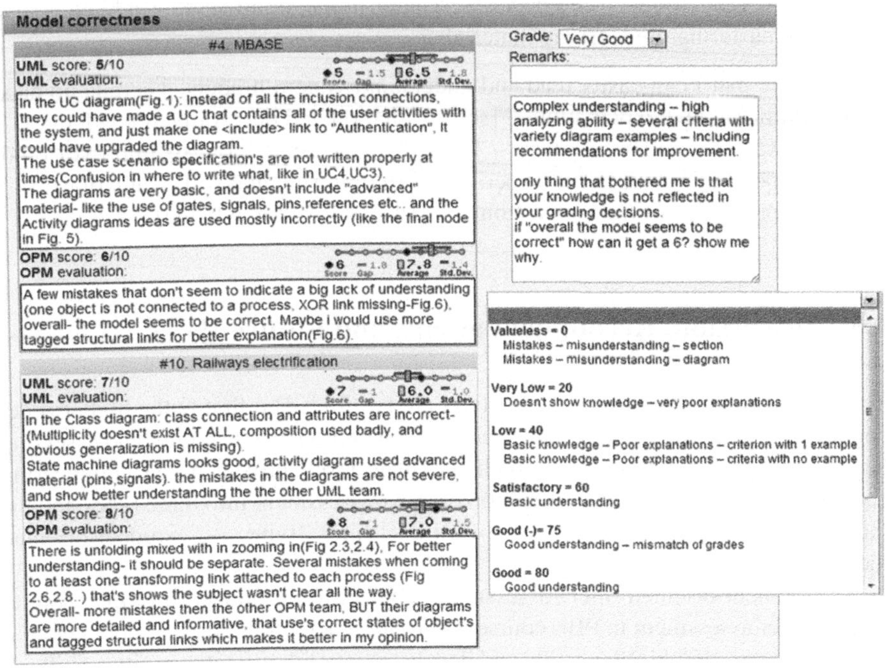

Fig. 9.4 A screenshot of the Web-based meta-assessment system – a team course interface

Students' reflection on teamwork in the small-scale graduate course revolves around teaming experience and skills:

- "Each team member's role in the project was well-defined, and each member took complete ownership of his contribution." [planning and monitoring the activities]
- "Each one had to…check the ego at the door." [procedural knowledge]
- "As a group we had to…recognize strengths and weaknesses of each other." [declarative and conditional knowledge]

The Peer Assessment Task Students' reflections on the peer assessment in the large-scale undergraduate course stressed the problem of the limited time allotted for this comparison task and uncertainties about its requirements, as well as reflection on one's own project compared to those that were assessed:

- "The time is too short for a thorough examination of two projects." [resource managing]
- "The peer assessment made me think about our project execution; we could make an effort to be more specific." [self-evaluation]

Students' reflections on the peer assessment in the small-scale graduate course related to mastering the learning materials:

- "I realize that I can easily read and understand system models. There is no need for additional text explanations." [self-evaluation and declarative knowledge]
- "[…it was] A challenging task – I could recommend [how to improve] other teams' projects based on my experience." [monitoring and self-evaluation]
- "Good practicing of the course content." [declarative knowledge]

9.5 Discussion, Recommendations, and Future Work

The theoretical contributions of our study are twofold. The first is the higher order cognitive- and metacognitive-based pedagogical approach to teaching and assessing large undergraduate-level and small graduate-level project-based engineering courses. The second is the classification of meta-assessment into three types and the introduction and development of the third type – student-oriented meta-assessment.

The student-oriented meta-assessment approach introduced in this chapter extends peer assessment in PBL courses. This approach promotes students' application of higher order thinking skills and development of their metacognitive skills. It also provides for assessing individuals' learning outcomes, enabling differentiation between the performance of the team and the individual participating as a member in the team.

At the practical level, we have developed a new approach to teaching PBL courses, implemented on the Web in an intricately detailed structured scheme. Our approach incorporates peer assessment of projects other teams carried out and (student-oriented) meta-assessment. In the large undergraduate course, each student individually assessed four models – one OPM model and one UML model in each of the two projects each student assessed. This way, each model had 12 independent scores, each of which had to be accompanied with convincing arguments made by the scoring student. We decided that only projects whose score had a large standard deviation would be examined by a course team member, but this turned out to be unnecessary.

We employes the peer assessment to assess teamwork. All the students in the same team received the same grade for the project. The meta-assessment provided the individual component of the students' assessment, accounting for one third of the final course grade. Thus, the peer assessment had three purposes. One was to get the projects assessed at the team level without having to spend the prohibitively large amount of course team time. The second purpose was to train the future engineers in assessing projects, a task which they might encounter during their professional career. Lastly, the third purpose was to provide the course team with means to carry out the student-oriented meta-assessment and thus get the individual component that complements the team assessment component.

Both the peer assessment and the meta-assessment have distinctive cognitive and metacognitive elements. To do a good peer assessment job, the assessing student had to compare and rank models of two projects, using a structured rubric comprised of a list of criteria for each one of the following categories: (1) clarity and understandability, (2) completeness, (3) correctness, and (4) documentation. Moreover, the student had to provide convincing arguments for the ranking with actual examples from the models. This requires a high level of course material mastery, as well as analysis and synthesis capabilities. These are all bona fide higher order *cognitive* skills.

On top of this, since the student knew that her or his assessment was going to be the basis for his own individual assessment by the course team – the student-oriented meta-assessment – the student had to write the assessment while considering whether what she writes will be understandable and acceptable to the course team assessor, such that her individual grade would be maximized. This is clearly a *metacognitive* task, as the student explicitly or implicitly asks himself/herself: "How do I write the assessment of these models in these projects so they are grounded in facts with examples, such that I will be best understood, seem intelligent, and also be fair to the team whose project I am assessing?" In addition, as we saw from students' reflections: "How was I and my team doing in comparison to the project that I am now assessing?"

A unique new element in our research is the development of the Web-based meta-assessment tool. Indeed, without relying on Web technology, it would be next to impossible to collect and analyze the thousands of data items, both numeric and textual, that were collected over the duration of the course.

Our *student-oriented meta-assessment* is summative in nature. It focuses on peer assessment and enables the teaching team to use it for evaluating the individual student's learning outcomes, based on the quality of that student's peer assessment. In subsequent semesters of this course, which followed the one in which this research was carried out, we have added elements of formative assessment, aimed primarily at ensuring that students will attend the class sessions and not procrastinate in making progress with their projects till almost the deadline.

As our literature review has shown, there are engineering courses that apply the PBL approach in several variants, such as CDIO (Crawley et al. 2008, 2011). The educational literature has also reported about using peer assessment as a means to evaluate students. The uniqueness of our method is the combination of PBL with peer assessment of graduate and undergraduate engineering students and the addition of student-oriented meta-assessment – assessing the assessments of students by the course team in order to evaluate each individual student skills as reflected by his ability to assess and compare models in different projects and in two different modeling languages on the basis of what had been learned. The student-oriented meta-assessment adds an individual dimension to the assessment of the team as a whole, enabling differentiation of individuals within each team based on their performance of high-level cognitive and metacognitive tasks beyond the scope of their team project.

The peer assessment categories in our meta-assessment tool are adapted to a project-based learning environment and to model-based systems engineering subject matter. Therefore, using our meta-assessment tool in other fields of study requires adaptation of the categories and criteria for rubrics that provide for structured comparison among projects.

While the particular meta-assessment tool requires adaptation, our new approach for assessing the student's outcomes can be used as a complementary method to project-based learning courses. Instructors in project-based learning courses usually assess the quality of the project and assign the same grade to all the students in the project team, regardless of their individual contribution. Cheng and Warren (2000) proposed to require group members to report about each other's contribution to the project. While this method can help differentiate individual contributions, it does not directly assess the individual learning outcomes and is likely to be affected by comradeship. In contrast, our meta-assessment approach enables individual summative assessment of each student's learning outcomes, because high-quality assessment mandates that the assessing student be proficient with the subject matter in order to provide meaningful arguments and examples that are in line with and justify the score that the student assigned to the model in question. As argued, for a student to be assessed highly, she or he must demonstrate not only higher order cognitive skills, such as critical thinking and argumentation, but also a sufficiently high level of metacognitive skills (Avargil et al. 2018; Kohen and Kramarski 2018). The metacognitive skills required for our approach include skills such as being able to monitor student's own understanding, evaluate his/her peers' understanding, and tell whether the assessment is clear and easily communicable to the receiving end – the course team meta-assessor. We call this particular metacognitive skill "meta-understanding" – the ability of an individual to determine the extent to which his (written or verbal) input would be understood by the other side. Meta-understanding is a metacognitive skill that is intimately related to the concept of transactional distances (Wengrowicz et al. 2014), and relations between the two are an intriguing subject for future research.

Although we found the meta-assessment tool to be effective for measuring the individual, students' reflections on their peer assessment and the course as a whole (see Appendix) have raised two main issues that require attention and might be considered as limitations of this study. The first relates to the time allotted for the peer assessment task. A number of students wrote that "this task is too long and therefore exhausting," while others claimed that "the time is too short for thorough examination of two projects" or "the time given to this task is intolerable." Liu and Carless (2006), who emphasized the potential of peer assessment to enhance students' learning, also mentioned time as a limitation and a source of students' resistance to peer assessment tasks. Since the peer assessment is supposed to be a positive learning experience in and of itself, and the grade for this assessment should reflect the knowledge and assessment skills of each student, a follow-up research should examine the effect of time on the validity and reliability of our meta-assessment tool. In the semesters that followed this study, we have indeed increased the time allotted for the peer assessment of the same undergraduate course. We plan to investigate the effect of this change in a follow-up study. Arguably, giving too much time to perform a peer assessment task can have other adverse effects, so an optimal solution should be sought.

A second limitation that came up from students' comments was students' lack of experience with the peer assessment task, as reflected in their comments, such as "This is the first time I had to performs such a task," "we do not have adequate tools to analyze our peer projects," and "I'm not sure that I have classified my findings to the appropriate categories since I have never done it before." As noted, the students had been exposed in a class presentation to the categories and their related criteria during the course prior to performing the peer assessment, but that is evidently not good enough. Based on the students' reflections, not just lecture presentation but also "hands-on" training in peer assessment should be conducted before requiring them to do the "real" assignment. Indeed, several researchers (McDonald 2010; Sluijsmans et al. 2002; Topping 2010; Van Zundert et al. 2010) pointed out that training and practice improve both the reliability and the validity of peer assessment and have a positive effect on students' attitudes. Therefore, we incorporated peer assessment training during the large-scale course in the subsequent semester, as part of the midsemester project report – a "dry run" prior to the "wet" peer assessment at the end of the course. This provides students with the opportunity to practice this technique, use the Web-based tools we developed, and apply criteria accompanied with concrete examples for meaningful assessment. However, we are also aware of the fact that this seemingly constructive activity will put further burden in terms of time and effort requirements of the course, which is already overloaded with time-consuming assignments for both the students and the course teaching team. Here too, one must find a delicate balance between these conflicting time and training level factors. It is an ongoing trial-and-error effort to achieve this balance.

Finally we provide recommendations for instructors, who are interested in experimenting with our new approach for developing their students' higher-order thinking and metacognitive skills. More specifically, what we provide below is a

five-step guideline for adapting the peer assessment and meta-assessment method in a PBL learning environment for engineering courses.

- Design your course as a collaborative PBL course, placing team projects as the focus of the course.
- Design the peer assessment rubric as a list of categories and related criteria for assessing the projects.
- Task each student with assessing and comparing two different peer projects based on the rubric.
- Carry out meta-assessment by evaluating each student's assessment skills based on her or his peer assessment and especially how she or he justifies their ranking.
- Ask your students to write an individual reflection report on the course, on their team work, and on the peer assessment process. In this chapter, we presented the advantages of incorporating metacognition in general and meta-assessment in particular into engineering education and using it for enhancing students' metacognitive skills. Formative assessment was made possible by providing feedback to the teams as they were engaged in the project-based learning. Our meta-assessment approach enables individual summative assessment of each student's learning outcomes, whereas each team project served as a basis for the collective team's summative assessment.

Appendix: Students' Reflections on the Course as a Whole

Students' reflections on the course as a whole in the large-scale undergraduate course revolved around the high demands of the course and the effort they had to spend on the one hand and the lack of familiarity with this kind of tasks on the other hand:

- "The course requires a lot of work."
- "Where is the peace of mind? I invested in this course more than any other course of the semester."
- "Takes up a lot of time."
- "The course is a time and effort consumer."
- "The course is not an easy course; it required a very large investment relative to other courses I've taken to date."
- "Busy course relative to other courses."
- "High level of time and effort investment than any other course."
- "This is the first time I had to perform such a task."
- "We do not have adequate tools to analyze our peer projects."
- "I'm not sure that I have classified my findings to the appropriate categories since I have never done it before."

Reflection on the course as a whole in the small-scale graduate course revolved around the learning of the two modeling languages in parallel. Students emphasized that this method helped them to better understand the uniqueness of each language:

- "I thoroughly enjoyed this class. … not only the syntax of both modeling languages, but also how they compare to each other."
- "I like presenting both OPM and SysML in the class, not necessarily so I can efficiently use both, but so I could understand their differences, strengths and weaknesses."
- "I was able to develop a good understanding of the various types of modeling language over the course."
- "Very useful, and helped solidify understanding … when translating between the two modeling languages."
- "The hands-on session and converting other standard diagrams with class discussion are awesome experiences."
- "The exercises of converting other diagrams are really good ways to understand other diagrams and at the same time improve students' understanding and skill."

References

ABET. (2014). Criteria for accrediting engineering programs, 2015–2016. Retrieved 1 Jan 2015, from http://www.abet.org/eac-criteria-2015-2016/

Akoka, J., Comyn-Wattiau, I., & Cherfi, S.S.S. (2008). Quality of conceptual schemas an experimental comparison. In *2008 Second International Conference on Research Challenges in Information Science* (pp. 197–208). IEEE. doi:10.1109/RCIS.2008.4632108

Anderson, L.W., Krathwohl, D.R., Airasian, P.W., Cruikshank, K.A., Mayer, R.E., Pintrich, P.R., … Wittrock, M.C. (2001). A taxonomy for learning, teaching, and assessing: A revision of bloom's taxonomy of educational objectives, abridged edition. White Plains: Longman.

Avargil, S., Lavi, R., & Dori, Y. J. (2018). Students' metacognition and metacognitive strategies in science education. In Y. J. Dori, Z. Mevareach, & D. Bake (Eds.), *Cognition, metacognition and culture in STEM education*. Springer.

Bedford, S., & Legg, S. (2007). Formative peer and self feedback as a catalyst for change within science teaching. *Chemistry Education Research and Practice, 8*(1), 80. doi:10.1039/b6rp90022d.

Boud, D. (1990). Assessment and the promotion of academic values. *Studies in Higher Education, 15*(1), 101–111. doi:10.1080/03075079012331377621.

Bransford, J. D., & Stein, B. S. (1993). *The ideal problem solver: A guide to improving thinking, learning, and creativity* (2nd ed.). New York: Freeman.

Brodeur, D.R., Young, P.W., & Blair, K.B. (2002). Problem-based learning in aerospace engineering education. In *Proceedings of the 2002 American society for engineering education annual conference and exposition* Montreal, Canada (pp. 16–19).

Brown, S., & Knight, P. (1994). *Assessing learners in higher education*. London: Kogan Page.

Carr, R.L., & Strobel, J. (2012). Work in progress: Development of a metacognition scaffold in STEM/P-6 engineering context: MCinEDP. In *Frontiers in education conference (FIE), 2012* (pp. 1–2). IEEE.

Cheng, W., & Warren, M. (2000). Making a difference: Using peers to assess individual students' contributions to a group project. *Teaching in Higher Education, 5*(2), 243–255. doi:10.1080/135625100114885.

Covert, S. (2012). OMG's unified modeling language (UML) celebrates 15th anniversary. Retrieved from http://www.omg.org/news/releases/pr2012/08-01-12-a.htm

Crawley, E. F., Brodeur, D. R., & Soderholm, D. H. (2008). The education of euture aeronautical engineers: Conceiving, designing, implementing and operating. *Journal of Science Education and Technology, 17*(2), 138–151. doi:10.1007/s10956-008-9088-4.

Crawley, E. F., Malmqvist, J., Lucas, W. A., & Brodeur, D. R. (2011). The CDIO syllabus v2.0. an updated statement of goals for engineering education. In *Proceedings of 7th international CDIO conference*. Denmark. Retrieved from http://publications.lib.chalmers.se/records/fulltext/local_143186.pdf

Cross, D. R., & Paris, S. G. (1988). Developmental and instructional analyses of children's metacognition and reading comprehension. *Journal of Educational Psychology, 80*(2), 131.

Cruz-Lemus, J. A., Genero, M., Manso, M. E. E., Morasca, S., & Piattini, M. (2009). Assessing the understandability of UML statechart diagrams with composite states – A family of empirical studies. *Empirical Software Engineering, 14*(6), 685–719. doi:10.1007/s10664-009-9106-z.

Cruz-Lemus, J. A., Maes, A., Genero, M., Poels, G., & Piattini, M. (2010). The impact of structural complexity on the understandability of UML statechart diagrams. *Information Sciences, 180*(11), 2209–2220. doi:10.1016/j.ins.2010.01.026.

De Graaff, E., & Christensen, H. P. (2004). Editorial: Theme issue on active learning in engineering education. *European Journal of Engineering Education, 29*(4), 461–463.

Dewey, J. (1934). *Art as experience*. New York: Minton, Balch and Company. Retrieved from http://dcg.mit.edu/wp-content/uploads/2011/10/DEWEY_expressiveObject.pdf.

Dori, D. (2002a). *Object-process methodology*. Berlin: Berlin/Heidelberg. doi:10.1007/978-3-642-56209-9.

Dori, D. (2002b). Why significant UML change is unlikely. *Communications of the ACM, 45*(11), 82–85. doi:10.1145/581571.581599.

Dori, Y. J. (2003). From nationwide standardized testing to school-based alternative embedded assessment in Israel: Students' performance in the matriculation 2000 project. *Journal of Research in Science Teaching, 40*(1), 34–52.

Dori, Y. J., & Sasson, I. (2013). A three-attribute transfer skills framework–part I: Establishing the model and its relation to chemical education. *Chemistry Education Research and Practice, 14*(4), 363–375.

Dym, C. L., Agogino, A. M., Eris, O., Frey, D. D., & Leifer, L. J. (2005). Engineering design thinking, teaching, and learning. *Journal of Engineering Education, 94*(1), 103–120.

Flavell, J. H., & Wellman, H. M. (1977). Metamemory. In R. V. Kail & J. W. Hagen (Eds.), *Perspectives on the development of memory and cognition* (pp. 3–33). Hillsdale: Erlbaum.

Ford, C. L., & Yore, L. D. (2012). Toward convergence of critical thinking, metacognition, and reflection: Illustrations from natural and social sciences, teacher education, and classroom practice. In A. Zohar & Y. J. Dori (Eds.), *Metacognition in science education* (pp. 251–271). Dordrecht: Springer-Verlag.

Fulcher, K. H., & Good, M. R. (2013). The surprisingly useful practice of meta-assessment to title. [Web log post]. *National Institute for Learning Outcomes Assessment.* Retrieved from http://illinois.edu/blog/view/915/99344

Fulcher, K. H., Swain, M., & Orem, C. D. (2012). Expectations for assessment reports: A descriptive analysis. *Assessment Update, 24*(1), 1–16. doi:http://doi.wiley.com/10.1002/au.241.

Hadim, H. A., & Esche, S. K. (2002). Enhancing the engineering curriculum through project-based learning. In *Frontiers in education, 2002. FIE 2002. 32nd Annual* (Vol. 2, pp. F3F–1). IEEE.

Herscovitz, O., Kaberman, Z., Saar, L., & Dori, Y. J. (2012). The relationship between metacognition and the ability to pose questions in chemical education. In A. Zohar & Y. J. Dori (Eds.), *Metacognition in science education* (pp. 165–195). Dordrecht: Springer-Verlag.

ISO. (2015). ISO/PAS 19450 –Automation systems and integration – Object-process methodology. Retrieved July 27, 2015, from http://www.iso.org/iso/catalogue_detail.htm?csnumber=62274/

Johri, A., & Olds, B. M. (2011). Situated engineering learning: Bridging engineering education research and the learning sciences. *Journal of Engineering Education, 100*(1), 151–185.

Kohen, Z., & Kramarski, B. (2018). Promoting mathematics teachers' metacognition. In Y. J. Dori, Z. Mevareach, & D. Bake (Eds.), *Cognition, metacognition and culture in STEM education.* Springer.

Kollar, I., & Fischer, F. (2010). Peer assessment as collaborative learning: A cognitive perspective. *Learning and Instruction, 20*(4), 344–348. doi:10.1016/j.learninstruc.2009.08.005.

Kuhn, D. (2000). Metacognitive development. *Current Directions in Psychological Science, 9*(5), 178–181.

Lawanto, O. (2009). Metacognition changes during an engineering design project. In *Frontiers in education conference, 2009. FIE'09. 39th IEEE* (pp. 1–5). IEEE.

Lewis, P., Aldridge, D., & Swamidass, P. M. (1998). Assessing teaming skills acquisition on undergraduate project teams. *Journal of Engineering Education, 87*(2), 149–155.

Lin, X. (2001). Designing metacognitive activities. *Educational Technology Research and Development, 49*(2), 23–40.

Lindland, O. I., Sindre, G., & Solvberg, A. (1994). Understanding quality in conceptual modeling. *IEEE Software, 11*(2), 42–49. doi:10.1109/52.268955.

Liu, N.-F., & Carless, D. (2006). Peer feedback: The learning element of peer assessment. *Teaching in Higher Education, 11*(3), 279–290. doi:10.1080/13562510600680582.

McDonald, B. (2010). Improving learning through meta assessment. *Active Learning in Higher Education, 11*(2), 119–129. doi:10.1177/1469787410365651.

Mills, J. E., & Treagust, D. F. (2003). Engineering education – Is problem-based or project-based learning the answer? *Australasian Journal of Engineering Education, 3*, 2–16.

MIT SDM. (2015). MIT system design & management (SDM). Retrieved July 27, 2015, from https://sdm.mit.edu/

Mohagheghi, P., & Aagedal, J. (2007). Evaluating quality in model-driven engineering. In *International workshop on modeling in software engineering (MISE'07: ICSE Workshop 2007)* (pp. 6–6). IEEE. doi:10.1109/MISE.2007.6.

Newell, J., Dahm, K., Harvey, R., & Newell, H. (2004). Developing metacognitive engineering teams. *Chemical Engineering Education, 38*(4), 316–320.

NSF. (1998). *The action agenda for systemic engineering education reform – NSF 98–27.* Retrieved from http://www.nsf.gov/pubs/1998/nsf9827/nsf9827.htm

Olds, B. M., Moskal, B. M., & Miller, R. L. (2005). Assessment in engineering education: Evolution, approaches and future collaborations. *Journal of Engineering Education, 94*(1), 13–25. doi:10.1002/j.2168-9830.2005.tb00826.x.

OMG SysML. (2015). Documents associated with systems modeling language (SysML), Version 1.3. Retrieved 27 Jul 2015, from http://www.omg.org/spec/SysML/1.3/

OMG UML. (2015). Unified modeling language™ (UML®) Version 2.5. Retrieved 27 Jul 2015, from http://www.omg.org/spec/UML/

Orem, C. D. (2012). *Demonstrating validity evidence of meta-assessment scores using generalizability theory*. Harrisonburg: James Madison University.

Ory, J. C. (1992). Meta-assessment: Evaluating assessment activities. *Research in Higher Education, 33*(4), 467–481.

Peleg, M., & Dori, D. (2000). The model multiplicity problem: Experimenting with real-time specification methods. *Software Engineering, IEEE Transactions, 26*(6), 742–759.

Pintrich, P. R. (1999). The role of motivation in promoting and sustaining self-regulated learning. *International Journal of Educational Research, 31*(6), 459–470.

Popham, W. J. (2004). Curriculum, instruction, and assessment: Amiable allies or phony friends? *The Teachers College Record, 106*(3), 417–428.

Reinhartz-Berger, I., & Dori, D. (2005). OPM vs. UML: Experimenting with comprehension and construction of web application models. *Empirical Software Engineering, 10*(1), 57–80.

Resnick, L. B. (1987). *Education and learning to think*. Washington, DC: National Academy Press.

Rugarcia, A., Felder, R. M., Woods, D. R., & Stice, J. E. (2000). The future of engineering education I. A vision for a new century. *Chemical Engineering Education, 34*(1), 16–25.

Schraw, G., Crippen, K. J., & Hartley, K. (2006). Promoting self-regulation in science education: Metacognition as part of a broader perspective on learning. *Research in Science Education, 36*(1–2), 111–139.

Schraw, G., Olafson, L., Weibel, M., & Sewing, D. (2012). Metacognitive knowledge and field-based science learning in an outdoor environmental education program. In A. Zohar & Y. J. Dori (Eds.), *Metacognition in science education* (pp. 57–88). Dordrecht: Springer-Verlag.

Schunk, D. H., & Zimmerman, B. J. (2003). Self regulation and learning. In W. M. Reynolds & G. E. Miller (Eds.), *Handbook of psychology – volume 7* (pp. 59–78). Hoboken: Wiley.

Selic, B. (2003). The pragmatics of model-driven development. *Software, IEEE, 20*(5), 19–25.

Sluijsmans, D. M. A., Brand-Gruwel, S., & van Merriënboer, J. J. G. (2002). Peer assessment training in teacher education: Effects on performance and perceptions. *Assessment & Evaluation in Higher Education, 27*(5), 443–454. doi:10.1080/0260293022000009311.

Thomas, D. (2004). MDA: Revenge of the modelers or UML utopia? *IEEE Software, 21*(3), 15–17.

Topping, K. J. (1998). Peer assessment between students in colleges and universities. *Review of Educational Research, 68*(3), 249–276.

Topping, K. J. (2010). Methodological quandaries in studying process and outcomes in peer assessment. *Learning and Instruction, 20*(4), 339–343. doi:10.1016/j.learninstruc.2009.08.003.

Van Zundert, M., Sluijsmans, D., & van Merriënboer, J. (2010). Effective peer assessment processes: Research findings and future directions. *Learning and Instruction, 20*(4), 270–279. doi:10.1016/j.learninstruc.2009.08.004.

Veenman, M. V. J. (2012). Metacognition in science education: Definitions, constituents, and their intricate relation with cognition. In A. Zohar & Y. J. Dori (Eds.), *Metacognition in science education* (pp. 21–36). Dordrecht: Springer-Verlag.

Vos, H., & De Graaff, E. (2004). Developing metacognition: A basis for active learning. *European Journal of Engineering Education, 29*(4), 543–548.

Vrugt, A., & Oort, F. J. (2008). Metacognition, achievement goals, study strategies and academic achievement: Pathways to achievement. *Metacognition and Learning, 3*(2), 123–146.

Wengrowicz, N., Dori, Y. J., & Dori, D. (2014). Transactional distance in an undergraduate project-based systems modeling course. *Knowledge-Based Systems, 71*(6), 41–51. doi:10.1016/j.knosys.2014.05.022.

Zohar, A., & Dori, Y. J. (2003). Higher order thinking skills and low-achieving students: Are they mutually exclusive? *The Journal of the Learning Sciences, 12*(2), 145–181.

Zugal, S., Pinggera, J., Weber, B., Mendling, J., & Reijers, H. A. (2012). Assessing the impact of hierarchy on model understandability – A cognitive perspective. In J. Kienzle (Ed.), *Models in software engineering*. Berlin: Springer. doi:10.1007/978-3-642-29645-1_14.

Chapter 10
The Impact of Culture on Engineering and Engineering Education

Adam R. Carberry and Dale R. Baker

10.1 Culture and Engineering

A culture is the result of symbolic elements shaped by the given system within which they are created, distributed, evaluated, taught, and preserved (Peterson and Anand 2004). The discussion of culture in this chapter holds many different meanings based on various lens used to discuss cultural impacts. Whether intentional or not, the actions of those in the field of engineering have established a culture that differentiates engineers by how they think, do, relate with others and the environment, accept difference, and identify as being an engineer (Godfrey and Parker 2010). Practice of these cultural dimensions is a major influencer of how the field is perceived to those looking in from the outside. The perception of an engineering culture is connected to the discipline and how engineering institutions and industries conduct business. This lens discusses how individuals perceive the field and how they see themselves fitting in with the established culture. The perception of the field and how individuals view themselves as fitting in is especially germane to efforts that increase the participation of women in engineering, which has historically been a male-dominated field with its own brand of masculine culture. Perception of the field also influences how engineering is taught and how western engineers work in non-western cultures. Engineering is also a discipline that aims to serve society. As such, the established culture of the given society being served has impacts on how engineers go about solving problems. This lens recognizes the

A.R. Carberry (✉)
Ira A. Fulton Schools of Engineering, The Polytechnic School, Arizona State University, Mesa, AZ, USA
e-mail: Adam.Carberry@asu.edu

D.R. Baker
Mary Lou Fulton Teachers College, Educational Leadership and Innovation, Arizona State University, Tempe, AZ, USA

Y.J. Dori et al. (eds.), *Cognition, Metacognition, and Culture in STEM Education*, Innovations in Science Education and Technology 24,
https://doi.org/10.1007/978-3-319-66659-4_10

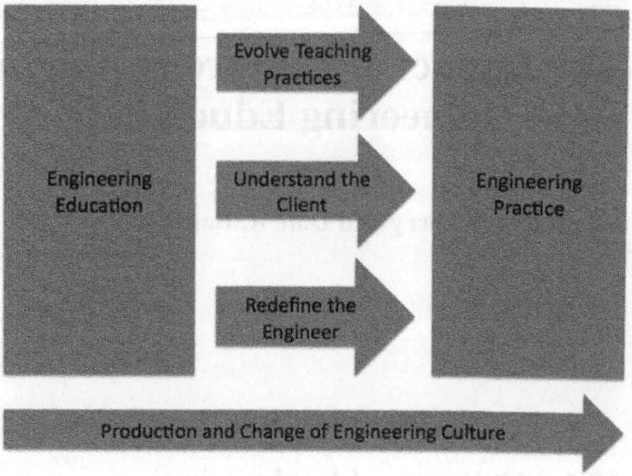

Fig. 10.1 Schematic overview of the impact of factors discussed in the chapter

importance of the user/client and how their culture can impact the design of a solution.

The broad scope used in this discussion is intended to provide a general description of how culture impacts the engineering field as a whole. The following sections will break down this discussion to investigate perception, production, education, society, enculturation, and the implications of these factors on engineering and society to provide recommendations for educators, learners, and practitioners. Figure 10.1 depicts how engineering education must evolve teaching practices, stress the need to deeply understand clients, and redefine who engineers are and what they do to produce and change the engineering culture that exists in practice and is perceived by society.

10.2 Perception and Production of Engineering Culture

A key indicator of an established culture is public perception. A society's perception of a given context reflects their understanding as a summation of experiences and interactions with the given context. Encounters can range from firsthand mastery experiences to simple word-of-mouth information. Engineering is a body of knowledge that the majority of the public has never had the opportunity to formally learn or experience. This is evident by evaluations that clearly suggest an overall lack of awareness and understanding about engineering and what engineers do. Marshall et al.'s (2007) assessment of public attitudes and perceptions of engineering and engineers in Great Britain revealed a degree of confusion and uncertainty about the discipline. Respondents associated engineering only with fixing things, providing things people rely on in their everyday lives, and causing key problems in society (e.g., climate change). The National Academy of Engineers' (1998; later update in 2002) assessment of public perceptions in the United States added that Americans were generally uninformed about engineering and viewed engineering as simply the application of science. Engineering as a discipline doesn't have to search any further than itself as being a major culprit to blame for the public's lack in understanding about the field. Cultural norms established by the field from Australia and New Zealand to the United States and throughout Europe portray engineering as boring and masculine (Hansen and Godfrey 1997; McLean et al. 1997; Sagebiel and Dahmen 2006; Tonso 1996).

Western society has adopted a culturally influenced notion that engineering drives innovation and technology and fosters entrepreneurship (Nazan and Bogers 2015; Vickers et al. 2001) through ABET-accredited programs that educate students in applied sciences, computing, engineering, and engineering technology. A major contributor to the notion that engineering is boring and only for men is the established reputation of engineering being a highly complex field fit only for those who excel in mathematics and the "hard" sciences (e.g., physics, chemistry, and biology). This notion is propagated through media outlets that push a pervasive image of what engineering and engineers look like. For example, automobile company commercials often have engineering in their slogans (e.g., Ford: "Engineered that lasts" or Audi: "Truth in Engineering") leading to the perception that complex machines such as automobiles are equivalent to engineering. Engineering program websites also tend to project a certain image of what an engineering student should look like. Television shows portray how people view the field both positively and negatively (Tang 2013). For example, shows like *Design Squad, Engineering Marvels, Extreme Engineering, Epic Engineering, How it's Made, Engineering Disasters*, and *MythBusters* portray these fields in a positive light; however, many may find these shows boring to watch or even frightening to think about the catastrophic failure that can result from poor engineering. Additionally, shows like the *Big Bang Theory* make it appear as though only nerdy super geniuses can be successful in science and engineering. These media representations can be very influential on public opinion and interest in engineering as demonstrated by research asking children to "Draw an Engineer" (Capobianco et al. 2011; Ganesh 2011; Knight and Cunningham 2004). Students think about engineers as using tools to build and fix car engines, designing things such as buildings or machines, or someone who drives and/or works with trains. Drawings also indicate that students think engineers are mostly men. The established masculine culture of engineering has helped to propagate these notions resulting in a perpetual cycle that recruits and retains only those who fit the established cultural mold.

The large quantitative survey analyses (Davis and Gibbin 2002; Marshall, McClymont and Joyce, 2007) and complementary qualitative workshop polling (Marshall, McClymont and Joyce, 2007) provided a broad view of the general public's perception of engineering culture. Additional qualitative analysis of young children's drawings (Capobianco et al. 2011; Ganesh 2011; Knight and Cunningham 2004) provided a full picture of how society – youth to adult – perceive engineering. These findings should be alarming to those in the field of engineering and those who seek to recruit and retain a diverse population of future students. The broad conclusions suggest that efforts being made to change public perception need to either be rethought or expanded to reach a greater percentage of the general public. The solution to a well-informed society is through improved education. Citizens need to be educated on what engineering is and what engineers actually do as early as middle school to grow interest and understanding (Zunker 1994). Current educational practices are clearly falling short on projecting an accurate depiction of the field, which is heavily influenced by the masculine culture established within traditional engineering education programs.

10.3 Engineering Education Culture

10.3.1 Traditional Engineering Programs

The culture of engineering schools is reflected in instructional approaches that influence learning, metacognition, interest, and teaching. Research by Nelson Laird, Shoup, Kuh, and Schwarz (2008) found that engineering faculty were less likely to engage in instructional practices that encouraged deep learning than in what they called soft applied fields. They attribute this difference to disciplinary socialization and a culture of consensus in the content and methods of inquiry, which they state do not exist for soft fields. Brint et al. (2008) also conclude that the academic culture in engineering may discourage the development and implementation of experiences that promote the use of deep approaches to learning. Brint, Cantwell, and Hanneman describe the culture of engineering as one that rewards:

> *…industrious, but unimaginative students who perform technical tasks competently but express little initiative outside of required activities and little interest in connecting ideas or interacting with their professors. Interaction between students and faculty and participation in class are minimal, and interest in jobs seems to greatly outweigh the inspiration of ideas.* (p. 398)

Brint, Cantwell, and Hanneman did not expect to see many changes to engineering in the future. The established culture of engagement, where students participate in class and are interested in ideas, is perceived by faculty to be more appropriate for majors in the arts, social sciences, and humanities rather than science and engineering. It is no surprise then that Finelli et al. (2014) found 60 % of the engineering classes they observed to lack any form of active learning.

Boiarsky (2004) describes the culture of engineering education as narrowly focused on content that does not teach students how to learn-to-learn. Bucciarelli, Einstein, Terenzini, and Walser (2000) also have an unflattering description of the predominant engineering culture. They describe it as "… based on compartmentalization of knowledge, individual specialization, and a wholly research-based reward structure" (p.141). The lecture format also creates a barrier between professors and students that results in lower self-efficacy, academic confidence, and GPA among students (Blinkenstaff 2005; Vogt 2008.). Students in large lecture format classes are also more likely to rate instruction poorer than students in smaller classes (Johnson et al. 2013).

A case study of engineering culture in a high ranking engineering school in New Zealand found faculty describing their teaching as learning the hard way to cover material through traditional lecture-based courses. Problem solving was taught using a reductionist method with an emphasis on mathematics (Godfrey and Parker 2010). Godfrey and Parker emphasize the mathematical and learning the hard way culture of engineering by quoting a student who repeated a well-known joke to them. He said "You know you're an engineer if you haven't got a life and can prove it mathematically" (p. 10).

Montfort et al. (2014) feel that faculty epistemological beliefs are at the heart of the difficulties in bringing about reform and educational innovations in teaching engineering. For example, the belief that the natural world is too large and complex results in an absence of real-world examples and de-contextualization of concepts. Personal epistemologies are also relied upon to determine which questions, issues, or opinions to address in courses. These choices by faculty impact the beliefs students bring to their future studies and further on down the road in their careers (Carberry 2014). Montfort, Brown, and Shinew concluded that questions and issues that are unaddressed could have an impact on students' continuing interest and retention in engineering.

10.3.2 Teaching Methods in Engineering Education

Although many engineering professors are aware of and respect the research on learner-centered teaching, they are reluctant to adopt these instructional strategies because their institutional culture rewards research productivity and high-level professional activities (King 2012), while discouraging high levels of effort to improve teaching (Crawley et al. 2007). Other faculty members do not embrace pedagogical reforms presented with strong evidence of effectiveness because they are unwilling to invest the time to teach the course using new techniques to replace teacher-centered approaches. The primary reason for this reluctance is that the time commitment to learn and use innovative pedagogies is greater than for traditional lectures (Fairweather 2008). Reluctance to adopt pedagogical reforms may also be due to little or no training in how to teach. One junior faculty interviewed by Godfrey and Parker (2010) said in reference to teaching, "… you are just dumped into the job – there is no real preparation beforehand" (p. 13). Faculty lack the education and, as Graham (2009) found of faculty in the United Kingdom, confidence to design and use assessments to evaluate learner-centered practices such as project-based learning activities.

New engineering faculty are essentially "well intentioned gifted amateurs" who need to develop expertise, which requires commitment, time, focused resources, and recognition in the institutional reward structure (Ambrose and Norman 2006). Fairweather (2008) notes that the more time a faculty member spends on teaching, the lower their salary, while the more time spent in research and publications, the higher their salary. Institutions of higher learning valuing teaching can therefore be viewed as merely rhetoric. King (2012) sees the problem differently suggesting poor alignment between those who would benefit from changes in engineering curricula and instructional pedagogy (e.g., students and the public) and those who have influence over whether change will take place (e.g., faculty). It may actually be engineering professors who do not foster change who are contributing to retention problems in engineering majors.

This notion is supported by students who transfer out of engineering majors citing poor teaching and advising as a primary cause (Marra et al. 2012). Students in the Marra et al. (2012) study put it thusly: "…and the professors didn't seem to care at all whether or not people did well in their class" and "The advising system was very poor. I was a number not a name. The first two years are when students most need advising…we had advisors who basically told you to just follow the rubric in the engineering manual" (p. 18).

Institutional culture, the culture of the university in which the engineering program resides, also has an impact on engineering students. Seymour and Hewitt (1977) in their landmark study found that institutional culture influenced students' decision to switch from an engineering major to other majors. Tonso (2007), studying a reform curriculum of design in engineering, found that the masculine campus culture of an engineering school made women feel invisible and like outsiders negating the effects of the experiential curriculum design to bring women into engineering. Marra et al. (2012) also found that one of the factors leading to dropping out of an engineering major was an engineering culture that made students feel like outsiders.

It is not just a culture that rewards research activities over teaching that has a negative effect on students. Women, and in particular women of color, find the competitive culture of engineering detrimental to their success (Godfrey and Parker 2010; Johnson 2007). Cultural change is needed in order for the discipline to evolve and grow.

10.3.3 Changing the Culture of Engineering Education

Understanding the existing culture established within engineering or perceived by the public is essential to informing change. It is clear that engineering needs to be more engaging, relevant, and welcoming (Clough 2004) and that such change must be driven by engineering faculty and administrators (Jamieson and Lohmann 2012); but change is difficult for both people and institutions. Change in engineering, according to Graham (2012), comes about only when there is a shared purpose among faculty and agreement that change is imperative. McKenna et al. (2014) note that considerable work in reform has taken place at the local level by individuals or teams to change pedagogy or curriculum, but that these reforms have not had an impact on engineering culture at large. Furthermore, despite many reports about what engineering education should look like, there is little information about how change in the engineering culture can come about (Besterfield-Sacre et al. 2014). Besterfield-Sacre et al.'s analysis of the data from *Innovation with Impact* (Jamieson and Lohmann 2012) indicated that faculty, chairs, and deans felt that transformative change could come about through developing and disseminating innovative pedagogy, support for the scholarship of teaching, and implementing policies that supported and rewarded innovative pedagogy. Conspicuously absent in terms of the mechanisms to bring about reform was developing a shared vision and strategic planning for changing the culture of engineering. Godfrey (2014) suggests "…that change at the levels of curricula, structures, and behaviors is not sufficient for sustained cultural change. Cultural change requires transformation – forming new collective understandings and creating new beliefs about what is valued in engineering education" (p. 452). So what then can drive such a transformation? Graham (2012) found that motivation for reform, most often project-based learning, came from a school or college's position in the marketplace (70–80 % of the time). In contrast, project-based learning was implemented only 5–10 % of the time, even in schools and departments where a culture of innovation already existed.

There are better and stronger arguments for using a variety of student-centered experiential pedagogies than market place positioning. Engineering programs and curricula that reflect a culture that has embraced experienced-based teaching methodologies and student engagement are more likely to result in students using deeper approaches to learning (Chen et al. 2008; Shawcross and Ridgman 2012) in addition to traditional reading and studying (Kuh et al. 2004). Faced with numerous choices, faculty members are more likely to use just one research-based instructional strategy than they are to use two or more. From among the many effective pedagogical strategies, faculty are most likely to use case-based teaching, just-in-time teaching, and inquiry strategies (Borrego et al. 2013). Moderate levels of strategy use were found for think-aloud-paired problem solving, cooperative learning, collaborative learning, problem-based learning, and think-pair-share. The lowest level of strategy use was found for peer instruction and service learning. Strategy use by faculty contrasts the most commonly used student-centered instructional strategies of design projects and service found in most engineering curricula. Fisher et al. (2005) found that instructional reforms in engineering service courses improved ABET-related student learning outcomes in problem solving and analysis of complex problems. These strategies support the development of engineering expertise, but have not yet been rigorously tested for impact on learning (Litzinger et al. 2011).

10.3.4 Methods and Tools for Investigating Engineering Education

The studies cited describing engineering education culture used a variety of methods that allowed us to come to conclusions, make recommendations, and identify implications. These studies included:

- Thought-provoking pieces or position papers grounded in broad and interdisciplinary research studies (Ambrose and Norman 2006; Choresh et al. 2009; Clough 2004; Jamieson and Lohmann 2012; King 2012)
- Large-scale literature reviews with a synthesis of the findings and recommendations derived from the synthesis (Blinkenstaff 2005; Fairweather 2008; Litzinger et al. 2011; McKenna et al. 2014)
- Recommendations for reform drawn from existing literature and firsthand personal teaching (Boiarsky 2004)
- Information gained through a workshop (Bucciarelli et al. 2000) used
- A developed model for conceptualizing student engagement in engineering (Chen et al. 2008)

The referenced studies used a variety of quantitative and qualitative research methods. Most quantitative studies used survey data from a variety of instruments (e.g., National Survey of Student Satisfaction, Faculty Survey of Student Satisfaction, self-efficacy assessments, self-reports, and opinions) with statistical analysis of the data (Besterfield-Sacre et al. 2014; Borrego et al. 2013; Brint et al. 2008; Kuh et al. 2004; Marra et al. 2009; Nelson Laird et al. 2008). For example, Johnson et al. (2013) used student evaluations of the course and instructor and subjected them to statistical analysis to identify issues with teaching. Qualitative studies used case studies (e.g., Graham 2009), large-scale ethnographies (e.g., Godfrey & Parker, 2010; Seymour and Hewitt 1977; Tonso 2007), or interviews (e.g., Graham 2012; Johnson 2007; Montfort et al. 2014). There were also a number of studies that used a mixed methods approach. Examples include the RTOP based on classroom observations and factor analysis (Piburn et al. 2000; Sawada et al. 2002), institutional change plans and identified student learning outcomes using focus groups and surveys (Finelli et al. 2014; Fisher et al. 2005), and qualitative and quantitative techniques used to examine skill development in an engineering master's program (Shawcross and Ridgman 2012).

The various studies and approaches to analyze engineering education have shown us that an environment where students are engaged in deep thinking will require professors to change the way they teach, interact with students, and revise the curriculum (Brint et al. 2008). First and foremost, there should be more active learning. This includes (1) more class discussions about readings and ideas encountered in class and readings, (2) group work on projects in class and outside of class, (3) community-based projects, and (4) opportunities to tutor other students about course material. Professors will have to provide more, prompt formative feedback during the semester for all active learning activities. Other strategies to increase deep learning include internships, senior capstone projects, electronic discussions boards, and learning communities.

Coursework should be academically challenging and require effort. Students should be required to write papers that exceed 20 pages and engage in other activities that require analysis, synthesis, and the application of knowledge to novel problems and situations. The nature of teacher–student interactions will have to become more personal so that students can discuss career options, grades, and assignments with professors as well as engage in research projects with their professors.

The leadership of engineering schools needs to communicate how important good teaching and student relationships are for student success and that efforts in this regard will be valued and rewarded. Teaching must be evaluated in a more rigorous and systematic way with items that reflect changes in teaching using instruments such as the Reformed Teaching Observation Protocol (RTOP) (Piburn et al. 2000; Sawada et al. 2002). Leadership must facilitate changes by providing opportunities for professors to learn about effective pedagogy as well as opportunities to learn and discuss the research stemming from the learning sciences about the nature of knowledge and how individuals create it.

For all teaching situations, time, effort, and money should be put into smaller class sizes that allow discussion, professor–student interactions, and a studio model rather than a lecture model of instruction. Interdisciplinary courses should be developed that reflect real-world and interesting contexts. Students should also be provided with better advising, and student academic support should be provided so that students feel cared for as individuals and where they can get help without fear of negative consequences from professors or peers.

As a new field of study intent on improving the preparation of engineers, it is not surprising that the citations mentioned and the results presented in this section of the chapter reflect a synthesis of research from many disciplines that form the basis of reform recommendations. Engineering education scholars are also engaged in their own studies of engineering using a full complement of inquiry tools that will enrich our understanding of engineering education as a discipline and contribute to our understanding of how best to educate future engineers.

10.4 Societal Culture and Engineering: Beyond the Western Culture

According to Bernard Amadei, "Engineers have a collective responsibility to improve the lives of people living around the world" (Amadei 2004, p. 24). He notes that technical aspects of an engineering project are less important to success than cultural, social, economic, environmental, and ethical considerations. He also states that engineering schools are not adequately preparing engineers to think beyond the technical; except in rare instances where service learning is integrated throughout a curriculum (Duffy et al. 2011), faculty from different countries collaborate (Dori and Silva 2010), or pedagogical approaches like product archaeology are used (Ulrich and Pearson 1998; Lewis et al. 2010, 2011). Consequently, engineering graduates from western universities are often unaware of the cultural factors that have an impact on the transfer of technologies to non-western societies and the developing world. Many assume that it is as simple as providing the technology or borrowing technology and expecting that it will be successfully used anywhere (Kedia and Bhagat 1988).

Technology transfers are influenced by societal culture, organizational culture, and strategic management processes. Culture is least important to successful technology transfer when the technology transfer is from one industrialized nation to another and most important to success when the technology transfer is from an industrialized nation to a developing nation (Kedia and Bhagat 1988). Many factors influence technology transfer and adoption, which makes it problematic. The first issue to consider is whether the technology is appropriate. The topic of technology appropriateness to a given population and culture is not currently addressed in most engineering curricula or research because it is perceived as low tech and unimportant (Amadei 2004). Engineers must develop the skill of identifying when a technology is appropriate by learning how to assess benefit, resources, and knowledge to sustain technology, local conditions impacting success, user needs, government and other agency support, and cultural beliefs (Bhatia 1990; Sas 2011). In addition to these considerations, cultural factors such as avoidance of uncertainty, power distance, individualism versus collectivism, masculinity versus femininity, and abstractive versus associative characteristics must be taken into account (Kedia and Bhagat 1988). Engineers must develop listening skills to address these considerations because, according to Parsons (1996), engineers make contributions to developing countries "when the engineer truly listens to the desires of those he/she is attempting to serve" (p. 170).

10.4.1 Examples of Success and Failure of Engineering Projects

There are a plethora of wonderful examples illustrating the success and failure of engineering projects and curricula beyond Western civilization (Sas 2011). The data about engineering in developing countries was limited both in scope and methods. Unlike some other areas we explored in this chapter, there were fewer articles in engineering education journals and a greater concern for technology diffusion than the education of future engineers.

In India, the choice of which reusable energy technology to introduce depends upon the circumstances of the farmer. A study of the introduction of renewable energy sources found that a biogas engine for farming worked best for a relatively large farm where the farmer raises two crops a year and has capital and livestock to run a biogas plant. The farmer must also have knowledge of operating and repairing diesel engines and electric motors. In contrast, wind power or solar power requires less technical knowledge making it more successful and appropriate for a marginal farmer who brings produce to a local market (Bhatia 1990).

Improvements in agriculture in Africa have been difficult due to cultural perceptions of the role of agricultural engineers. Locally trained agricultural engineers have had limited success in increasing food production because of the perception that agricultural engineers are either farmers or tractor mechanics (Mafe 2005). The misperception of what agricultural engineers do and the lack of interest in agricultural engineering as a career choice are, in part, due to the curriculum in African universities. The curriculum has been adopted wholesale from developed industrialized countries, and students are not being prepared to create endogenous technologies that reflect the local needs (Adewunmi 2008).

In Pakistan, culture influences the education of engineers as seen in teamwork dynamics and the team roles individuals prefer to undertake. In particular, students take roles that resemble traditional Pakistani family dynamics. The discomfort with change and comfort with traditional practices limit the number of students willing to undertake team roles that foster creativity and stifle creativity. The university curriculum reinforces this problem by limiting course work to basic science rather than courses that foster problem solving and creativity (Hassan et al. 2014).

A project to construct houses and a water system in Nicaragua was deemed a failure because 2 years after construction the houses were in disrepair and the water system was not in use. The project failed for several reasons, including cultural factors such as (1) no money or expertise to fix the broken water pump and no means to transport it to be repaired, (2) no input to the project from members of the community about their desires or needs, and (3) high illiteracy rates and lack of knowledge among the community members about how to govern themselves as members of a cooperative overseeing the water system. In contrast, a similar project to bring drinking water to people in Nepal was successful because technical advice, a budget for skilled labor, and materials were provided by a local committee that managed the project. The local committee also provided the unskilled labor and was responsible for maintaining the system and buying spare parts available in the local market as needed (Parsons 1996).

The case of information technology transfer in Arab countries such as Saudi Arabia and Kuwait is another interesting case. They are developing countries, but money and education are not constraints. Cultural beliefs have been seen to be predictors of IT transference. For example, Arabs prefer to deal with people face-to-face, build consensus, build family-like environments in the workplace, and have a more relaxed sense of time. These cultural factors mitigate against technologies such as email and the use of online meeting places (Straub et al. 2002).

10.4.2 *Know Your Client*

The literature that supports the conclusions about societal culture and engineering came from a variety of sources using a limited number of methods. Three authors used literature reviews to advocate for a particular position and make recommendations for the developing world (Adewunmi 2008; Amadei 2004; Mafe 2005), while two other authors used literature reviews to develop a conceptual model of technology transfer (Kedia and Bhagat 1988). Two authors took the case study approach to examine engineering projects in developing countries (Bhatia 1990; Sas 2011). Parsons (1996) used the research literature to inform her interviews and then used qualitative analysis of the interviews and the literature review to support her recommendations about engineering in the context of the developing world. Straub et al. (2002) used both qualitative and quantitative analyses to examine the transfer of technology in the Arab world, while Hassan et al. (2014) used simple percentages to analyze survey data about national and engineering culture on team role selection.

These examples provide context for the reality that a community's culture has grave influences on whether or not a developed solution, particularly one involving technology, has the potential for success. An ideal solution for one community may not have success in another. For example, the rocket cookstove designed to reduce smoke and subsequent smoke-related health issues is for all intents and purposes an ideal solution; yet some villages that have been provided with this solution still do not use the product due to adverse reactions by tribes to technology. The term ideal becomes relative to location of use and culture. Engineers must obtain feedback from their potential users to identify what factors may influence a design. Only a perspective from these users will provide the necessary information they need to produce an ideal solution for that society.

Cultural considerations have major implications on how engineers approach design problems. This implies that user feedback is essential to the design process and cannot be assumed or guessed. Engineers must understand their clients to ensure they satisfy the needs of all stakeholders.

10.5 Enculturation: Becoming an Engineer

Engineering is more than simply looking, talking, and acting within a masculine culture. To become an engineer, one must traverse across the novice–expert continuum to master disciplinary knowledge, problem solving and problem identification, and understanding and engagement with data (Stevens et al. 2008). By engaging in engineering activities, engineers come to see themselves as part of a culture defined by technology because they are producers of technology and use technology to solve problems.

10.5.1 Underrepresented Groups

Becoming part of the engineering culture by developing an identity as an engineer is critical to persistence in an engineering major; this relationship is particularly strong for women (Jones, Ruff, & Peretti, Jones et al. 2012). The engineering culture may make developing an engineering identity difficult for women (Jorgenson 2002). In contrast to women, men report having an engineering identity long before beginning formal engineering training at the higher education level. They often report that they have always wanted to be an engineer, while most women do not consider engineering as a major or career until they start to apply for college. Women who do develop an early engineering identity are more likely to stay in the field (Bieri Buschor et al. 2014; Godwin et al. 2016); however, many women find engineering to be incompatible with their gender identity, which can lead to stress, questioning of ability, and poor achievement expectations and performance (Ancis and Phillips 1996; Rosenthal et al. 2011). Many women who decide to major in engineering typically have little knowledge of what engineers do because they often do not have the tinkering experience so characteristic of males; however, tinkering experiences are becoming less regular with advancements in technology. The choice of engineering, for women, is based on wanting to do something useful with their strong math and science background (Du 2006) and a desire to help people (Miller et al. 2000).

A strong math background and a desire to help people are often not enough to interest many women to study engineering. According to Ceci and Williams (2010), sex differences in rates of participation in math-intensive fields reflect career preferences, lifestyle choices, and gender inequity in engineering. This conclusion is reinforced by the data concerning the lack of sex differences in math achievement worldwide, but does not account for other barriers to engineering careers placed on women by some countries' culture. In their meta-analysis of TIMMS (Trends in International Mathematics and Science Study and the Programme for International Student Assessment) data, Else-Quest et al. (2010) concluded that girls do as well as boys in math even though they report less confidence in their mathematical abilities. The meta-analysis also found that boys were more motivated to succeed. It is hypothesized that a lack of confidence, rather than a lack of ability, is the reason why girls are less likely than boys to pursue careers in science, technology, engineering, or mathematics. Factors in the TIMSS data responsible for small differences were education, curriculum, and the value that schools, teachers, and families placed on girls learning mathematics. When male–female differences in mathematics achievement are found, they are correlated with gender inequity. The more inequity, the larger the performance gap favoring males (Guiso et al. 2008); however, a state-by-state comparison of mathematics achievement in the United States found that girls and boys do equally well on state standardized math tests from elementary through high school (Hyde et al. 2008). Girls and boys need to develop a strong engineering identity to strengthen interest and reduce the likelihood that they will transfer out of an engineering major. This is a statistically significant relationship.

Jones et al. (2012) reported a correlation of −0.43 with engineering identity and changing major for men and −0.69 for women. For both men and women, successful enculturation into the world of engineering means internalizing an engineering identity, adapting to the culture of engineering by adopting the norms and values of engineering, and establishing solidarity with others in the profession (Drybaugh 1999).

Engineering is a discipline that serves people, but has long been a profession predominantly made up of white, privileged, males (Bix 2004; Drybaugh 1999; Frehill 2004). To be an engineer one must look, talk, and act like an engineer (McIlwee and Robinson 1992). As a consequence, women engineering students often feel like outsiders who do not belong and are not part of the culture of engineering (Foor et al. 2007) prompting them to change majors (Marra et al. 2009). Minority of women in engineering feel particularly excluded as demonstrated by lower feelings of inclusion the longer African American women stay in an engineering program (Marra et al. 2009). This feeling is exacerbated if women of color are poor and lack the cultural capital of their white, female counterparts studying engineering (Foor et al. 2007).

Some aspects of the environment contribute to the masculine nature of the culture. Du (2006) found that project spaces were often strewn with beer bottles, pornography was on the walls, and male students engaged in swearing, aggressive behavior in discussion, and jokes using a technical vocabulary. In addition, male engineering students often held and transmitted negative stereotypes about women's abilities in engineering (Jones et al. 2012). Discourse patterns in whole class discussions and teams also reflect a masculine way of doing engineering (Tonso 1996).

It is a culture where hands-on work is valued and there is a fascination with tinkering and/or making; however, the farther removed engineers are from the production of technology, the less respect they receive from other engineers (Robinson and McIlwee 1991). One female student who was having a difficult time fitting in to the technical culture of engineering put it this way:

> *I did not know that there is such a high demand in the technical part. I felt so stupid because there was something they [males] knew before and I did not know. I am taking some training courses in my spare time, and I think I will reach the same level as them in one or two more semesters.* (Du 2006, p. 38)

Despite studies that document the negative effects of engineering culture on women, there are studies that find that women are not affected by stereotyping of engineering as a male domain. It is unclear whether these findings are due to a change in the culture of engineering, changes in the way engineering is taught, or changes in the culture at large (Beasley and Fischer 2012; Crisp et al. 2009; Jones et al. 2012). Initial hints suggest that good teaching may be responsible for this cultural change. One key feature of engineering programs is problem-based learning. This approach provides students with experience solving problems and working in teams. It is a strong socializer for males and provides them with a professional engineering identity. It also works for females but the impact isn't as strong (Du

2006). Another way that an engineering identity is acquired is through opportunities to work and learn in industry. Workplace experience can help individuals make the transformation from engineering students to students of engineering (Dehing et al. 2013). This is especially true when the workplace supervisor perceives the student as an engineer. Students come away from the workplace experience with a better understanding of what their professional future will be like.

A more metacognitive approach also strengthens professional identity. Creating professional portfolios helped engineering students become more aware of their own values and interests in engineering and was equally successful with males as with females. The process of putting a portfolio together helped engineering students define themselves as engineers (Elliot and Turns 2011). Metacognitive activities such as the creation of a professional portfolio strengthen group membership because metacognition is not generic, but rather reflects a specific discipline (Bransford et al. 2000). In other words, metacognition supports thinking like an engineer, and when one thinks like an engineer, one is an engineer. The impact of metacognition on professional development is not limited to engineering. It has been found to occur in fields as diverse as teacher education (Graham and Phelps 2003) and law (Fruehwald 2015).

Doctoral students are enculturated into the world of academic engineering through grantsmanship to become academic capitalists. Being a successful engineer/engineering researcher is equated with obtaining external funds. In this culture, students and faculty who receive multi-year funding are perceived as being better engineers than less successful grant writers (Szelenyi 2013).

10.6 Redefining Engineers

Data about how individuals become enculturated into the world of engineering came from studies that predominantly used surveys and questionnaires that could be analyzed statistically. These surveys looked at identity (Dehing et al. 2013; Elliot and Turns 2011), identity and stereotypes (Jones et al. 2012), or self-esteem and self-efficacy (Crisp et al. 2009; Marra et al. 2009). Other survey studies addressed attrition from the major using the National Longitudinal Study of Freshman (Beasley and Fischer 20120) or an instrument design to measure belonging in engineering (Marra et al. 2012).

Historical studies relied on document analysis of archived materials, newspaper reports, and research literature (Bix 2004; Frehill 2004). All but one of the remaining studies used some form of qualitative analysis such as large-scale ethnographies (McIlwee and Robinson 1992; Stevens et al. 2008; Tonso 1996) or an ethnography of the particular to tell one person's story (Foor et al. 2007). Drybaugh (1999) used three qualitative techniques (i.e., observations, interviews, and focus groups) to examine enculturation into engineering, while Du (2006) used comparative case studies to examine constructing an engineering identity. To look at the socialization of graduate students in engineering, Szelenyi (2013) conducted interviews. Only one study used both quantitative and qualitative data sources and analyses to examine the culture of engineering for men and women (Robinson and McIlwee 1991).

10.6.1 The Need to Broaden the Definition of Engineering and Engineering Education

From the cited research, we can see that the definition of an engineer should be broadened especially in regard to what an engineer looks like and does. This process should be started before students decide to enter a university program in engineering. The new K-12 Next Generation Science and on Engineering Standards (NGSES) are a good beginning, but we will have to expand upon them. Curricula is needed at the K-12 level that goes beyond the current NGSES focus of understanding the design process and how engineering and science are the same and different. Additions to the curricula should focus on engineering that includes opportunities to both tinker and develop an engineering identity. Engineering activities in the curricula should be contextualized in such a way as to make clear how engineering contributes to social good and how expanding who becomes an engineer is part of contributing to the social good. Early experiences of this kind would increase both women and minorities' interest in engineering majors.

Change has to also take place in university settings. Professional societies focused on women (e.g., Society of Women Engineers, Association for Women in Computing, and Women in Science and Engineering) and minorities (e.g., National Society of Black Engineers, American Indian Science and Engineering Society, and Society of Hispanic Professional Engineers) as well as successful gatherings such as the Grace Hopper Celebration of Women in Computing (http://gracehopper.org) and the Richard Tapia Celebration of Diversity (http://tapiaconference.org) are good examples that aim to gather and recruit underrepresented groups. These efforts help universities and engineering programs toward increasing diversity of their students and the engineering profession, but unintentional biases may still exist. Engineering instructors need to become aware of inclusive pedagogical practices and how some traditional practices may create what Hall and Sandler (1982) called a chilly climate for women. Hiring more female and minority engineering professors who can serve as role models can send a message that engineering is for everyone. Increasing the visibility and support of organizations such as the Society of Women Engineers (SWE), Women in Science and Engineering (WISE), National Society of Black Engineers (NSBE), and American Indian Science and Engineering Society (AISES) also indicates that diversity is valued. Professors can also provide opportunities for women and minorities to take leadership positions in team projects and ensuring that the team respects and follows them as leaders. Finally, there should be an institutional mechanism for reporting incidents, activities, and behaviors, whether initiated by professors or fellow students, that create a hostile environment, exclude women and minorities, and reinforce stereotyping of who is an engineer.

10.7 Conclusions

Changing current practices can be a daunting task, especially within institutions that have longstanding traditions and entrenched individuals who live by the notion that "if it ain't broke, don't fix it." The issue is that traditional systems don't appear broken to those who have found it to be successful for themselves. Improvements are merely an inconvenience that will cause an inordinate amount of unnecessary time to implement. It is this approach that perpetuates the misconception that engineering is boring, difficult, and only for males who are good at math and science. This misconception is held beyond the United States, as supported by *Chapter: Engineering Education in Higher Education in Europe* by Corlu et al. (2018), and expands our discussion to include the impact that history, innovation, and instructional best practices can also have on engineering education. Changing these aspects is highly influenced by the current culture established in engineering education institutions. It is important that we address these aspects to create a better-informed society interested in pursuing engineering careers. What must be undertaken is massive organizational change, both within higher education and industry. Further studies like the ethnographic study of an engineering division within a large American high-tech corporation conducted by Kunda (2009) will help inform this change. The culture of engineering education and engineering industries must provide a platform for a new generation of learners and future workers that will provide diverse perspectives to address twenty-first century grand challenges. This will require a focus on the relationship between engineering organizational culture, engineering identity, and the perception/image projected by engineering to the general public (Hatch and Schultz 1997). Additionally, *Chapter 7: Technology, Culture, and Values – Implications for Enactment of Technological Tools in Precollege Science Classrooms* by Waight and Abd-El-Khalick (2018) reminds us that technology and engineering go hand-in-hand. Technology can make life better, but the use of technology in the early years of education has the potential for negative consequences on student perceptions. Technology is taken for granted in university-level engineering education, but can negatively impact perceptions of engineering if the technology is problematic.

10.8 Recommendations for Engineering Educators, Learners, and Practitioners

This review has made several suggestions for recommended changes in engineering education and practice that can help the evolution of the engineering discipline. These recommendations include:

- Improving education and awareness of engineering within the general public. The public's perception is a key indicator of how engineering culture is viewed from outside the profession. Diversifying the field starts by increasing interest beyond those who have already been enculturated.

- Increasing educational environments that engage students in deep thinking through active learning approaches. Modifying engineering education to be more hands-on and less teacher centered will provide positive experiences to retain students.
- Making the student–teacher relationship more personal to provide students with opportunities to become more deeply acculturated in the field. These opportunities could include teaching and research experiences.
- Greater support by administrators for the scholarship of teaching engineering. A greater emphasis on teaching will allow faculty to expend more effort in turn providing better experiences for students.
- Increased opportunities for engineering students and engineers to know their customers. Designing a product should be for the sole purpose of solving a client's need. Knowing your client is essential in ensuring the solution is appropriate.
- Diversifying engineering and broadening the definition of what an engineer looks like and does. The notion of engineering being strictly for white males good at the hard sciences is a misconception that has been perpetuated and needs to be dispelled.

Enacting these recommendations should prove to positively impact the perceived and existing culture of engineering.

References

Adewunmi, B. A. (2008). Engineering education for agricultural and rural development in Africa. *European Journal of Engineering Education, 33*(3), 321–330.

Amadei, B. (2004). Engineering for the developing world. *The Bridge, 34*(2), 24–31.

Ambrose, S. A., & Norman, M. (2006). Preparing engineering faculty as educators. *The Bridge, 36*(2), 25–32.

Ancis, J. R., & Phillips, S. D. (1996). Academic gender bias and women's behavioral agency self-efficacy. *Journal of Counseling & Development, 75*(2), 131–137. doi:10.1002/j.1556-6676.1996.tb02323.x.

Beasley, M. A., & Fischer, M. I. (2012). Why they leave: The impact of stereotype threat on the attrition of women from science, mathematics and engineering majors. *Social Psychology of Education, 15*, 427–448.

Besterfield-Sacre, M., Cox, M., Borrego, M., Beddoes, K., & Zhu, J. (2014). Changing engineering education: Views of U.S. faculty, chairs, and deans. *Journal of Engineering Education, 103*(2), 193–219.

Bhatia, R. (1990). Diffusion of renewable energy technologies sin developing countries: A case study of biogas engines in India. *World Development, 18*(4), 575–590.

Bieri Buschor, C., Berweger, S., Keck Frei, A., & Kapper, C. (2014). Majoring in STEM – What accounts for women's career decision making? A mixed methods study. *Journal of Educational Research, 107*(3), 167–176. doi:10.1080/00220671.2013.788989.

Bix, A. S. (2004). From "engineeresses" to "girl engineers" to "good engineers": A history of women's U.S. engineering education. *National Women's Studies Association Journal, 16*(1), 27–73.

Blinkenstaff, J. C. (2005). Women in science careers: Leaky pipeline or gender filter? *Gender and Education, 17*(4), 369–386.

Boiarsky, C. (2004). Teaching engineering students to communicate effectively: A metacognitive approach. *International Journal of Engineering Education, 20*(2), 251–260.

Borrego, M., Cutler, S., Prince, M., Henderson, C., & Froyd, J. (2013). Fidelity of implementation of research-based instructional strategies (RBIS) in engineering science courses. *Journal of Engineering Education, 102*(3), 394–425. doi:10.1002/jee.20020.

Bransford, J. D., Brown, A. L., & Cocking, R. R. (2000). *How people learn: Brain, mind, experience, and school*. Washington, DC: National Academy Press.

Brint, S., Cantwell, A. M., & Hanneman, R. A. (2008). The two cultures of undergraduate academic engagement. *Research in Higher Education, 49*(5), 383–402.

Bucciarelli, L., Einstein, H. H., Terenzini, P. T., & Walser, A. D. (2000). ECSEL/MIT engineering education workshop '99: A report with recommendations. *Journal of Engineering Education, 89*(2), 141–150. doi:10.1002/j.2168-9830.2000.tb00508.x.

Capobianco, B. M., Diefes-dux, H. A., Mena, I., & Weller, J. (2011). What is an engineer? Implications of elementary school student conceptions for engineering education. *Journal of Engineering Education, 100*(2), 304–328. doi:10.1002/j.2168-9830.2011.tb00015.x.

Carberry, A. (2014). Engineering epistemological beliefs. In J. Heywood & A. Cheville (Eds.), *Philosophical perspectives on engineering and technological literacy* (pp. 58–69). Dublin: Original Writing Ltd.

Ceci, S., & Williams, W. (2010). Sex differences in math intensive fields. *Current Directions in Psychological Science, 19*(5), 275–279.

Chen, H., Lattuca, L., & Hamilton, R. (2008). Conceptualizing engagement: Contributions of faculty to student engagement in engineering. *Journal of Engineering Education, 97*(3), 339–353. doi:10.1002/j.2168-9830.2008.tb00983.x.

Choresh, C., Mevarech, Z. R., & Frank, M. (2009). Developing argumentation ability as a way to promote technological literacy. *International Journal of Educational Research, 48*(4), 225–234.

Clough, G. W. (2004). *The engineer of 2020: Visions of engineering in the new century*. Washington, DC: National Academy of Engineering.

Corlu, M. S., Svidt, K., Gnaur, D., Lavi, R., Borga, O., & Corlu, M. A. (2018). Engineering education in higher education in Europe. In J. Dori, Z. Mevareach, & D. Baker (Eds.), *Cognition, metacognition and culture in STEM education* (pp. 241–260). Cham: Springer.

Crawley, E., Malmqvist, J., Östlund, S., Brodeur, D., & Edström, K. (2007). *Rethinking engineering education, the CDIO approach*. Cham: Springer.

Crisp, R. J., Bache, L. M., & Maitner, A. T. (2009). Dynamics of social comparison in counter-stereotypic domains: Stereotype boost, not stereotype threat for women engineers. *Social Influence, 4*(3), 171–184.

Davis, L. A., & Gibbin, R. D. (Eds.). (2002). *Raising public awareness of engineering*. Washington, DC: National Academies Press.

Dehing, A., Jochems, W., & Baartman, L. (2013). The development of engineering students' professional development during workplace learning in industry: A study in Dutch bachelor education. *Engineering Education, 8*(1), 41–64.

Dori, Y. J., & Silva, A. (2010). Assessing international product design & development graduate courses: The MIT-Portugal program. *Advances in Engineering Education, 2*(2), 1–30.

Drybaugh, H. (1999). Work hard, play hard: Women and professionalization in engineering- adapting to the culture. *Gender and Society, 13*(5), 664–682.

Du, X.-Y. (2006). Gendered practices of constructing an engineering identity in a problem-based learning environment. *European Journal of Engineering Education, 31*(1), 35–42.

Duffy, J., Barrington, L., West, C., Heredia, M., & Barry, C. (2011). Service-learning integrated throughout a college of engineering (SLICE). *Advances in Engineering Education, 2*(4), 1–23.

Elliot, M., & Turns, J. (2011). Constructing professional portfolios: Sense-making and professions identity development for engineering undergraduates. *Journal of Engineering Education, 100*(4), 630–654. doi:10.1002/j.2168-9830.2011.tb00030.x.

Else-Quest, M., Hyde, J., & Linn, M. (2010). Cross-national patterns of gender differences in mathematics: A meta-analysis. *Psychological Bulletin, 136*(1), 21–38.

Fairweather, J. (2008). Linking evidence and promising practices in science, technology, engineering and mathematics (STEM) undergraduate education. A status report for the National Academies National Research Council Board of Science Education. https://www.nsf.gov/attachments/117803/public/Xc--Linking_Evidence--Fairweather.pdf. Accessed 11 Nov 2014.

Finelli, C. J., Daly, S. R., & Richardson, K. M. (2014). Bridging the research to practice gap: Designing an institutional change plan using local evidence. *Journal of Engineering Education, 103*(2), 331–361. doi:10.1002/jee.20042.

Fisher, P., Zeligman, D., & Fairweather, J. (2005). Self-assessed student learning outcomes in an engineering service course. *International Journal of Engineering Education, 21*, 446–456.

Foor, C. E., Walden, S. E., & Trytten, D. A. (2007). "I wish that I belonged more in this whole engineering group:" Achieving individual diversity. *Journal of Engineering Education, 96*(2), 103–113. doi:10.1002/j.2168-9830.2007.tb00921.x.

Frehill, L. M. (2004). The gendered construction of the engineering profession in the United States, 1893-1920. *Men and Masculinities, 6*(4), 383–403.

Fruehwald, E.S. (2015). Developing law students' professional identities university of La Verne Law Review. https://ssrn.com/abstract=2599057.

Ganesh, T. G. (2011). Children-produced drawings: An interpretive and analytic tool for researchers. In E. Margolis & L. Pauwels (Eds.), *The Sage handbook of visual research methods* (pp. 214–240). London: Sage.

Godfrey, E. (2014). Understanding disciplinary cultures: The first step to cultural change. In A. Johri & B. M. Olds (Eds.), *Cambridge handbook of engineering education research*. New York: Cambridge University Press.

Godfrey, E., & Parker, L. (2010). Mapping the cultural landscape in engineering education. *Journal of Engineering Education, 99*(2), 5–22.

Godwin, A., Potvin, G., Hazari, Z., & Lock, R. (2016). Identity, critical agency, and engineering: An affective model for predicting engineering as a career choice. *Journal of Engineering Education, 105*(2), 312–340. doi:10.1002/jee.20118.

Graham, R. (2009). UK approaches to engineering project-based learning. White Paper sponsored by the Bernard M. Gordon-MIT Engineering Leadership Program. http://www.rhgraham.org/RHG/Recent_publications_files/MIT%20White%20Paper%20-%20UK%20PjBL%20April%202010.pdf. Accessed 18 Nov 2014.

Graham, R. (2012). *Achieving excellence in engineering education: Ingredients of successful change*. London: Royal Academy of Engineering. http://www.ifees.net/pdf/RuthGrahamEducational-change-summary-report.pdf. Accessed 14 Sept 2014.

Graham, A., & Phelps, R. (2003). Being a teacher: Developing teacher identity and enhancing practice though metacognitive and reflective learning. *Australian Journal of Teacher Education, 27*(2), 1–14.

Guiso, L., Monte, F., Sapienza, P., & Zingales, L. (2008). Culture, gender and math. *Science, 320*(5880), 1164–1165.

Hall, R., & Sandler, B. (1982). *The classroom climate: A chilly one for women?* Washington, DC: Project on the Status of Women, Association of American Colleges.

Hansen, N., & Godfrey, E. (1997). *The Auckland University Engineering Society: Whose fault is it anyway?* Paper presented at the 4th Australasian Women in Engineering Forum: Taking a Place and Making it your Own, Ballarat, VIC.

Hassan, I. E., Talib, N. A., Riaz, A., & Iqbal, J. (2014). Influence of national and engineering culture on team role selection. *International Journal of Technology Design Education, 24*, 91–105.

Hatch, M. J., & Schultz, M. (1997). Relations between organizational culture, identity and image. *European Journal of Marketing, 31*(5/6), 356–365.

Hyde, J., Lindberg, S., Ellis, A., Linn, M., & Williams, C. (2008). Gender similarities characterize standardized test performance. *Science, 321*(5888), 494–495.

Jamieson, L., & Lohmann, J. (2012). *Innovation with impact: Creating a culture for scholarly and systematic innovation in engineering education*. Washington, DC: American Society for Engineering Education. http://www.asee.org/member-resources/reports/Innovation-with-Impact/Innovation-With-Impact-Report.pdf. Accessed 5 Sept 2014.

Johnson, A. C. (2007). Unintended consequences: How science professors discourage women of color. *Science Education, 91*(5), 805–821.

Johnson, M., Narayanan, A., & Sawaya, W. (2013). Effects of course and instructor on student evaluation of teaching across a college of engineering. *Journal of Engineering Education, 102*(2), 289–318. doi:10.1002/jee.20013.

Jones, B. D., Ruff, C., & Paretti, M. C. (2012). The impact of engineering identification and stereotypes on undergraduate women's achievement and persistence in engineering. *Social Psychology of Education, 16*, 471–493.

Jorgenson, J. (2002). Engineering selves: Negotiating gender and identity in technical work. *Management Communication Quarterly, 15*(3), 350–380. doi:10.1177/0893318902153002.

Kedia, B. L., & Bhagat, R. S. (1988). Cultural constraints on transfer of technology across nations: Implications for research in international and comparative management. *The Academy of Management Review, 13*(4), 559–571.

King, C. J. (2012). Restructuring engineering education: Why, how and when? *Journal of Engineering Education, 101*(1), 1–5. doi:10.1002/j.2168-9830.2012.tb00038.x.

Knight, M., & Cunningham, C. (2004). *Draw an engineer test (DAET): Development of a tool to investigate students' ideas about engineers and engineering*. Paper presented in American Society for Engineering Education Annual Conference and Exposition, Salt Lake City, UT.

Kuh, G., Nelson Laird, T., & Umbach, P. (2004). Aligning faculty activities and student behavior: Realizing the promise of greater expectations. *Liberal Education, 40*(4), 24–31.

Kunda, G. (2009). *Engineering culture: Control and commitment in a high-tech corporation*. Philadelphia: Temple University Press.

Lewis, K., Simpson, T.W., Chen, W., Kremer, G. & McKenna, A. (2010). *From product dissection to product archaeology: Understanding the global, economic, environmental, and societal foundations of engineering design*. Paper presented at *the* ASME Design Engineering Technical Conference, Montreal, CA.

Lewis, K., Moore-Russo, D., Ashour, O., Kremer, G., Simpson, T. W., Neumeyer, X., McKenna, A. & Chen, W. (2011). *Teaching the global, economic, environmental, and societal foundations of engineering design through product archaeology*. Paper presented in American Society for Engineering Education Annual Conference and Exposition, Vancouver, CA.

Litzinger, T., Lattuca, L., Hadgraft, R., & Newstetter, W. (2011). Engineering education and the development of expertise. *Journal of Engineering Education, 100*(1), 123–150. doi:10.1002/j.2168-9830.2011.tb00006.x.

Mafe, O. A. (2005). Refocusing engineering curricula in developing countries for endogenous technology development and entrepreneurship. *West Indian Journal of Engineering, 28*(1), 1–12.

Marra, R., Rodgers, K., Shen, D., & Bogue, B. (2009). Women engineering students and self-efficacy: A multi-year, multi-institution study of women engineering student self-efficacy. *Journal of Engineering Education, 98*(1), 27–38. doi:10.1002/j.2168-9830.2009.tb01003.x.

Marra, R., Rodgers, K., Shen, D., & Bogue, B. (2012). Leaving engineering: A multi-year single institution study. *Journal of Engineering Education, 101*(1), 6–27. doi:10.1002/j.2168-9830.2012.tb00039.x.

Marshall, H., McClymont, L., & Joyce, L. (2007). *Public attitudes to and perceptions of engineering and engineers*. London: Royal Academy of Engineering. http://www.raeng.org.uk/events/pdf/Public_Attitude_Perceptions_Engineering_Engineers_2007.pdf. Accessed 10 Sept 2014.

McIlwee, J., & Robinson, J. G. (1992). *Women in engineering: Gender, power and workplace culture*. Albany: State University of New York Press.

McKenna, A., Froyd, J., & Litzinger, T. (2014). The complexities of transforming engineering higher education: Preparing for next steps. *Journal of Engineering Education, 103*(2), 188–192. doi:10.1002/jee.20039.

McLean, C., Lewis, S., Copeland, J., Lintern, S., & O'Neill, B. (1997). Masculinity and the culture of engineering. *Australasian Journal of Engineering Education, 7*(2), 143–156.
Miller, P. H., Rosser, S. V., Benigno, J. P., & Zieseniss, M. L. (2000). A desire to help others: Goals of high-achieving female science undergraduates. *Women's Studies Quarterly*, 128–142.
Montfort, D., Brown, S., & Shinew, D. (2014). The personal epistemologies of civil engineering faculty. *Journal of Engineering Education, 103*(3), 388–416. doi:10.1002/jee.20050.
Nazan, E., & Bogers, M. (2015). Open for entrepreneurship: How open innovation can foster new venture creation. *Journal of Engineering Education, 24*(4), 574–584. doi:10.1111/caim.12146.
Nelson Laird, T. F., Shoup, R., Kuh, G. D., & Schwarz, M. J. (2008). The effects of discipline on deep approaches to student learning and college outcomes. *Research in Higher Education, 49*(6), 469–494.
Parsons, L. B. (1996). Engineering in context: Engineering in developing countries. *The Journal of Professional Issues in Engineering Education and Practice, 122*(4), 170–176.
Peterson, R., & Anand, N. (2004). The production of culture perspective. *Annual Review of Sociology, 30*, 311–334.
Piburn, M., Sawada, D., Turley, J., Falconer, K., Benford, R., Bloom, I., & Judson, E. (2000). *Reformed teaching observation protocol: Reference manual (Technical report no. INOO-3)*. Tempe: Arizona Collaborative for Excellence in the Preparation of Teachers. https://mathed.asu.edu/instruments/rtop/RTOP_Reference_Manual.pdf. Accessed 23 Oct 2014.
Robinson, J. G., & McIlwee, J. (1991). Men, women and the culture of engineering. *Sociological Quarterly, 32*(3), 403–421.
Rosenthal, L., London, B., Levy, S. R., & Lobel, M. (2011). The roles of perceived identity compatibility and social support for women in a single-sex program at a co-educational university. *Sex Roles, 65*(9–10), 725–736. doi:10.1007/s11199-011-9945-0.
Sagebiel, F., & Dahmen, J. (2006). Masculinities in organizational cultures in engineering education in Europe: Results of the European Union Project WomEng. *European Journal of Engineering Education, 31*(01), 5–14.
Sas, L. (2011). Appropriate technology for the development of the "third world". *Totem: The University of Western Ontario Journal of Anthropology, 12*(1), 73–78.
Sawada, D., Piburn, M., Judson, E., Turley, J., Falconer, K., Benford, R., & Bloom, I. (2002). Measuring reform practices in science and mathematics classrooms: The reformed teaching observation protocol. *School Science and Mathematics, 102*(6), 245–253.
Seymour, E., & Hewitt, N. (1977). *Talking about leaving: Why undergraduates leave the sciences*. Boulder: Westview Press.
Shawcross, A., & Ridgman, T. (2012). Manufacturing excellent engineers: Skill development in a masters programme. *Engineering Education, 7*(2), 38–50.
Stevens, R., O'Connor, K., Garrison, L., Jocuns, A., & Amos, D. M. (2008). Becoming an engineer: Toward a three dimensional view of engineering learning. *Journal of Engineering Education, 97*(3), 355–368. doi:10.1002/j.2168-9830.2008.tb00984.x.
Straub, D. W., Loch, K. D., & Hill, C. E. (2002). Transfer of information technology to the Arab world: A test of cultural influence. In M. Dadashzadeh (Ed.), *Information technology management in developing countries* (pp. 92–134). Hershey: IRM Press.
Szelenyi, K. (2013). The meaning of money in the socialization of science and engineering doctoral students: Nurturing the next generation of academic capitalists. *Journal of Higher Education, 84*(2), 266–294.
Tang, K.-S. (2013). Out-of-school media representations of science and technology and their relevance for engineering learning. *Journal of Engineering Education, 102*(1), 51–76. doi:10.1002/jee.20007.
Tonso, K. L. (1996). The impact of cultural norms on women. *Journal of Engineering Education, 85*(3), 217–225. doi:10.1002/j.2168-9830.1996.tb00236.x.
Tonso, K. (2007). *On the outskirts of engineering*. Rotterdam: Sense Publishers.
Ulrich, K. T., & Pearson, S. (1998). Assessing the importance of design through product archaeology. *Management Science, 44*(3), 352–369.

Vickers, K., Salmo, G., Loewer, O., & Ahlen, J. (2001). Creation of an entrepreneurial university culture, The University of Arkansas case study. *Journal of Engineering Education, 90*(2), 617–622. doi:10.1002/j.2168-9830.2001.tb00648.x.

Vogt, C. M. (2008). Faculty as a critical juncture in student retention and performance in engineering programs. *Journal of Engineering Education, 97*(1), 27–36. doi:10.1002/j.2168-9830.2008.tb00951.x.

Waight, N., & Abd-El-Khalick, F. (2018). Technology, culture, and values – Implications for enactment of technological tools in precollege science classrooms. In J. Dori, Z. Mevareach, & D. Baker (Eds.), *Cognition, metacognition and culture in STEM education* (pp. 139–165). Cham: Springer.

Zunker, V. G. (1994). *Career counseling: Applied concepts of life planning* (4th ed.). Belmont: Brooks/Cole Publishing.

Chapter 11
Engineering Education in Higher Education in Europe

M. Sencer Corlu, Kjeld Svidt, Dorina Gnaur, Rea Lavi, Oğuz Borat, and M. Ali Çorlu

11.1 Introduction

This chapter concerns the historical development of engineering education in higher education for two European countries, Denmark and Turkey, in the light of innovation and knowledge society. The aim of this chapter is to demonstrate an approach to viewing engineering education through the lens of history and culture. We begin by providing an overview of innovation in Europe and different countries' innovation status; we continue by providing brief overviews of engineering education in Europe in general and programs in Denmark and Turkey in particular, two countries at opposite ends of the innovation scale. Next, we provide a summary and conclusion of our overview, and, finally, we provide recommendations for policy makers, educators, and researchers.

M.S. Corlu
Graduate School of Education, Bilkent University, Ankara, Turkey

K. Svidt
Department of Civil Engineering, Aalborg University, Aalborg, Denmark

D. Gnaur
Department of Learning and Philosophy, Aalborg University, Aalborg, Denmark

R. Lavi (✉)
Faulty of Education in Science and Technology, Technion, Haifa, Israel
e-mail: realavi@tx.technion.ac.il

O. Borat
Department of Industrial Engineering, Istanbul Commerce University, Istanbul, Turkey

M.A. Çorlu
Department of Mathematics, Istanbul Commerce University, Istanbul, Turkey

Y.J. Dori et al. (eds.), *Cognition, Metacognition, and Culture in STEM Education*, Innovations in Science Education and Technology 24,
https://doi.org/10.1007/978-3-319-66659-4_11

11.2 European Innovation and Knowledge Society

Innovation is the generation, utilization, and circulation of new knowledge; it transcends the isolated understanding of engineering. There is a general consensus among policy makers in Europe that innovation is critical for countries to stay competitive in the twenty-first century, as an innovative culture can only flourish in a *knowledge society*. The economic goals of European innovation strategies, such as *Europe 2020*, primarily focus on overcoming the difficulties that have arisen due to the aging population in Western Europe and decreasing the disparities in particular European regions. Among the proposed actions are creating new jobs for the unemployed, improving energy efficiency, and enhancing business-research cooperation. However, these goals will not be met without change from a traditional educational approach to a more interdisciplinary educational approach, especially in engineering education.

The European Commission's (2014) Innovation Union Scoreboard refers to different European countries' respective public and private investment in innovation, innovation partnerships among companies and academia, educational basis for innovation, and innovative research. The innovation performances of European countries in 2014, displayed in Fig. 11.1, show that the lower end of the figure is mostly populated by countries in Eastern Europe, Bulgaria (BG), Turkey (TR), and Romania (RO), whereas the upper end is populated by countries in Western Europe, Switzerland (CH), Sweden (SE), and Denmark (DK). While Germany (DE) is one of the innovation leaders in Europe, France (FR) is just above the European Union (EU) average.

With regard to education, the European goals of innovation include increasing educational attainment, attracting talent to prominent fields, and connecting subject

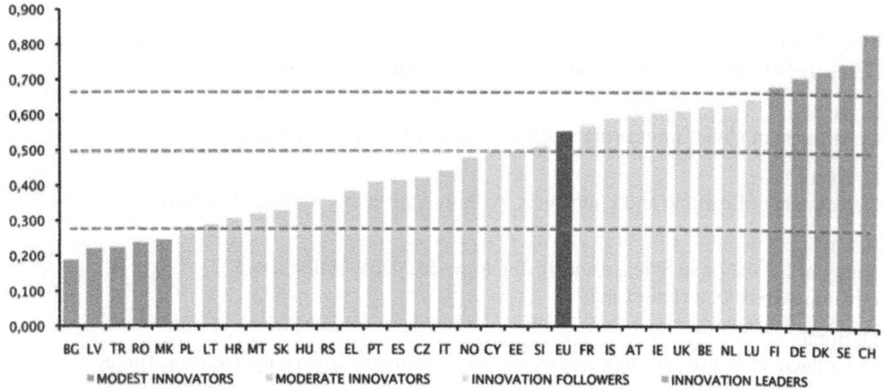

Fig. 11.1 Innovation Union Scoreboard 2014, showing the relative positions of Turkey (TR), 0.2 (low score for innovation); the UK, 0.6; and Denmark (DK), 0.7, with respect to the EU average (Reproduced from European Commission 2014)

matter and knowledge-based life (European Commission 2014). Thus, the process of innovation in European education includes three major characteristics:

(a) *Permeable disciplinarity*—Each discipline has developed a unique knowledge base and certain habits of mind over their long history. While this knowledge base is recognized and appreciated, innovation takes this to a new level. Permeable disciplinarity requires active exchanges of information across and specific specializations within disciplines.
(b) *Coexisting communities of practice*—Coexisting communities of practice foster the circulation of new knowledge through mutual engagement, joint enterprise, and shared repertoire (Wenger 1998) that requires free movement of knowledge within academia and between academia and business.
(c) *Knowledge-based life*—Knowledge-based life reinterprets the concept of real life according to the changing conditions of the twenty-first century. This requires a consideration of the complexities of the knowledge society and lifelong learning principle of the Europe 2020 vision.

11.3 Historical Development of Engineering Education in Europe

11.3.1 A Brief Overview of Engineering Education in Europe

Understanding the dynamics of engineering education in the European context depends on cognizing the historical divide between Continental European and Anglo-American traditions. While highly bureaucratic states, such as France or Germany, represent the former, a more local self-government model, such as the one in England, represents the latter. During the first Industrial Revolution (1760–1840s), there were two main differences between these traditions of engineering education. The first difference is engineers in Continental Europe were mainly public servants (e.g., military or civil engineers), whereas engineers in the Anglo-American tradition held roles as entrepreneurs or freelance professionals. The second difference is engineers in Continental Europe were equipped with advanced mathematics and science knowledge that they learned in school; early examples of such schools were the *École Nationale des Ponts et Chaussés* in France (1747) and *Genie-Akademie* in Austria (1778). In contrast, Anglo-American engineers trained on the job acquire practical skills.

The difference between the two traditions continued until the second Industrial Revolution (1850s, World War I). With the rise of science-based industries in the late nineteenth century, it became a necessity for both traditions to scale up the academic training of engineers in mathematics and science. Additionally, the importance of business in the application of science to practical ends was recognized. Yet, classical universities of Europe with roots in pure knowledge had difficulty embracing the practical side of engineering. It was the German dual approach during the postwar period that introduced the concept of technical schools specializing in educating engineers for practical skills (Lundgreen 2006).

Today, engineering is placed at the center of the knowledge triangle that combines research, education, and innovation (see European Federation of National Engineering Associations—*Fédération Européenne d'Associations Nationales d'Ingénieurs* [FEANI] 2010; European Society for Engineering Education—*Société Européenne pour la Formation des Ingénieurs* [SEFI] 2013). This knowledge triangle is steadily materialized after the 1950s from the transition of tradition-specific education to a more integrated form of engineering science and practical application. Even more, with the increased mobility of business, engineers across European countries necessitated a more continental, even global, approach to engineering education, which in turn sparked the question: "What is the relationship between preparing engineers to serve their home country and preparing engineers to serve Europe?" (Downey et al. 2008, p. 437).

Although not uniform, there is more consistency between prominent engineering programs and curriculum in European engineering education institutions. The introduction of the European Higher Education Area with the Bologna Declaration in 1999 was what made this consistency possible, by allowing students and academia to be mobile across institutions. Additional efforts for a more consistent program were followed by the establishment of a Pan-European authority for accreditation—the European Accredited Engineer (EUR-ACE, European Accreditation of Engineering Programs 2008; SEFI 2012).

11.3.2 Engineering Education in Denmark

11.3.2.1 Overview of the Structure and Content of Engineering Education Programs

In light of the technological, societal, and environmental challenges in the contemporary world, Danish engineering education aims at providing top qualified engineers and engineering scientists who are equipped to respond in informed and innovative ways to the changing demands of their profession. Danish engineering education programs differ in content depending on their specialization and professional orientation. The Danish system offers two basic degree programs in engineering: a practical, professionally oriented, 3.5-year undergraduate program that results in a bachelor's degree and a relatively theoretical and research-oriented 3- and 2-year program leading to a master's degree. After this, there is the possibility to embark on a 3-year program leading to a doctorate degree.

The 3.5-year undergraduate program develops a well-rounded engineer, leading to a more practical, hands-on engineering career, e.g., a technical engineer (diploma engineer), and is offered at both universities and engineering colleges. It can, however, be extended with a specially designed 2-year graduate program offered by universities. The university offers a 3- and 2-year program, which yields a formal bachelor's degree, with an intent to continue at the graduate level in engineering or a compatible discipline, allowing for a broader spectrum. Insofar, students who work toward a theoretical bachelor's degree and continue to earn a master's degree will hold the other type of professional qualification, which is a more speculative and research-minded profession, e.g., a civil engineer. The two types of engineering programs supplement each other, as they cover the entire range of problems solved by engineering professionals—both task-oriented problems and complex ones requiring higher theoretical insight and technological competence. In this way, engineering education continues to attract a broader selection of students.

The Danish government owns, finances, and regulates engineering in higher education institutions, which are subject to ministerial guidelines regarding quality assurance and accreditation systems; however, the government affords institutions certain degrees of freedom in their educational structure, such as teaching philosophies. Furthermore, engineering education is free of charge, offering students state grants when enrolled in a program. To gain admission, students generally need to complete a 3-year upper secondary science-oriented program and have a relatively high performance in subjects such as mathematics, physics, and chemistry. In the undergraduate program, students can combine vocational training with a supplementary study program to achieve a level similar to upper secondary education (Jensen 2000). Additionally, in 2014, the Danish government introduced a new type of education that combines vocational training with an upper secondary level education termed EUX. The intent of the program is to give academically strong students a better opportunity to achieve an upper secondary level education in connection with their vocational courses and training (Danish Ministry of Education 2014).

Engineering programs in Denmark address the need to include high proficiency level in the technological and scientific areas; professional elements such as communication, foreign languages, ethics, project management, and business knowledge; and a general understanding of societal, environmental, and global challenges that affect the professional environment and the responsibility placed on the engineering profession (Danish Society of Engineers [IDA] 2009). From an industrial standpoint, both types of engineers are valued and in demand, as this dual orientation provides a satisfactory solution to the industry's need for both practical professionals and technical specializations.

As an important stakeholder in the design of the engineering programs, IDA encourages the maintenance and the further consolidation of the two educational profiles (IDA 2009). The recommendations for the professional program include the affiliation with related research environments in order to ensure state-of-the art exposure, while also maintaining the main emphasis on the practical qualifications. This provides a broad spectrum of opportunities to interact with professionals in the field and suggests a preference for recruiting professionally active teaching staff at the university colleges. Similarly, IDA recommends that the civil engineering programs continue their engineering science profile through research and development, including high degrees of specialization and front-edge technologies. To this end, IDA recommends emphasizing the consolidation of the various programs within their respective research environments and a requirement for the 3 + 2-year education program to be exclusively research based.

The research and development orientation is the essential element in various doctoral programs, in which the candidates are qualified to conduct research and development activities independently in an international environment. The research programs include the Industrial PhD program, where the government and a private company co-jointly support a university-hosted PhD project. The government capitalizes on the long-term benefits of the high-level research and development program and, thereby, strengthens the technological and scientific base of engineering education. This makes it a worthwhile investment toward securing continual growth and welfare in the Danish society.

11.3.2.2 Tradition and Renewal in Danish Engineering Education

The dual orientation within the Danish engineering education sector—the specialized, practical, professional strand and the theoretical, scientific strand—originates from the alternate movement between theoretical hegemony and practice. Historically, there have been various designs in Denmark, as well as in the rest of Europe and the USA, with either a practical, skill-based technique or a theoretical, science-based method. The former was the product of technical schools, whereas the latter was embraced at technical universities and engineering schools. The practical and theoretical orientations led to the formation of two distinct institutional traditions, that of technical engineering and that of engineering education, which still exist in the present engineering landscape. The Danish solution to this divide resembles the northern European development, as it embraces the two models: the technical school tradition of educating students with practical engineering skills for industries with technical and mathematical foci and a more academic-oriented engineering education program that is the focus of technical universities, producing consultant engineers valued for their specialized theoretical and technical qualifications.

As in most parts of Europe, the youth protest and democratization movement in the late 1960s transformed the backdrop of higher education in Denmark from being an opportunity reserved for only an elite group of people into a mass phenomenon with an ever-increasing intake of students from various backgrounds. Not only did engineering education become easily accessible, but it also moved in a more scholarly direction as the content of the technical and theoretical programs grew.

In the 1990s, Denmark, along with the rest of Europe, had increasingly questioned its relatively scientific-dominated engineering programs. Teaching less practical skills decreased these programs' relevance for the industry. At the same time, a predominantly technical focus overshadowed analytical qualifications, and this proved insufficient in fostering innovative capacity. It did not deliver the type of creative design that engineers needed to be capable of managing, nor did it keep up with the rapid technological developments. The critique came from the technical universities, as they were against this narrow, science-focused program and emphasized the need to respond to societal and industrial demands of engineering skills and competencies.

This led to a renewed focus of engineering education design, which culminated with the expansion of project- and problem-based learning approaches that had been the landmark of only two progressive universities, specifically Aalborg University (AAU) and Roskilde University Center (Barnes 2010; Kjersdam and Enemark 1994). The philosophy of the project and of the problem-solving approach had its root in the main concern of engineering: problem solving in a real-life context as well as interdisciplinary settings through activation and integration of relevant, practical, theoretical, and analytical skills necessary to a career in engineering. Furthermore, team-working skills, such as communication and cooperation; interpersonal competencies, such as flexibility and autonomy; and project management skills were all integrated into the learning space.

The increasing diversity of technologies and the complex problems posed by globalization constitute a new set of challenges that called for a unified front and the need to reform engineering education once more (Jørgensen 2007). The Danish strategy of acknowledging and supporting each of the two orientations in their own right for their distinct contributions to the disciplinary and professional domain implies that the traditional tension is no longer the focal point. Meanwhile, the efforts to reform the educational setup in engineering are visible in creating strong, competent, and knowledge-based learning and in developing environments to cope with modern-day challenges. The tendency to associate scientific and technological competencies is widely visible in recent developments of merging university colleges into universities that offer both types of engineering education. Ramskov (2012) pointed out there is only one independent university college left on the Danish map of engineering educations, namely, VIA University College in Horsens. Other university colleges, along with their historical programs of early professional engineering (technical and academy engineers), have merged with four existing universities. These are AAU, with campuses in Aalborg, Esbjerg, and Copenhagen; Aarhus University (AU), which had absorbed the former university colleges in Herning and Aarhus; the University of Southern Denmark (SDU), including former technical institutions of Odense and Sønderborg; and the Danish Technical University (DTU), located in the Copenhagen area.

It is questionable, though, whether the massive mergers were initiated out of a wish to reconcile the two traditions or whether it is proof of the final victory of the blend of the theoretical, science-based school and the scholastics of professional life, complying with the national strategy for creating a knowledge society. However, the latest development does increase competition among the universities attracting students since they all offer the same range of programs. This will arguably make the quality of teaching and the specific subject combinations an important competitive parameter.

Nevertheless, one problem remains unsettled, and it pertains to the expanding range of topics offered within various engineering programs. Disciplinary congestion can occur when attempting to cover all of the skills and competencies demanded of future engineers (Jørgensen 2007). According to a comprehensive report by the Danish Technological Institute, Centre for Policy and Business Analysis (Teknologisk Institut-Center for Analyse og Erhvervsfremme [DTI] 2013), on future job functions and competency requirements for engineers, the Danish companies are in need of engineers with in-depth science and technology expertise, as well as the ability to apply this expertise in practice. There exists an expectation from engineers to collaborate in interdisciplinary settings, in order to attend to complex tasks in ways that are both efficient and innovative. As development and innovation processes increasingly involve customers and suppliers in often globally distributed networks, organization and advanced business understanding, as well as communication and personal aptitude, is expected. Most of all, it will be the successful combination of all of these various competencies that will prove decisive for future engineers' ability to identify new market opportunities for delivering new, hi-tech knowledge and thorough products and services. The complex competency profile expected of modern engineers puts high demands on education. The way educational institutions have dealt with this so far has been with a renewed focus on scientific knowledge, as this ensures the necessary preparation in dealing with the increasing complexity of technology. They also have created a whole set of new courses within the *human dimensions* in response to the new demands of the engineering profession (Christensen et al. 2009).

Curriculum design remains a major area of attention as it has an effect on learning outcomes and competency development. Rather than adding yet more isolated courses to cover the extended fields in engineering to an already heavy curriculum, the solution sought presently involves different approaches to teaching and learning. The project- and problem-based learning methods (PBL) in use at Roskilde University Centre and AAU have gained renewed attention (Graff and Kolmos 2007). Arguably, they contain the potential for more integrative forms of education that will include social, environmental, and economic considerations into the problems that are the central drivers of learning in PBL. By facilitating learning based on identifying and working with real-life engineering problems in their socioeconomic context, students learn to manage increasingly complex systems of scientific knowledge, practical skills, and contextual awareness in an organic, integrated fashion (Jørgensen 2007). One critical component of PBL is to scaffold a network of increasing problem complexity, spanning from reproductive learning in the form of routine to complex problem solving in contexts involving the complexity of real-life settings, in order to promote creative learning. Ellström (1997) described PBL progression as having degrees of freedom. The lowest degree of freedom in PBL is when students are given both the problem and the problem-solving method. A higher degree of freedom is achieved by giving students a problem left open (e.g., a case study) and the method for solving it. The highest degree of freedom is achieved by leaving a problem for students to solve, in such a way as to support the attainment of the learning goals (Biggs and Tang 2009).

Another similar approach to teaching and learning of engineering is the conceive-design-implement-operate (CDIO) methodology that has similarly arisen out of the need to rethink engineering education (Crawley et al. 2013). The objective of this approach is to prepare engineers that are ready to engineer. The Swedish government originally developed this system, but it has won increasing attention in Denmark as well, as an inspiration or a framework for curricular planning and outcome-based assessment. Nearly all Danish engineering schools are members of the CDIO network, and some like DTU have based some of their study programs on the CDIO concept (e.g., CDIO Projects in DTU's Bachelor of Engineering in Information Technology Study Program, http://orbit.dtu.dk/). These four terms, *conceive, design, implement, and operate*, and the activities and outcomes of the four phases are applicable to a wide range of engineering disciplines, in which engineers lead or are involved in all phases of a product, process, and system life cycle. The *conceive* stage is about defining customer needs, taking into account the technological parameter, the enterprise strategy, and various regulations, and finally developing the conceptual, technical, and business plans. The *design* stage involves the whole design process including the plans, drawings, and algorithms that describe the product, process, and/or system. The *implement* stage focuses on transforming the design into the product, and this encompasses the hardware production, software programming, testing, and validation. The final stage, *operate*, uses the implemented design to produce the intended value, including maintaining, evolving, recycling, and closing down the system. The CDIO approach to curriculum design builds on stakeholder input to identify the learning needs and then sequence a path of integrated scholarship experiences to meet those needs. In so doing, the CDIO contribution to engineering education promotes in-depth learning of foundational engineering knowledge, skills, and principles that engineers contribute to society.

An important criterion in ensuring international recognition of Danish engineering education programs is the quality assurance, which "…includes all planned and systematically executed proceedings and procedures to ensure that quality/performance indicators defined for a product or service, an educational program be performed in a complete manner" (Borat 2010, p. 40). In alignment with international standards, Denmark too has implemented the European Qualifications Framework of the Bologna Working Group, supplemented by the Dublin Descriptors, which offers general statements of the typical expectations of achievements and abilities associated with qualifications within the particular disciplines, representing the end of each Bologna cycle. Referring specifically to engineering education, the accreditation and quality assurance are in accordance with the internationally acknowledged accreditation system, EUR-ACE, which identifies high-quality engineering degree programs in Europe and is managed by the Danish Evaluation Centre, and the Danish Accreditation Institution, ACE.

11.3.3 *Engineering Education in Turkey*

11.3.3.1 Foundations of Turkish Engineering Education

In the eighteenth century, the Ottoman Empire conducted educational reforms based on similar reforms in Europe. The year 1734 marked the establishment of the first secular school to educate military engineers, and 1773 marked the establishment of the first formal engineering school, with the mission to educate potential teachers of engineering rather than to produce large numbers of engineers. These schools and many others belonged to the Continental European model of engineering education, with a mix of French and Ottoman teaching staff and a curriculum that included science (mechanics-astronomy), technology (technical drawing and design of military equipment), and mathematics (geometry, algebra, and logarithms). Constraints such as the extended period of schooling (15 years), restrictions regarding the number of enrolled students, and the expectation of the students to support themselves during their long period of education, hindered the production of engineers compared to other European countries (Kaçar 2007).

By the end of the nineteenth century, most of the teachers at engineering schools in the country were Turkish. The practical aspects of engineering gained importance for the development of economy during the second Industrial Revolution. German-inspired engineering schools were providing a more theory-based engineering education, and several technical schools were founded, following a more practical curriculum when compared to the academic-oriented engineering schools (Kaçar et al. 2012). Through the educational reforms after the eighteenth century, engineering schools in Turkey had broader impacts on society and contributed to the development of scientific knowledge in other STEM fields. For the first time, engineering schools taught subjects such as differential and integral calculus, physics, chemistry, hydraulics, mechanics, and electricity. Additionally, these schools published several books in Turkish on these subjects, and students in other schools used these books too. In addition, some prominent teachers of engineering contributed to the field with their work at these schools, including Karl von Terzaghi (1883–1963) and Fritz Georg Arndt (1885–1969).

Despite the influence of Continental European tradition on Turkish engineering education, Robert College engineering school (1912) distinguished itself as an example of the Anglo-American tradition in Turkey. In contrast to their peers at other engineering schools, students at Robert College were provided with abundant opportunities to work with machinery (Freely 2013; Nurdoğan 2009). However, even some of the existing faculty members at Robert College did not welcome such a practice-based education nor it became as popular as the other engineering schools in the country at that time (Scipio 1955 as cited in Nurdoğan 2009).

The founder of modern Turkey in 1923—Mustafa Kemal Atatürk—outlined in his vision the need to "...establish and operate small and large industries … …to reach the ideal of a developed and prosperous Turkey" (Tantekin-Ersolmaz et al. 2012, p. 31). Graduates of the Ottoman-era engineering schools responded to

Ataturk's vision by planning and constructing roads, bridges, dams, factories, buildings, energy plants, communication networks, villages, and cities (Tantekin-Ersolmaz et al. 2012). Other important developments after the declaration of the Republic related to Turkey's membership in the North Atlantic Treaty Organization (NATO) from 1949 and the OECD in 1960, both of which enabled a closer relationship with the USA and thus the Anglo-American tradition in engineering education, gained popularity among policy makers following World War II. The adoption of the Anglo-American model with an increased emphasis on practical skills was a perceived solution to the increasing workforce requirements of a growing postwar economy and the increased demand for access to engineering education (c.f. democratization movement as mentioned in 10.2.2).

Several universities, including the Middle East Technical University (METU) in Ankara, Turkey, were founded during this postwar period. However, the open-market economy era in the 1980s was when the country experienced an exponential increase in the number of engineering faculties (Tantekin-Ersolmaz et al. 2012). Bilkent University, founded in 1984, was the first nonprofit foundation university in the country, which opened a pathway to the private sector to invest in higher education.

11.3.3.2 The Current Situation of Engineering Education in Turkey

Today, the number of engineering faculties has reached almost 130 situated in more than 170 public and nonprofit foundation universities. This is partially a result of the current government's ambitious goal of founding a university in every city in the country. So far, the total number of engineering students exceeds 250,000, approximately 29% of whom are female students. Over 9000 faculty members manage this massive number of students (Measuring, Selection and Placement Center—Ölçme, Seçme ve Yerleştirme Merkezi [ÖSYM] 2013). Early examples of engineering institutions such as İTU, Boğaziçi University (formerly known as Robert College), METU, and Bilkent University continue to be the most popular engineering education faculties in the country. The majority of the graduates of these universities prefer to work in the private sector in high-paying jobs. Public service is a popular choice for the majority of engineers who graduate from other universities (Chamber of Mechanical Engineers 2013). While these numbers show the high interest among Turkish youth for engineering education, in recent years, there is an increasing interest in emerging fields of engineering such as mechatronics (Akpinar 2006). However, with an economic-growth model based on construction, it is not surprising that civil engineering has maintained its position over the years as one of the most popular engineering fields in the country. See Table 11.1 for the most popular engineering departments in Turkey in terms of the number of applications to engineering programs.

However, this high interest toward engineering in the country comes with a price. Some claim that the government favors the funding of engineering projects with practical outcomes over scientific research projects (Nesin 2014). In addition, some

Table 11.1 Number of applications submitted to the most popular engineering education fields in 2008 and 2012 (NTVMSNBC 2013)

	2008	2012	Change in percentage
Civil engineering	185,349	312,536	69
Mechanical engineering	190,806	260,491	37
Electrical and electronic engineering	89,889	182,250	103
Computer engineering	81,411	128,828	58
Food engineering	68,432	98,089	43
Industrial engineering	65,363	96,256	47
Environmental engineering	61,453	58,567	−5
Geomatic engineering	32,807	39,638	21
Mechatronics engineering	4660	39,611	750
Metallurgical and materials engineering	31,957	38,122	19
Chemical engineering	41,963	35,336	−16
Biomedical engineering	5151	18,403	257

Note. Departments ranked in terms of the number of applications in 2012

universities are closing their science faculties due to low student interest in a career in pure science or mathematics. In order to revive student interest for careers in pure sciences, faculties of science are desperately trying to grant their students eligibility for teaching upon graduation through quick-fix certification programs. However, there has been an oversupply of engineering teachers in the country, and such alternative teaching certification programs are negatively affecting the graduates of faculties of education in finding jobs at public schools, as well as the quality of engineering teaching in the country (Çorlu 2013; Corlu et al. 2014).

Engineering students need to be educated as innovators with a global perspective, who can work in multilingual and multidisciplinary environments and find solutions to the problems of a knowledge society (İnan 2005). However, engineering students in Turkey generally express negative opinions about their education with regard to fostering their engineering innovation. Students at engineering faculties believe their education does not foster the necessary skills for being globally employable. In one of the few studies on engineering education in Turkey, engineering students expressed their perception of their instructors as the sole expert of theoretical knowledge rather than someone they collaborate with in finding solutions to the problems of the society. One conclusion of this study was that the engineering education in Turkey does not encourage interdisciplinary approaches, communication-based teamwork, risk taking, or entrepreneurship (Sevindik and Akpınar 2007; c.f. Çakır and Yelmen 2011).

In another study, Gençoğlu and Gençoğlu (2005) found engineering education programs at Turkish universities to be uniform, providing a bachelor's degree at the end of a 4-year study with almost identical coursework across universities. Evidently, Turkish engineering education had disregarded the diverse needs of different regions of the country, and the programs have not changed much over the past 50 years.

There is evidence, however, that externally accredited engineering faculties in Turkey were likely to be providing a more interdisciplinary learning environment, particularly when compared to faculties of science or mathematics (Corlu 2013). In addition, some of the universities have recently opened faculties of engineering with a more interdisciplinary approach. The sharp increase in interest toward engineering fields, which foster an interdisciplinary approach, such as mechatronics engineering and biomedical engineering, provides evidence that Turkish students are well aware of the changing conditions of the twenty-first century, and some engineering faculties in the country are showing an effort to adapt to these conditions.

Today, women engineers in Turkey make up the 27% of the engineering workforce. In fact, Turkey has one of the highest percentages of women engineers in the world, compared to 14% in the USA and 6% in the UK (Smith and Dengiz 2010; Tantekin-Ersolmaz et al. 2012; U.S. Congress Joint Economic Committee 2012; Women's Engineering Society 2014). Women academicians reported fair and equal treatment in the academic world, which seemed to be a valid argument for women engineers in professional life (Acar 1991, as cited in Zengin-Arslan 2002; Zengin 2000). However, one study elaborated on the challenges that women engineering students face during their education:

> …more covert forms of discrimination still occur in the educational institutions of Turkey, such as the tendency to guide female graduate students into those fields of engineering which are viewed as more convenient for women, jokes made by the professors about women's incompetence in engineering and the marginalizing attitudes of male classmates towards female students. (Zengin 2000, p. 407)

Such challenges are unlikely to be specific only to Turkish women engineers.

As the supreme body responsible for the supervision of universities, the Council of Higher Education (Ayvaz et al. 2016) considers the Bologna Process as a planned and systematic roadmap to ensure that Turkish universities function according to the European standards. Within the higher education institutions in Turkey, internal quality assurance is progressing well (Borat 2009; Borat 2010; Varol et al. 2013). Each university in Turkey prepares its strategic plan—in accordance with the Public Financial Management and Controlling Law. For the external auditing, a license has been given to the Association for Evaluation and Accreditation of Engineering Programs—*Mühendislik Eğitim Programları Değerlendirme ve Akreditasyon Derneği* (MÜDEK)—on November 2007. Initially, MÜDEK adopted the criteria of some other popular accreditation organizations in the USA and Europe, including the Accreditation Board for Engineering and Technology (ABET), Washington Accord Graduate Attributes, and EUR-ACE Framework Standards. All these activities and expected developments are gathered in the framework of the Turkish Qualifications Framework (Borat 2014).

Today, MÜDEK accredits only the undergraduate engineering programs in Turkey (MÜDEK 2013), and their list of accredited institutions is not exhaustive (see the list of accredited programs as of 2014 July at http://www.mudek.org.tr/en/akredit/akredite2014.shtm). Some of the most popular engineering faculties in the country, including İTÜ, Boğaziçi University, METU, or Bilkent University, are accredited by ABET, but not by MÜDEK.[1]

[1] See the list of accredited programs as of 2014, October, here: http://main.abet.org/aps/Accreditedprogramsearch.aspx

METU, ITU, Gazi University, Ege U/niversity, YTU, 9Eylul University, Bogazici University, Firat University, Erciyes University, Selçuk University, KTU, Hacettepe University and Bilkent University are first 13 universities according University Ranking by Academic Performence (URAS) in engineering area and their engineering programs are accredited by ABET or by MÜDEK (URAS 2017).1 Turkish Higher Education Quality Board having public legal personality with administrative and financial autonomy was established to carry out evaluations according to national and international quality standards of quality of education and research activities and administrative services of higher education institutions and to carry out internal and external quality assurance, accreditation processes and authorization processes of independent external evaluation institutions (Quality Board 2017). It is expected that the autonomous Quality Board will influence the developments and quality of engineering education and research activities in Turkey positively.

11.3.4 Engineering Education in the UK

11.3.4.1 The History of Engineering Education in the UK

Engineering education in the UK represents the Anglo-American tradition of engineering education. Historically, the public and economic position of engineers in the UK has been low compared with other high-skill professions. Engineering qualifications in the UK were provided by professional organizations rather than academic institutions, and they involved apprenticeships rather than degrees. Indeed, until the end of the nineteenth century, no examination was required to obtain any engineering qualification. Throughout the first part of the twentieth century, the number of qualified engineers was only several thousands, and by the mid-century, a skill shortage of technical engineers was evident. It was not until 1956 that the Ministry of Education, led by Prime Minister Sir Winston Churchill, in response to the rise of technical education in the USSR, produced a white paper, which argued for a complete change in technical education throughout the UK. The implementation of the policy outlined in this paper led to a large, fast increase in the number of technical colleges growing into full-fledged universities. However, despite these changes, the number of engineering graduates, as well as their public status and pay relative to other highly skilled professions, still remained low throughout the 1960s and 1970s, compared with competing developed countries, and the lack of vocational training during graduate studies remained, with academic studies detached from the reality of engineering work (Albu 1980).

11.3.4.2 Current Status of Engineering Education in the UK

Presently, engineers in the UK do not require a license to practice their profession, but one can obtain such licenses, including engineering technicians, incorporated engineers, chartered engineers, and information and communication technology technicians, from various institutions under the devolved accreditation of the UK Engineering Council. Registering as a chartered engineer is called initial professional development and normally takes 4–8 years following the completion of the first engineering degree. It depends on peer review and sometimes further exams. Engineering undergraduate studies may or may not include 1 year of on-the-job experience (termed "sandwich courses"), and further qualifications of more competences named continuing professional development are available following registration as a licensed engineer (Engineering Council 2017). More practical aspects of engineering have been introduced into academic degrees, particularly project-based learning, which is especially diverse in UK engineering education (Graham 2010; Graham and Crawley 2010). However, historical problems of low public perception of the engineering profession, as well as skill shortages in some engineering sectors, still linger. The introduction of institutes of technology, proposed by the Royal Academy of Engineering (2017), aims to help ameliorate some of these problems.

11.4 Comparison of Engineering Education in Denmark, Turkey, and the UK

Table 11.2 presents a comparative summary of present-day engineering education in Denmark, Turkey, and the UK in higher education (Barlex 2011; c.f. Çakır and Yelmen 2011; Chamber of Mechanical Engineers 2013; Graham and Crawley 2010;

Table 11.2 Comparison of engineering education in Denmark, Turkey, and the UK

Country	Innovation score[a]	Prevalent teaching methods	Advantages	Challenges
Denmark	0.7	Mostly student-centered (PBL and CDIO)	Theoretical and practical programs produce graduates who cover an entire range of problems faced by present-day engineers	Expanding range of topics offered within engineering programs can lead to disciplinary congestion
Turkey	0.2	Mostly traditional (teacher-centered)	Produces many engineers; private and public sectors employ local engineers	Low student interest in pure science
				Most programs do not encourage interdisciplinarity, team communication, risk taking, or entrepreneurship
UK	0.6	Both student- and teacher-centered	Characterized by a wider variety of approaches to PBL than typically found in other countries	Resource and expertise constraints can hinder PBL adoption

[a]According to Innovation Union Scoreboard (European Commission 2014)

Jørgensen 2007; Sevindik and Akpınar 2007). Denmark's prevalent student-centered teaching methods contrast with Turkey's prevalent teacher-centered teaching methods. While Denmark's challenges for engineering education center on limiting and optimizing the curriculum, Turkey's challenges center on expanding the curriculum to include more innovative teaching methods, as well as skills and abilities, that are not traditionally part of engineering education but required in the present-day globalized world. The UK's challenge lies mainly in training educators how to implement PBL and in making PBL cost-effective.

11.5 Conclusion

We provided a brief overview of engineering education in three European countries with disparate innovation levels: Denmark and the UK with a high level of innovation and Turkey with a low level of innovation (European Commission 2014). The purpose of these country-specific examples was to understand the development of engineering education in Europe and its relation to European society and culture. The narrative in each case approached the issue of engineering education in the European context from two distinct but complementary perspectives. We compared these countries' respective engineering education systems with each other. We conclude that Turkey's low innovation score and Denmark and the UK's relatively high innovation scores (see Fig. 11.1) could stem from differences in their educational programs and teaching methods: diverse, interdisciplinary, and practically based in Denmark and the UK and uniform and traditional in Turkey. Finally, we conclude that differences in innovation levels between European countries may be explained, in part, by differences pertaining to teaching methods and advantages and challenges to engineering education.

In Chap. 4 of this book, J. Sjöström and I. Eilks (2018) present three types of visions for scientific literacy: the first concerns scientific content and processes for later application; the second focuses on the usefulness of science to life through learning in meaningful context; and the third vision, belonging to the authors themselves, is more critical and more concerned with global and social issues than the two other, prevalent visions. It would seem from Sjöström and Eilks' description of the three visions for scientific literacy that Turkey mainly follows the first vision described, while Denmark and the UK mainly follow the second vision described. This difference in vision of scientific literacy may also serve to explain the difference in innovation scores discussed above.

In Chap. 10 of this book, A. R. Carberry and D. R. Baker (2018) recommend for engineering education to become more hands-on and less teacher-centered, focusing instead on active learning by students. Their recommendation echoes our own summary of the differences between highly innovative countries (Denmark and the UK) and less innovative countries (Turkey), where we found student-centered pedagogy for the former versus teacher-centered pedagogy for the latter.

11.6 Recommendations

Based on our review of engineering education research and praxis in Europe, we can make the following recommendations:

- For educational policy makers and educators looking to improve their country's innovation performance, we suggest introducing more student-centered teaching methods into engineering education programs while training teachers and allocating resources for successful implementation. While doing so, policy makers and educators should be wary of and mitigate for interdisciplinary congestion.
- For researchers, we suggest adding more European countries, from both the Anglo-American and Continental traditions, into the comparison summarized in Table 11.2. This addition would enable categorization of engineering education systems and help identify common attributes in engineering education among low- and high-innovation performance countries.

References

Akpinar, B. (2006). Mechatronics education in Turkey. *Mechatronics, 16*(2006), 185–192.

Albu, A. (1980). British attitudes to engineering education: a historical perspective. In *Technical innovation and British economic performance* (pp. 67–87). London: Palgrave Macmillan.

Ayvaz, B., Kusakci, A. O., & Borat, O. (2016). Quality Assurance and Accreditation Processes / Kalite Güvencesi ve Akreditasyon Süreçleri. *New Turkish Journal/Yeni Türkiye Dergisi, Sayı, 88*, s 921–928.

Barlex, D. (2011). The STEM programme in England. In *Positioning technology education in the curriculum* (pp. 63–74). Rotterdam: Sense Publishers.

Barnes, S. (2010). *Principles of problem and project based learning: The Aalborg model*. Retrieved from http://www.en.aau.dk/about-aau/aalborg-model-problem-based-learning/. Accessed 13 Jan 2015.

Biggs, J., & Tang, C. (2009). *Teaching for quality learning at university*. Berkshire: McGraw-Hill.

Borat, O. (2009). *Challenges facing higher education in the globalized world*. Bahru: Invited presentation at the University President Forum for Global Academic Network. Bahru, Malaysia.

Borat, O. (2010). Improvements expected in education sector within the scope of lifelong learning. *Journal of Projects, 4*(4), 32–44.

Borat, O. (2014). HR Development within the Turkish qualifications framework. New Turkish Journal, 528–539/Türkiye Yeterlilikler Çerçevesi Kapsamında İnsan Kaynaklarının Gelişimi, Yeni Türkiye Dergisi, s.528–539, Yeni Türkiye Stratejik Araştırma Merkezi, Ankara.

Çakır, M. T., Yelmen, B. (2011). Engineering education in Turkey. *Proceedings of 2nd International Conference on new trends in education and their implications*. Ankara: Siyasal Kitabevi. Retrieved from http://iconte.org/FileUpload/ks59689/File/094..pdf. Accessed 13 Jan 2015.

Carberry, A. R., & Baker, D. R. (2018). The impact of culture on engineering and engineering education. In Y. J. Dori, Z. Mevarech, & D. Baker (Eds.), *Cognition, metacognition, and culture in STEM education* (pp. 217–239). Springer.

Chamber of Mechanical Engineers. (2013). *Üye profili araştırma raporu* [A report on member profiles] [White paper]. Ankara: Author.

Christensen, S. H., Delahousse, B., & Meganck, M. (2009). *Engineering in context*. Aarhus: Academica.

Corlu, M. S. (2013). Insights into STEM education praxis: An assessment scheme for course syllabi. *Educational Sciences: Theory & Practice, 13*(4), 2477–2485.

Çorlu, M. A. (2013). *Uzman alan öğretmeni eğitimi modeli ve görüşler* [A white paper about teacher education theory and praxis]. Retrieved from http://tstem.com/. Accessed on 13 Jan 2015.

Corlu, M. S., Capraro, R. M., & Capraro, M. M. (2014). Introducing STEM education: Implications for educating our teachers in the age of innovation. *Education in Science, 39*(171), 74–85.

Danish Ministry of Education. (2014). *Improving vocational education and training – Overview of reform of the Danish vocational education system* [White paper]. Retrieved from http://www.uvm.dk/vocational. Accessed on 13 Jan 2015.

Danish Society of Engineers [IDA]. (2009). *Uddannelsespolitik for Ingeniørforeningen, IDA* [Educational policy of the Danish Society of Engineers, IDA]. Retrieved from http://ida.dk/sites/prod.ida.dk/files/IDAs%20uddannelsespolitik%20160609.pdf. Accessed on 13 Jan 2015.

Danish Technological Institute, Centre for Policy and Business Analysis [DTI]. (2013). *Den Danske Ingeniør 2020, Jobfunktioner og Kompetencekrav* [White paper]. Retrieved from http://ida.dk/sites/prod.ida.dk/files/den_danske_ingenioer_2020_jobfunktioner_og_kompetencekrav_final_loe_sidk_0.pdf/. Accessed on 13 Jan 2015.

Ellström, P. E. (1997). The many meanings of occupational competence and qualification. *Journal of European Industrial Training, 21*(6/7), 266–273.

Engineering Council. (2017). Retrieved from https://www.engc.org.uk/informationfor/students-apprentices-and-graduates/on 19th June, 2017.

European Accreditation of Engineering Programs. (2008). EUR-ACE *framework standards for the accreditation of engineering programs* [White paper]. Retrieved from http://www.enaee.eu. Accessed 13 Jan 2015.

European Commission. (2014). *Innovation union scoreboard 2014* [White paper]. Retrieved from http://ec.europa.eu. Accessed on 13 Jan 2015.

European Federation of National Engineering Associations [FEANI]. (2010). *Educational policy* [FEANI position paper]. Retrieved from http://www.feani.org/. Accessed on 13 Jan 2015.

European Society for Engineering Education [SEFI]. (2012). Engineering education and the Bologna Process [SEFI Position Paper]. Retrieved from http://www.enaee.eu/. Accessed on 13 Jan 2015.

European Society for Engineering Education [SEFI]. (2013). *The knowledge triangle: Re-inventing the future*. Finland: Author.

Freely, J. (2013). *A history of Robert College*. İstanbul: Yapı Kredi Yayınları.

Graff, E., & Kolmos, A. (2007). *Management of change: Implementation of problem-based and project-based learning in engineering*. Rotterdam: Sense Publishers.

Graham, R. (2010). United States. Bernard M. Gordon MIT Engineering Leadership Program. *UK approaches to engineering project-based learning*. Cambridge, MA.

Graham, R., & Crawley, E. (2010). Making projects work: A review of transferable best practice approaches to engineering project-based learning in the UK. *Engineering Education, 5*(2), 41–49.

İnan, K. (2005). *Mühendislik eğitiminde yeni yaklaşımlar ve Türkiye* [New approaches to engineering education and Turkey] [White paper]. Retrieved from http://www.emo.org.tr/ekler/4492cbe90fbdbf8_ek.doc. Accessed on 13 Jan 2015.

Jensen, H. P. (2000). Quality management: Danish engineering education. *International Journal of Engineering Education, 16*, 127–135.

Jørgensen, U. (2007). Historical accounts of engineering education. In E. Crawley, J. Malmqvist, S. Östlund, & R. Brodeur (Eds.), *Rethinking engineering education: The CDIO approach* (pp. 216–240). New York: Springer.

Kaçar, M. (2007). Tersâne Hendesehânesi'nden Bahriye Mektebi'ne Mühendishâne-i Bahrî-i Hümâyûn [The Ottoman School of Naval Engineering: From 'Geometry Room' to Naval Academy]. *Osmanlı Bilimi Araştırmaları: Studies in Ottoman Science, 9*(1–2), 51–77.

Kaçar, M., Zorlu, T., Barutçu, B., Bir, A., Ceyhan, C. O., & Neftçi, A. (2012). *İstanbul Teknik Üniversitesi ve mühendislik tarihimiz* [İstanbul Technical University and our history of engineering]. İstanbul: İstanbul Teknik Üniversitesi.

Kjersdam, F., & Enemark, S. (1994). *The Aalborg experiment: Project innovation in university education*. Aalborg: Aalborg University Press.

Lundgreen, P. (2006). Engineering education in Europe and the U.S.A, 1750-1930: The rise to dominance of school culture and the engineering professions. *Annals of Science, 47*(1), 33–75.

Measuring, Selection and Placement Center [ÖSYM]. (2013). *Öğretim yılı yükseköğretim istatistikleri* [Higher education statistics] [White paper]. Retrieved from http://www.osym.gov.tr. Accessed on 13 Jan 2015.

MÜDEK. (2013). Association for Evaluation and Accreditation of Engineering Programs. Application form for accreditation agencies –Authorisation to award the EUR-ACE® label [White Paper]. Retrieved from http://www.mudek.org.tr. Accessed 13 Jan 2015.

Nesin, A. (2014). *2023 için çok geç, belki 2073* [Too late for 2023; perhaps in 2073]. *TÜSİAD Görüş*, 85. Retrieved from http://www.stemtusiad.org. Accessed on 13 Jan 2015.

NTVMSNBC. (2013). *En popüler mühendislik inşaat mühendisliği* [Civil engineering is the most popular engineering]. Retrieved from http://www.ntv.com.tr/arsiv/id/25452921. Accessed on Jan 13, 2015.

Nurdoğan, A. M. (2009). Robert Koleji mühendislik okulu [Robert College engineering school]. *Divan Disiplinlerarası Çalışmalar Dergisi, 14*(26), 203–232.

Oxford English Dictionary. (2010). *Engineer*. Oxford: Oxford University Press.

Quality Board. (2017). Higher Education Quality Board. Law No.7033. Official Gazette date: 1 July 2017, No.: 30111, Ankara.

Ramskov, J., (2012). Sådan vandt universiteterne kampen om at uddanne ingeniører [This is how universities won the quest of educating engineers] [White paper]. Retrieved from http://ing.dk/artikel/sadan-vandt-universiteterne-kampen-om-uddanne-ingeniorer-128561. Accessed on 13 Jan 2015.

Royal Academy of Engineering. (2017). Engineering an academy that works for all. London.

Sevindik, T., & Akpınar, B. (2007). The effects of the changes in postmodern pedagogical paradigms on engineering education in Turkey. *European Journal of Engineering Education, 32*(5), 561–571.

Sjöström, J., & Eilks, I. (2018). Reconsidering different visions of scientific literacy and science education based on the concept of *Bildung*. In Y. J. Dori, Z. Mevarech, & D. Baker (Eds.), *Cognition, metacognition, and culture in STEM education* (pp. 65–88). Springer.

Smith, A. E., & Dengiz, B. (2010). Women in engineering in Turkey: A large scale quantitative and qualitative examination. *European Journal of Engineering Education, 35*(1), 45–57.

Tantekin-Ersolmaz, S. B., Ekinci, E., & Sağlamer, G. (2012). Engineering education and practice in Turkey. *Technology and Society Magazine, 25*(2), 26–35.

URAS. (2017). University Ranking by Academic Performence (URAS) in engineering area. Retrieved from http://tr.urapcenter.org/2011/2011_7_t8.php. Accessed on 25 Aug 2017.

U.S. Congress Joint Economic Committee. (2012). *STEM education: Preparing for the jobs of the future* [White paper]. Retrieved from http://www.jec.senate.gov/. Accessed on 13 Jan 2015.

Varol, K., Borat, O., & Ceviz, E. (2013). Challenges facing higher education in the globalized world. Istanbul Commerce University Journal of Sciences, 12(23) 47–53/Küreselleşen Dünyada Yükseköğrenimde Karşılaşılan Güçlükler. *İTİCÜ Fen Bilimleri Dergisi, 12*(23) 47–53.

Wenger, E. (1998). *Communities of practice: Learning, meaning, and identity*. Cambridge: Cambridge University Press.

Women's Engineering Society. (2014). *Statistics on women in engineering* [White paper]. Retrieved from http://www.wes.org.uk. Accessed on 13 Jan 2015.

Yetkiner, E. Z. (2010). *Achievement and opportunity gaps in mathematics education in turkey compared to european union countries*. Unpublished doctoral dissertation, Texas A&M University, College Station.

Zengin, B. (2000). Women engineers in Turkey: Gender, education and professional life: A case study on METU. Unpublished master's thesis, Middle East Technical University, Ankara, Turkey.

Zengin-Arslan, B. (2002). Women in engineering education in Turkey: Understanding the gendered distribution. *International Journal of Engineering Education, 18*(4), 400–408.

Chapter 12
Cognition, Metacognition, and Mathematics Literacy

Zemira R. Mevarech and Lianghuo Fan

12.1 Introdution

The twenty-first century is characterized as the information era. As such, it emphasizes the importance of enhancing students' literacy at all levels of education from kindergarten to the end of high school. This is in contrast to previous years in which skills and the carrying out of algorithms were considered as the main objectives of mathematics education.

Furthermore, in the last decade, the concept of "literacy" has been widely broaden from "ability to read and write" (Oxford English Dictionary 1995) to include specific literacy domains, such as mathematics, science, engineering, and technology (Programme for International Student Assessment [PISA] 2003).

The present chapter focuses on mathematics literacy and how it can be promoted from the perspectives of both Western and Eastern educational systems. The chapter opens with an introduction that includes the definition of mathematics literacy and how it relates to deeper learning, meaningful learning, and problem-solving. The second section briefly reviews cognitive-metacognitive pedagogies and their effects on mathematical reasoning as found in Western and Eastern countries. In the third part of the chapter, we offer examples of a learning environment that promotes students' math literacy. Finally, in the last section, we briefly discuss the implications of the research reviewed throughout the chapter for teachers, researchers, and policy makers.

Z.R. Mevarech (✉)
David Yellin Academic College of Education, Bar-Ilan University, Jerusalem, Israel
e-mail: zemiram@dyellin.ac.il

L. Fan
University of Southampton, Southampton, UK

Y.J. Dori et al. (eds.), *Cognition, Metacognition, and Culture in STEM Education*, Innovations in Science Education and Technology 24,
https://doi.org/10.1007/978-3-319-66659-4_12

12.2 Mathematics Literacy: Definition

Mathematics literacy was defined by PISA as "…the capacity to identify, understand and engage in mathematics as well as to make well-founded judgments about the role that mathematics plays in an individual's current and future life as a constructive, concerned, and reflective citizen" (PISA 2003, p. 20). In 2012, when mathematics literacy was PISA main domain, the concept of mathematics literacy was further elaborated as follows:

> Mathematical literacy is an individual's capacity to formulate, employ, and interpret mathematics in a variety of contexts. It includes reasoning mathematically and using mathematical concepts, procedures, facts and tools to describe, explain, and predict phenomena. It assists individuals to recognize the role that mathematics plays in the world and to make the well-founded judgments and decisions needed by constructive, engaged and reflective citizens. (PISA 2012, p. 25)

Thus, the concept of "mathematics literacy" revolves around the *use* of mathematics in people life, rather than rote recall of mathematical facts (e.g., recalling the multiplication table) or carrying out computations and ready-made algorithms (e.g., solving equations of degree 3). PISA (2003) further emphasizes that "mathematics literacy implies not only the ability to pose and solve mathematical problems in a variety of situations but also the inclination to do so, a quality that often relies on personal traits such as self-confidence and curiosity" (PISA 2003, p. 20).

Mathematics literacy closely relates to "deeper learning" (Pellegrino and Hilton 2012), "meaningful learning" (Novak 2002), and "problem-solving" that has been a central theme in mathematics education research over the last few decades (e.g., Foong 2009; Koichu 2014; Schoenfeld 1992, 2007; Stanic and Kilpatrick 1988). Pellegrino and Hilton (2012) define "deeper learning" as "…the process of developing durable, transferable knowledge that can be applied to new situations" (p. 69). Wikipedia elaborates on this definition, indicating "deeper learning is a term that describes a set of student educational outcomes including acquisition of robust core academic content, higher-order thinking skills, and learning dispositions. It is associated with a growing movement in US education that places special emphasis on the ability to apply knowledge to real-world circumstances and to solve novel problems. Deeper learning is based on the premise that the nature of work, civic, and everyday life is changing and therefore increasingly requires that formal education provides young people with mastery of skills like analytical reasoning, complex problem solving, and teamwork."

The concept "deeper learning" has been evolved from "meaningful learning" (Novak 2002) which refers to students' ability to actively seek and integrate new knowledge with knowledge already in their cognitive structure and apply critical thinking. Several instructional methods have been suggested for enhancing meaningful learning; among them are the construction of cognitive maps, the use of advanced organizers, and the implementation of metacognitive scaffolding. Pólya's well-known book *How to Solve It* indicates that a successful math problem-solving process involves four steps: understanding the problem, devising a plan, carrying out the plan, and looking back (Polya 1957). For Polya, and in most literature on this topic – a "real" problem means an unfamiliar situation which requires a solution or answer but with no readily available means to an individual who needs to solve it (Kilpatrick 1985; Zhu and Fan 2006).

It is interesting to note that also in Shanghai, whose students were well known for being higher achievers in mathematics (PISA 2009, 2012), "mathematics literacy," or *Shu Xue Su Yang* in Chinese, is defined in a similar way. Accordingly, mathematics literacy is "the sum of all basic mathematics knowledge, basic skills, mathematics ideas, and perceptions that people gained through mathematics education and their own practical and cognitive activities, and through which, the characteristics of mathematics thinking and problem solving abilities." The Shanghai school curriculum emphasizes that "mathematics is an important component of modern culture" and "mathematics literacy is an essential literacy for modern citizens," and for which, mathematics education needs to help students "learn how to learn and learn how to think" (Shanghai Municipal Education Commission 2005, p. 1).

Similar to reading and science literacy, also mathematics literacy generally refers to three broad dimensions: content, process, and context. Each of these dimensions takes into consideration the other two dimensions to provide a comprehensive understanding of what mathematics literacy is. The following sections elaborate these aspects with regard to Western and Eastern educational systems.

12.2.1 Mathematics Content Knowledge

What kinds of mathematical knowledge are important for citizens in the modern world? Obviously, it is impossible to include all mathematical aspects in the school curriculum. On the one hand, we do not want to miss any mathematical phenomena that underlie many occurring situations and in which the mathematical "big ideas" are intertwined. On the other hand, it is impossible to refer to all mathematical concepts not only because we have to take into consideration the student's traits but also because of lack of time. Since mathematics literacy focuses on the use of mathematics, the content refers to clusters of relevant connected mathematical concepts that appear in real situations and contexts. These include four strands that provide the basis for large-scale assessment (PISA 2003):

- Change and relationships
- Space and shape
- Quantity
- Uncertainty and data analysis

Change and Relationships Our world is characterized by discrete and continuous changes and multitude relationships between objects, domains, people, and circumstances. Thus, understanding the different types of changes and relationships, being able to construct mathematical models that describe them, and developing the capabilities to predict them are essential parts of mathematically literate people.

Space and Shape are encountered all over. We come across patterns, objects, shapes, and visual information, when they are static or dynamic. Often, we are required to navigate, read maps, construct figures, and be able to handle three-dimensional objects. Students and adults do not have only to identify and describe shapes by using different representations but also to understand their properties and apply geometrical reasoning in different contexts.

Quantity includes number sense, understanding of measurements, magnitudes, units, and indicators. It also involves multiple representations of numbers, computations, estimations, and assessments. While in ancient culture, quantity includes only three units, one-two-many, in our modern cultures, quantity refers to continuous numbers, infinity, positive and negative numbers, as well as fractions and whole numbers (PISA 2013). In fact, although there have been different views about the nature of mathematics (e.g., Harel 2008; Stenlund 2014), mathematics was for a long time commonly defined as the science of quantity (Lenhard 2004). One cannot imagine our modern culture without referring to quantity.

Uncertainty and Data Analysis are at the heart of the twenty-first century. We encounter uncertainty in everyday life, science, and many problem-solving situations. Frequently, our decisions are based on evidence and data that take into consideration variations, probabilities, and type I and type II errors. Data analyses also relate to the way we represent the findings and communicate the results. Statistics has become a common language in newspapers, sport, governmental reports, polls, medicine, sciences, technology, and survey findings. The use of this language is quite new among "ordinary" citizens who are not professionals in these areas.

12.2.2 Mathematical Processes

Since mathematical literacy refers to the *use* of mathematics in various situations, describing the mathematical contents has to be followed by the processes that are employed in solving the problems. According to PISA (2012), the processes include:

- *Formulating* situations mathematically, including identifying the mathematical aspects of phenomena and problems and translating it into formal mathematical language
- *Employing* mathematical concepts, facts, procedures, and reasoning in solving mathematical tasks, making generalizations, and reflecting on mathematical arguments, explaining them, and justifying one's mathematical reasoning
- *Interpreting*, applying, and evaluating mathematical outcomes, such as inferring mathematical results back into the real world, judging the reasonableness of a mathematical solution in the real world, evaluating the model used to solve a problem and identifying its strengths and limits (PISA 2012).

12.2.3 Contexts

Mathematics, the queen of science, refers to a broad range of contexts. The contexts can refer to personal, occupational, societal, scientific, engineering, and technological settings or circumstances in which the situations need to be understood, analyzed, or solved mathematically using relevant mathematical knowledge, thinking skills and processes, and problem-solving abilities.

Another way to analyze the "context" is by referring to the "distance" of the situations from individuals – from those relating directly to the individual (e.g., comparing achievement scores) to technology problems of more general interest. Contexts, especially those relating directly to the individuals, are often related to where the individuals live, work, and study.

12.2.4 Mathematics Literacy and CUN Tasks

At the core of mathematics literacy are complex, unfamiliar, and non-routine (CUN) tasks that are the essence for teaching mathematics in innovative societies (Mevarech and Kramarski 2014). Yet, similar to what the studies on problem-solving have revealed (e.g., Kilpatrick 1985; Powell et al. 2009), the broad range of CUN tasks also raises questions regarding its practical implications: what is complex to one student might be simple to another; similarly, what is unfamiliar in one context can become familiar in another one; and what is non-routine at a certain learning stage might become a routine procedure after some practice.

Below are some examples of CUN task vs. routine-textbook tasks that can help clarify our understanding of CUN tasks. The first examples are provided in the book *Critical Maths for Innovative Societies: The role of metacognitive pedagogy* by Mevarech and Kramarski (2014, p. 26), and the last one is cited from Fan (2011a, pp. 28–37).

Example 1 The Supermarket Task
Before the holiday, several supermarkets advertised that they are the cheapest supermarket in town. Please collect data and find out which advertisement is correct. Please prepare a 60-min TV show to present your findings.

Example 2 A Sale Task
In supermarket A, 1 kg of meat costs EUR 8 and 1 kg of poultry costs EUR 4. In supermarket B, 1 kg of meat costs EUR 7 and 1 kg of poultry costs EUR 5. Mr. Jonson wants to buy 3 kg of meat and 2 kg of poultry.

Which supermarket is cheaper?

Is the "supermarket" a math task even though it does not include any number in it? Is it a CUN task? Clearly, this task is based on mathematical skills. It can be administered to first graders as well as to business administration students. It is complex, unfamiliar, and non-routine because it might have different solutions depending on the items chosen to be included in the analyses. It has no ready-made algorithms for solution and requires in addition to various mathematical knowledge also acquaintance with the situation, communication skills, and math creativity. Undoubtedly, the "supermarket" task is a mathematics literacy task as defined by both PISA (2003, 2009, 2012) and Shanghai Municipality Education Commission (2005).

In contrast, Example 2, the "sale" task is a typical textbook problem. It includes all the needed information, and it is based on ready-made algorithms for solution. To solve the sale' task, students need to perform the calculations, compare the two prices, and decide which is smaller.

The following example is from Singapore Mathematics Assessment Project (Fan 2011a). It shows how the CUN ideas are similarly reflected in Singapore school mathematics and embedded in Singapore context.

A River Cruise Task

Mrs. Lim intends to take her students for a river cruise on the Singapore River. There are two types of charges: The big boat, which can carry 6 people, charges a fare of $10 per boat, and the small boat, which can carry 4 people, charges a fare of $8 per boat. If there are 50 students in Mrs. Lim's class, what are all the possible options of renting the boats under each of the following conditions: (i) with a minimum cost? (ii) with a minimum number of boats? and (iii) with a minimum number of empty seats? From your solution to the question above, choose an option of renting the boats that you think is the best. Give your reason clearly.

More detailed information about this task and how it could be implemented in classroom can be found in Fan (2011a, pp. 28–37).

Over the last decade or so, CUN tasks have been a prominent feature in Singapore classroom-based research and practice in mathematics, which emphasized more on real-life contexts, disciplinarily in mathematics, and metacognition and self-reflection. For more such tasks in Singapore contexts, readers can refer to Fan (2011a, b), Fan et al. (2010), Foong (2009), and Wong et al. (2012).

To summarize the value of CUN tasks in mathematics teaching and learning, we would like to cite Mevarech and Kramarski (2014):

> Dramatic changes in our understanding of the nature of learning over the past century have resulted in a shift in focus from "what" to "how", and in particular, how to enhance students' abilities to solve complex, unfamiliar, and non-routine tasks (CUN). While these CUN problems allow students to develop skills needed in societies driven by innovation, most mathematics textbooks and teaching still focus on problem based on application of ready-made algorithms. CUN problems should become more central to mathematic education and examines innovative instructional methods that enhance mathematics education, notably metacognitive and cooperative pedagogies. (p. 15)

12.3 Metacognition and Mathematics Education

Exposing students to CUN tasks is a necessary condition for helping students to become math literate, but it is not sufficient. In order to solve CUN tasks, one has to apply a "higher-order thinking program" that plans the solution, regulates it, and reflects on all the stages from the very beginning to the very end. Flavell (1979) coined this process metacognition to emphasize its "meta" nature, namely, it is "beyond" the cognition. The metacognitive "program" receives information from the object level, processes it, debugs errors (when errors are identified), evaluates the execution, and provides information back to the object level for further elaboration (Nelson and Narens 1990; Schoenfeld 1987). A good metaphor for metacognition is GPS: it plans the route, controls and monitors the drive, recalculates a new route when an error occurs, and leads the driver all the way until s/he reaches the final destination.

Tens of studies have indicated positive relationships between metacognition and school achievements in general and the solution of CUN tasks in particular (Mevarech and Kramarski 2014). The positive correlations have been reported at all levels of education from kindergarten (Schneider 1998; Shamir et al. 2009; Whitebread 1999; Whitebread and Coltman 2010) through elementary school (Mevarech et al. 2010), high school (Mevarech and Amrany 2008; Veenman 2013; Veenman and Spaans 2005), and higher education (Mevarech and Fridkin 2006; Schraw and Dennison 1994). This is not surprising: a solver who plans ahead, regulates the solution, and reflects on it is more likely to solve the task correctly than a solver who does not activate these processes. Interestingly, while the implementation of metacognition is crucial for solving CUN tasks, one can solve a simple-routine task "automatically," by applying a ready-made algorithm, without referring to metacognitive processes (Mevarech and Kramarski 2014; Stillman and Mevarech 2010).

12.3.1 *Metacognitive Pedagogy*

Given these studies led researchers to design metacognitive pedagogies that aim at enhancing students' ability to plan, regulate, and reflect on the tasks and its solution. A widely used metacognitive teaching method is IMPROVE (Mevarech and Kramarski 1997). It is routed in three paradigms: cognitive-metacognitive theories, cooperative learning, and feedback-corrective procedure. IMPROVE is an acronym of all the teaching steps:

Introducing the new materials to the whole class by modeling the metacognitive questioning
Metacognitive questioning in small groups
Practicing by using the metacognitive questioning

Reviewing by using the metacognitive questioning
Obtaining mastery on lower and higher cognitive processes
Verification
Enrichment and remedial

The metacognitive questioning is at the core of the method. It includes four self-guiding questions based on Polya (1957) and Schoenfeld (1985) studies:

(a) Comprehension: What is the problem all about?
(b) Bridging: How is the problem at hand similar to or different from the problems you have solved in the past? Please explain your reasoning.
(c) Strategies: What strategies are appropriate for solving the problem? Please explain your reasoning.
(d) Reflection: Does the solution make sense? Are you stuck? Why?

IMRPOVE is a generic method. It was designed for mathematics but quite often is used also in science (Mevarech et al. 2014; Zion et al. 2015; Zion et al. 2005) or technology (Choresh et al. 2009). In these cases, small modifications in the metacognitive questioning are needed in order to fit the method to the specific domain (for more details, the readers are referred to Mevarech and Kramarski 2014).

IMRPOVE has been implemented in K-12 and higher education and was found to be very effective, particularly for solving CUN tasks and enhancing math reasoning and creativity. One of the first studies that implemented IMPROVE was conducted by Mevarech and Kramarski (1997). In this study, a random sample of eighth graders was exposed to IMPROVE over a full academic year, and their mathematics achievement and reasoning were compared to a control group who studied in a traditional manner. The IMPROVE students significantly outperformed the control group not only on mathematics achievement but also on measures of creativity (e.g., flexibility and originality). Interestingly, the effects of IMPROVE were maintained also on a delayed test that was administered a year after the students were exposed to IMPROVE (Mevarech and Kramarski 2003) and in other contexts, such as being tested on the matriculation exams (Mevarech and Amrany 2008) or through online interactions (Zion et al. 2015). Furthermore, lower achievers benefitted from IMPROVE, but not at the expense of higher achievers (Mevarech 1999). Using a meta-analysis technique to evaluate the effects of metacognitive pedagogies, Dignath and Buettner (2008) indicate that these methods are more effective in mathematics than in reading and writing and other domains (effect sizes were 0.96, 0.44, and 0.64 standard deviations, respectively).

Several studies examined the conditions under which the metacognitive pedagogy is effective. In particular, students who studied in ICT (information, communication, and technology) environments seem to be in need for metacognitive guidance because the very nature of the medium enables or even encourages the use of "trial and error" without reflecting on the procedure. Comparing ICT environments in which metacognitive pedagogy was implemented to ICT with no metacognitive guidance showed the positive contributions of the metacognitive

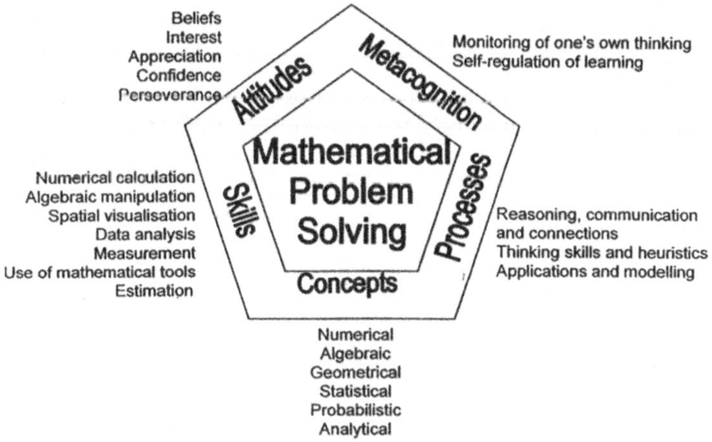

Fig. 12.1 The pentagon framework

scaffolding on schooling outcomes. Another issue relates to the relative efficiency of metacognition embedded in ICT vs. face to face. Zion et al. (2005) compared ICT to face-to-face (F2F) learning, each of which with or with no metacognitive scaffolding implemented via IMPROVE. This study indicates that students exposed to ICT with metacognitive scaffolding significantly outperformed all other groups and students who studied F2F with no metacognitive scaffolding received the lowest mean scores. Insignificant differences were found, however, between ICT with no metacognitive scaffolding and F2F with metacognitive scaffolding. Probably, the very fact that the interaction in ICT environments takes place mainly in writing led students to activate metacognitive processes spontaneously even though they were not guided to do so. For more information about IMPROVE and its impact, see Mevarech and Kramarski (2014).

In Singapore, IMPROVE principles are similarly reflected and implemented in the national mathematics curriculum. The following well-known pentagon framework illustrates the way mathematics is taught in this country (MOE 2012). Figure 12.1 presents the pentagon framework.

According to the Ministry of Education (MOE) of Singapore, this framework sets the direction and implemented teaching, learning, and assessment of mathematics at all levels of school education, and it also reflects the 21st competencies; stresses conceptual understanding, skills proficiency, and mathematics processes; and gives due emphasis to attitudes and metacognition (MOE 2012). Thus, teachers need to pay careful and explicit attention in their teaching in order to develop students' habit and skills of metacognition and self-reflection.

The following is a list of prompts for teachers in the Singapore Mathematics Assessment Project aforesaid to promote students' metacognition and self-reflection in different pedagogical scenarios, which teachers found helpful (Fan 2011b).

Prompts for Teaching Problem-Solving

Prompt 1. Where did you encounter difficulties? Why? (Scenario 1: after students did not know how to start or proceed in solving a problem)
Prompt 2. Where did you go wrong? Why? (Scenario 2: after students realized that he/she got a wrong solution or answer)
Prompt 3. Is the mistake a careless mistake? If not, why did you make the mistake? (Scenario 3: after students realized that he/she made a mistake)
Prompt 4. Are you sure your answer/solution is correct? Did you check? (Scenario 4: after students solved a problem or finished a task)
Prompt 5. Have you solved this kind of problems before? Does the problem look familiar to you? (Scenario 5: when students encounter difficulty in solving a problem, which appears to be essentially not new to him/her in his/her learning)
Prompt 6. What have you learned from solving this problem? (Scenario 6: after students have gone through an important or difficult problem)
Prompt 7. If you are given another problem like this, will you have confidence to solve it? (Scenario 7: after students solved a problem in a correct way)

Prompts for Other Teaching Scenarios

Prompt 8. What did you feel most difficult in learning this chapter (or topic, or lesson, or task, etc.)? Is it still difficult to you? (Scenario 8: after students finished learning a chapter, a topic, a lesson, a concept, etc.)
Prompt 9. Do you have any questions or difficulties to ask? (Scenario 9: when teacher prepares to close his/her teaching for chapter, a topic, a lesson, a section, a task, a problem, etc. and moves to next phase)
Prompt 10. How do you feel about your learning of mathematics recently? Do you think if you can improve your learning? How? (Scenario 10: when teacher realized that students might have problems recently in learning mathematics)

12.3.2 Reading Comprehension, Mathematics Literacy, and Metacognition

It is widely known that knowing how to read a math problem is fundamental to its solution. Given that math literacy tasks include a large portion of verbal descriptions, there is reason to suppose that without comprehending the text, the solvers would not be able to understand what the problem is all about and how to approach it. For example, if the task describes a shopping list and calls for adding the list of prices accurately, the solver would need to become proficient in using the addition procedure (with or without use of a calculator). For such a task, the solver would not need metacognitive scaffoldings that would allow the transfer of knowledge to new situations. Indeed, research has shown that metacognitive pedagogies are less needed in solving "routine tasks" (Mevarech and Kramarski 2014). Contrary to this, to solve CUN tasks, comprehending the texts is not sufficient. In these situations, the learner has to be engaged in planning (setting learning goals or prepare the solution outline), monitoring (keeping track of progress in solving the task), control (using, managing, or changing strategies to solve the task), and reflection (prior to, during, and after solving the task) (Pintrich 2004). By activating these processes, he or she may be able to regulate the solution rather than relying solely on ready-made algorithms. Thus, as Pellegrino and Hilton (2012) pointed out: "when the goal is to prepare students to be able to be successful in solving new problems and adapting to new situations, then deeper learning is called for" (p. 70). Acquiring metacognitive skills is essential for attaining this goal.

12.4 The Effects of IMPROVE on Students' Mathematics Literacy

To exemplify how metacognitive pedagogy could enhance mathematics literacy, we describe here a quasi-experimental study based on a quantitative method.

In this study, 71 tenth-grade students participated. The study was implemented in cooperation with Berger Irit and Madmony Yaniv, graduate students in the school of Education at bar-Ilan University, Israel. Participants were "typical" high-school students who studied in three classrooms with their regular math teachers. They all studied mathematics five times a week, out of which one period was devoted to the enhancement of mathematics literacy. In this period, the teachers provided "mathematics literacy tasks" similar to those used in PISA (2003) in order to familiarize the students with the new types of tasks to which they were not exposed before. In the other four periods, students solved "regular" math problems taken from the textbook. All teachers were certified math teachers and all had more than 10 years of experience in teaching mathematics.

For the purposes of the study, intact classrooms were randomly assigned into an experimental group ($N = 48$) and a control group ($N = 23$). Students in all classrooms, both in the experimental and control groups, solved the same math literacy tasks that were developed by the Israel Ministry of Education on the basis of PISA examinations.

In the experimental group, students were exposed to IMPROVE, i.e., they were trained to activate metacognitive processes. Students were provided with index cards on which the four metacognitive questions were printed. They were encouraged to use these questions while solving the tasks. At the beginning of the class, the teacher demonstrated how to solve such tasks by using the metacognitive questions; the teacher did so also at the end of the class during the review of the tasks' solutions. In the control group, students solved the same "math literacy" tasks. At the beginning of the class, the teacher demonstrated the solution of a task, and the students solved the "literacy" tasks as they used to solve the "regular" math tasks, without being explicitly exposed to metacognitive scaffolding; at the end of the class, the teacher reviewed the solutions with the students, as they did "regularly" without using the metacognitive scaffolding. To eliminate possible Hawthorne effects, both groups were told that they participate in a study in which mathematics literacy tasks are introduced for the first time. Since the study was conducted after the publication of PISA findings, all students and their teachers were aware of the importance of mathematics literacy and were motivated to participate in the study.

A ten-item examination was administered to all students prior to the beginning of the study. For the sake of simplicity, all scores are reported in terms of percent correct items. Factor analysis of the pretest indicated two factors, one included knowledge and computations (lower mental processes) and the other applications, synthesis, and analysis (higher mental processes) tasks. Multilevel analysis of variance (MANOVA) indicated no significant differences between the groups on the two pretest factors simultaneously ($F < 1.00$, $p > 0.05$) and on each factor separately: for lower mental processes, $M = 60$ and 62 and $SD = 27.5$ and 27.0 for the experimental and control group, respectively; for higher mental processes, $M = 69$ and 71 and $SD = 25$ and 29 for the experimental and control groups, respectively.

After the pretest, all classrooms started to solve mathematics literacy tasks, each classroom according to the intervention to which it was assigned. As indicated above, the experimental group studied via IMPROVE and the control group studied traditionally, without being explicitly exposed to metacognitive scaffolding.

At the end of the semester, all students were administered an eight-item posttest that covered both lower and higher mental processes. The lower mental processes (four items) referred to knowledge and computations, whereas the higher mental processes assessed application, analysis, and synthesis. Factor analysis revealed these two factors.

While no significant differences were found between the groups on the pretest, MANCOVA indicated significant differences on both factors simultaneously at the end of the study, even after controlling the pretest scores ($F = 7.92$, $p < 0.001$). The significant differences were found on both the lower mental processes factor ($F = 12.67$, $p < 0.001$) and the higher mental processes factor ($F = 6.75$, $p < 0.001$). The IMPROVE students outperformed the control group on lower mental processes ($M = 78.1$ and 46.5 and SD = 0.20 and 0.32, respectively) and on higher mental processes ($M = 69.5$ and 50.5 and $SD = 0.22$ and 0.26, respectively). Figures 12.2 and 12.3 present the mean scores by time and group on the total mathematics literacy task (Fig. 12.2) and on lower mental processes (LMP) and higher mental processes (HMP) (Fig. 12.3).

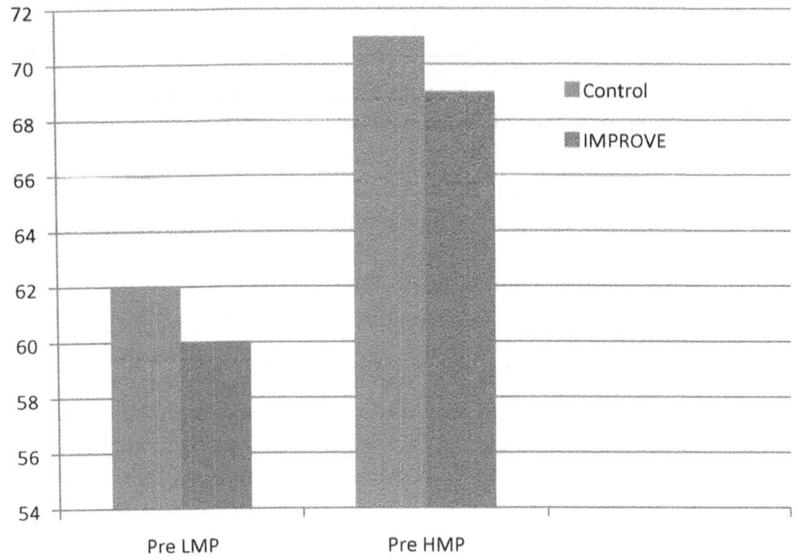

Fig. 12.2 Mean scores on total mathematics literacy (ML) scores by time and group

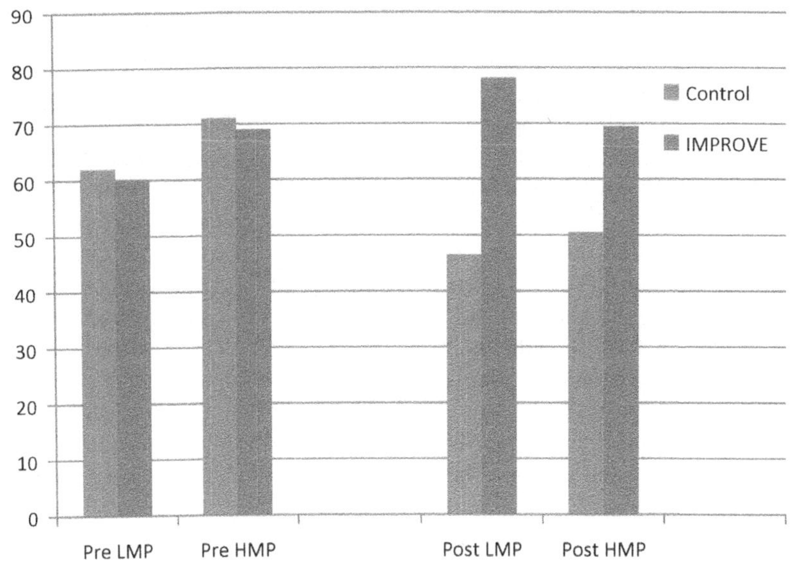

Fig. 12.3 Mean scores on lower mental processes (LMP) and higher mental processes (HMP) by time and group

12.5 Discussion

The studies reviewed above show that under certain conditions, metacognition is teachable. Research indicates that explicit teaching and intensive practicing of metacognitive processes are fundamental components in metacognitive pedagogies. Under these conditions, students at all levels of education, from kindergartens to higher education, benefit from metacognitive scaffolding implemented via IMPROVE or other metacognitive pedagogies (Mevarech and Kramarski 2014). The positive effects of metacognitive scaffolding are evident in Western and Eastern countries, in ICT and non-ICT environments, and on lower or higher achievers. In particular, IMPROVE advances students' math literacy as indicated in the quasi-experimental study reviewed above. Yet, classroom observations revealed that many teachers implement metacognitive methods implicitly and sporadically, a way that turned out to be ineffective. Similar conclusions have been found also in science education (Azevedo and Aleven 2013; Zohar and Dori 2012) and technology education (Choresh et al. 2009).

There are various ways for fostering metacognition in STEM education, including posing questions while reading scientific texts (Herscovits et al. 2012), embedding specific metacognitive instruction at different levels of the solution/science reading process (Kapa 2001; Michalsky et al. 2009), or providing students with the opportunities to cooperate with their peers during learning (Mevarech and Kramarski 2014). Also cyber-learning environments and the use of networked learning technologies have the potential to scaffold metacognition by allowing designers to create support for both individual and team metacognition (i.e., self-regulation and socially shared regulation, respectively) (Crippen and Antonenko, Chap. 5 in this book).

Yet, along the successes, there are some drawbacks. First, teachers who used metacognitive pedagogy for the first time complained that they would not be able to cover the curriculum because the implementation of the method takes more time than "regular" teaching. (It should be noted that after using the method for a while, the teachers realize that this complain is not valid anymore.) Others claim that mathematics has to deal with numbers and solutions per se, while under the metacognitive pedagogy, solvers have to verbalize their thinking. Still others declare that they use metacognitive cues in their teaching, although they do it implicitly and not systematically (see above). Nevertheless, our experience shows that most teachers, who were skeptical with regard to metacognitive pedagogy, changed their attitude after implementing the method and seeing its benefits for students' mathematical reasoning.

12.6 Recommendations

Recommendations emerging from the studies reviewed in this chapter have significant implications for mathematics education in the twenty-first century. The recommendations apply to researchers, teachers, and policy makers.

12.6.1 Recommendations for Researchers

- Implementing metacognitive pedagogies in mathematics classrooms enhances students' mathematics literacy which is at the core of mathematics education in Western and Eastern countries. Researchers are called to design math literacy tasks and study how students at different age levels approach this type of tasks.
- Singapore embedded metacognition in its national curriculum. This top-down approach proved to be efficient in enhancing students' mathematics literacy. Assessing the effects of metacognitive pedagogy on a macro level is definitely needed.
- Metacognition is not one entity. Its various components raise the necessity to develop teaching methods that would focus on specific components of metacognition rather than on the issue as a whole.
- Also mathematics is not one entity. The research on metacognition in mathematics has flourished, but mainly in the area of algebra. Very little is known at present on the effects of metacognitive pedagogies on advanced mathematics topics, such as calculus or topology.
- International collaborations on the development of mathematical literacy tasks and the appropriate metacognitive scaffolding might ease the change process that is indeed needed in mathematics education for the twenty-first century.

12.6.2 Recommendations for teachers and policy makers

- Given that "the quality of a country's educational system cannot exceed the quality of its teachers" (McKinsey Report 2007), there is a need for in-service and preservice professional development programs that focus on math literacy and metacognition. There are plenty of studies showing how to design such courses. For example, Kohen and Kramarski (Chap. 13 in this book, 2018) describe a case study showing the changes in pedagogical content knowledge of two student-teachers, one exposed IMPROVE and the other studied "traditionally." Generally, when the trainees get first-hand experience of the method and its effects, as in IMPROVE and other metacognitive pedagogies, it increases the likelihood of classroom implementations.
- An essential factor that affects mathematics teaching is the textbook. Embedding metacognitive scaffolding and mathematics literacy tasks in textbooks might be beneficial for encouraging teachers to implement metacognitive scaffolding in their attempts to develop mathematics literate students.
- Tests and assessments guide the teachers' work. Therefore, including metacognitive components in classroom assessments would enhance students' implementations of these processes. This recommendation is supported by Wengrowicz, Dori, and Dori (Chap. 9 in this book, 2018) who discuss the advantages of incorporating metacognition and meta-assessment in engineering education, particularly in large undergraduate and small graduate courses.

- Policy makers could develop national curriculum and stakeholder assessments that emphasize the use of metacognition, similarly to the Singapore model.
- Finally, globalization allows teachers to share ideas as in Erasmus projects. This opens new horizons for teachers and policy makers from Western and Eastern countries in attempts to promote students' mathematics literacy.

References

Azevedo, R., & Aleven, V. (Eds.). (2013). *International handbook of metacognition and learning technologies*. New York: Springer.

Choresh, C., Mevarech, Z. R., & Frank, M. (2009). Developing argumentation ability as a way to promote technological literacy. *International Journal of Educational Research, 48*(2), 225–234.

Dignath, C., & Buettner, G. (2008). Components of fostering self-regulated learning among students: A meta-analysis on intervention studies at primary and secondary school level. *Metacognition Learning, 3*(3), 231–264.

Fan, L. (Ed.). (2011a). *Performance assessment in mathematics: Concepts, methods, and examples from research and practice in Singapore classrooms*. Singapore: Pearson.

Fan, L. (2011b). Implementing self-assessment to develop reflective teaching and learning in mathematics. In B. Kaur & K. Y. Wong (Eds.), *Assessment in the mathematics classroom* (pp. 275–290). Singapore: World Scientific.

Fan, L., Zhao, D., Cheang, W. K., Teo, K. M., & Ling, P. Y. (2010). Developing disciplinary tasks to improve mathematics assessment and pedagogy: An exploratory study in Singapore schools. *Procedia – Social and Behavioral Sciences, 2*(2), 2000–2005.

Flavell, J. H. (1979). Metacognition and cognitive monitoring: A new area of cognitive development inquiry. *American Psychologist, 34*(10), 906–911.

Foong, P. Y. (2009). Review of research on mathematical problem solving in Singapore. In K. Y. Wong, P. Y. Lee, B. Kaur, P. Y. Foong, & S. F. Ng (Eds.), *Mathematics education: The Singapore journey* (pp. 263–300). Singapore: World Scientific.

Harel, G. (2008). What is mathematics? A pedagogical answer to a philosophical question. In B. Gold & R. Simons (Eds.), *Proof and other dilemmas: Mathematics and philosophy* (pp. 265–290). Washington, DC: Mathematical Association of America.

Herscovits, O., Kaberman, Z., Saar, L., & Dori, Y. J. (2012). The relationship between metacognition and the ability to pose questions in chemical education. In A. Zohar & Y. J. Dori (Eds.), *Metacognition in science education: Trends in current research* (pp. 165–195). Dordrecht: Springer.

Kapa, E. (2001). A metacognitive support during the process of problem solving in a computerized environment. *Educational Studies in Mathematics, 47*(3), 317–336.

Kilpatrick, J. (1985). A retrospective account of the past twenty-five years of research on teaching mathematical problem solving. In E. A. Silver (Ed.), *Teaching and learning mathematical problem solving: Multiple research perspectives* (pp. 1–15). Hillsdale: Lawrence Erlbaum Associates.

Kohen, Z., & Kramarski, B. (2018). Promoting mathematics' teachers pedagogical metacognition – A theoretical practical model and case study. In Y. J. Dori, Z. Mevareach, & D. Bake (Eds.), *Cognition, metacognition and culture in STEM education* (pp. 279–305). Springer.

Koichu, B. (2014) (with contributions by Gerald Goldin, Izzie Weinzweig, Shlomo Vinner and Roza Leikin). Reflections on problem solving. In M. N. Fried & T. Dreyfus (Eds.), *Mathematics and mathematics education: Searching for common ground. Advances in mathematics education* (pp. 113–135). Dordrecht: Springer.

Lenhard, J. (2004). Scepticism and mathematization: Pascal and Peirce on mathematical epistemology. *Philosophica, 74*(2), 85–102.

Mevarech, Z. R. (1999). Effects of metacognitive training embedded in cooperative settings on mathematical problem solving. *Journal of Educational Research, 92*(4), 195–205.

Mevarech, Z. R., & Amrany, C. (2008). Immediate and delayed effect of metacognitive instruction on regulation of cognition and mathematics achievement. *Metacognition Learning, 3*(2), 147–157.

Mevarech, Z. R., & Fridkin, S. (2006). The effects of IMPROVE on mathematical knowledge, mathematical reasoning and metacognition. *Metacognition Learning, 1*(1), 85–97.

Mevarech, Z. R., Halperin, C., Vaserman, S.. (2014). *Reading scientigic texts: the effects of metacognitive scaffolding on students' science literacy*. Paper presented at the sixth World Conference on Educational Sciences, Malta.

Mevarech, Z. R., & Kramarski, B. (1997). IMPROVE: A multidimensional method for teaching mathematics in heterogeneous classrooms. *American Educational Research Journal, 34*(2), 365–395.

Mevarech, Z. R., & Kramarski, B. (2003). The effects of worked-out examples vs. metacognitive training on students' mathematics reasoning. *British Journal of Educational Psychology, 73*(4), 449–471.

Mevarech, Z. R., & Kramarski, B. (2014). *Critical maths for innovative societies: The role of metacognitive pedagogies*. Paris: OECD Publishing. http://dx.doi/org/10/1787/9789264223561-en

Mevarech, Z. R., Terkieltaub, S., Vinberger, T., & Nevet, V. (2010). The effects of metacognitive instruction on third and sixth graders solving word problems. *ZDM International Journal on Mathematics Education, 42*(2), 195–203.

Michalsky, T., Mevarech, Z. R., & Haibi, L. (2009). Elementary school children reading scientific texts: Effects of metacognitive instruction. *The Journal of Educational Research, 102*(5), 363–374.

Ministry of Education (MOE). (2012). *Primary mathematics teaching and learning syllabus*. Singapore: Ministry of Education (MOE).

Nelson, T. O., & Narens, L. (1990). Metamemory: A theoretical framework and new findings. In G. H. Bower (Ed.), *The psychology of learning and motivation* (Vol. 26, pp. 1–45). New York: Academic.

Novak, J. D. (2002). Meaningful learning: The essential factor for conceptual change in limited or inappropriate propositional hierarchies leading to empowerment of learners. *Science Education, 86*(4), 548–571.

Oxford English Dictionary. (1995), England: Oxford University Press.

Pellegrino, J. W., Hilton, M. L. (2012). *Education for life and work: Developing transferable knowledge and skills in the 21st Century*. Committee on defining deeper learning and 21st century skills; Center for Education; Division on Behavioral and Social Sciences and Education; National Research Council.

Pintrich, P. R. (2004). A conceptual framework for assessing motivation and self-regulated learning in college students. *Educational Psychology Review, 16*(3), 385–407.

PISA. (2003). *First results from PISA 2003*. Paris: OECD Publishing.

PISA. (2009). *PISA 2009 results*. Paris: OECD Publishing.

PISA. (2012). *PISA 2012 results*. Paris: OECD Publishing.

PISA. (2013). *Assessment and analytical framework*. Paris: OECD Publishing.

Polya, G. (1957). *How to solve it?* (2nd ed.). Princeton: Princeton University Press.

Powell, A. B., Borge, I. C., Floriti, G. I., Kondratieva, M., Koublanova, E., & Sukthankar, N. (2009). Challenging tasks and mathematics learning. In E. J. Barbeau & P. J. Taylor (Eds.), *Challenging mathematics in and beyond the classroom, the 16th ICMI study* (pp. 133–170). New York: Springer.

Report, M. K. (2007). *How the world's best performing school system comes out on top*. London: McKinsey and Company.

Schneider, W. (1998). Performance prediction in young children: Effects of skill, metacognition and wishful thinking. *Developmental Science, 1*(2), 291–297.

Schoenfeld, A. H. (1985). *Mathematical problem solving*. New York: Academic.
Schoenfeld, A. H. (1987). What's all the fuss about metacognition? In A. H. Schoenfeld (Ed.), *Cognitive science and mathematics education* (pp. 189–215). Hillsdale: Lawrence Erlbaum Associates.
Schoenfeld, A. H. (1992). Learning to think mathematically: Problem solving, metacognition, and sense-making in mathematics. In D. Grouws (Ed.), *Handbook for research on mathematics teaching and learning* (pp. 334–370). New York: MacMillan.
Schoenfeld, A. (2007). Problem solving in the United States, 1970–2008: Research and theory, practice and politics. *ZDM – The International Journal on Mathematics Education, 39*, 537–551.
Schraw, G., & Dennison, R. S. (1994). Assessing metacognition awareness. *Contemporary Educational Psychology, 19*(4), 460–475.
Shamir, A., Mevarech, Z. R., & Gida, C. (2009). The assessment of young children's metacognition in different contexts: Individualized vs. peer assisted learning. *Metacognition learning, 4*(1), 47–61.
Shanghai Municipal Education Commission. (2005). *Mathematics curriculum standards for primary and secondary schools*. Shanghai: Shanghai Education Publishing House.
Stanic, G., & Kilpatrick, J. (1988). Historical perspective on problem solving in the mathematics curriculum. In R. Charles & E. Silver (Eds.), *The teaching and assessing of mathematical problem solving* (pp. 1–22). Reston: National Council of Teachers of Mathematics.
Stenlund, S. (2014). *The origin of symbolic mathematics and the end of the science of quantity*. Sweden: Department of Philosophy, Uppsala University. Retrieved from http://cas.uchicago.edu/workshops/wittgenstein/files/2014/05/Stenlund-The-Origin-of-Symb-Math-and-the-End-of-the-Science-of-Quantity.pdf
Stillman, G., Mevarech, Z. R. (2010). Metacognition research in mathematics education. *ZDM International Journal on Mathematics Education, 42*(2), Special Issue.
Veenman, M. V. J. (2013). *Metacognition and learning: Conceptual ad methodological considerations revisited. What have we learned during the last decade?* Keynote speech, 15th Bienniual EARLI conference, Munich.
Veenman, M. V. J., & Spaans, M. A. (2005). Relation between intellectual and metacognitive skills: Age and task differences. *Learning and Individual Differences, 15*(2), 159–176.
Wengrowicz, N., Dori, Y. J., & Dori, D. (2018). Metacognition and meta-assessment in engineering education. In Y. J. Dori, Z. Mevareach, & D. Bake (Eds.), *Cognition, metacognition and culture in STEM education* (pp. 191–216). Springer.
Whitebread, D. (1999). Interactions between children's metacognition abilities, working memory capacity, strategies and performance during problem solving. *European Journal of Psychology of Education, 14*(4), 489–507.
Whitebread, D., & Coltman, P. (2010). Aspects of pedagogy supporting metacognition and self-regulation in mathematics learning of young children: Evidence from observation study. *ZDM International Journal on Mathematics Education, 42*(2), 163–178.
Wong, K. Y., Zhao, D., Cheng, W. K., Teo, K. M., Lee, P. Y., Yen, Y. P., & So, H.-J. (2012). *Real life mathematics tasks: A Singapore experience*. Singapore: National Institute of Education, Nanyang Technological University.
Zhu, Y., & Fan, L. (2006). Focus on the representation of problem types in intended curriculum: A comparison of selected mathematics textbooks from Mainland China and the United States. *International Journal of Science and Mathematics Education, 4*(4), 609–626.
Zion, M., Adler, I., & Mevarech, Z. R. (2015). The effects of individualized and social metacognitive support on students' metacognitive performances in online discussions of environmental literacy. *Journal of Research in Science Teaching, 53*(4), 620–663.
Zion, M., Michalski, T., & Mevarech, Z. R. (2005). The effects of metacognitive instruction embedded within an asynchronous learning network on scientific inquiry skills. *International Journal of Science Education, 27*(8), 957–958.
Zohar, A., & Dori, Y. J. (2012). *Metacognition in science education: Trends in current research* (pp. 165–195). Dordrecht: Springer.

Chapter 13
Promoting Mathematics Teachers' Pedagogical Metacognition: A Theoretical-Practical Model and Case Study

Zehavit Kohen and Bracha Kramarski

13.1 Introduction

The importance of engaging students in meaningful learning as part of a coherent curriculum for developing problem-solving and mathematical reasoning has been emphasized in mathematics education reforms (National Council of Teachers of Mathematics [NCTM] 2000; Program for International Student Assessment [PISA] 2003), thus raising challenges for teachers' training goals concerning their pedagogical knowledge (e.g., Borko et al. 2015; Hill et al. 2005; NCTM 2000; Kramarski and Revach 2009). In essence, these goals maintain that teachers must cope with the complex, dynamic process of constructing/developing mathematical knowledge used to carry out the work of teaching mathematics. Examples of this work of teaching include "explaining terms and concepts to students, interpreting students' statements and solutions and providing students with examples of mathematical concepts, algorithms, or proofs" (Hill et al. 2005, p. 373). Moreover, it is suggested that training should challenge teachers to shift toward student-centered teaching that encourages knowledge construction through metacognition and self-regulation.

Metacognitive learners in mathematics and in other learning domains such as science, technology, and reading literacy are active participants in their own learning. They are metacognitive when they plan, set goals, select strategies, organize, self-monitor, and self-evaluate at various points during the process of acquisition. It is not only about the strategies that students use but also about students'

Z. Kohen (✉)
Faculty of Education in Science and Technology, Technion, Haifa, Israel
e-mail: zehavitk@ed.technion.ac.il

B. Kramarski
School of Education, Faculty of Social Studies, Bar-Ilan University, Ramat-Gan, Israel

Y.J. Dori et al. (eds.), *Cognition, Metacognition, and Culture in STEM Education*, Innovations in Science Education and Technology 24,
https://doi.org/10.1007/978-3-319-66659-4_13

considerations on *when*, *how*, and *why* to use them (e.g., Schoenfeld 1992; Schraw 1998; Zimmerman 2008).

Research has shown that teachers in high-metacognitive classrooms encourage student-centered learning, in which knowledge typically develops out of students' needs and interests (Perry et al. 2006; Randi 2004). In the context of mathematical classrooms, students are challenged to solve the tasks, to conceptualize their own opinions, and most important to adapt strategies to task demands (Dignath-van Ewijk et al. 2013; Kistner et al. 2010).

Educators and researchers claim that the ability to produce students who are metacognitive self-regulators of their planning, monitoring, and evaluation processes is tied to the teacher's own metacognition in two ways. First, teachers must be able to achieve metacognition for themselves. Second, teachers must be able to help their students achieve metacognition. We would therefore suggest starting with preservice mathematics teachers who will become teachers and, as an initial goal, teach them to become more effective learners. The second goal is to teach them to be more effective teachers with regard to pedagogical metacognition as learners and as teachers (Artzt and Armour-Thomas 1998; Kramarski and Michalsky 2009, 2010, 2015; Peeters et al. 2013).

The current study has three main goals: (a) building a theoretical-practical model of *pedagogical metacognition in teaching instruction* designed for preservice mathematics teachers, (b) applying this model in a microteaching course embedded with web-based learning supported by reflection, and (c) exploring the implementation of the model using a case study methodology analysis of two preservice mathematics teachers. Next, we elaborate on metacognition, which is the foundation of our theoretical-practical model.

13.2 Metacognition: Theoretical Framework

Metacognition is a person's knowledge about the cognitive processes necessary for understanding and learning (Flavell 1979). Metacognition is described as second-order cognitions: thoughts about thoughts, knowledge about knowledge, or reflections about actions. These definitions refer to all types of contexts and domains, like mathematics and other subject matter. Cognition and metacognition differ in their functions. The function of cognition is to solve problems, to bring cognitive enterprises to a successful conclusion. The function of metacognition is to regulate a person's cognitive operation in solving a problem or executing a task (Flavell 1979). For example, *presenting data in a graph is a cognitive function, whereas reflecting on the answer and realizing that the graph fits the givens are part of the metacognitive process.*

Many theoretical models of metacognition were built over the years (e.g., Brown 1987; Flavell 1979; Pintrich 2000; Schraw 1998; Zimmerman 2008). Schraw's and Zimmerman's models of metacognition serve as the theoretical framework for this study. Schraw (1998) explicitly differentiates between cognition and metacognition:

cognition refers to the use of *simple* strategies like memorization, information processing, and higher-level strategies such as *problem-solving* and *critical* thinking. Metacognition involves two strategic components: *knowledge of cognition* (KC) and *regulation of cognition* (RC). The KC involves three kinds of knowledge and students' *considerations* of strategy implementation: declarative knowledge refers to knowing about *what* strategy to use, procedural knowledge refers to knowing *how* to use the strategy, and conditional knowledge refers to knowing the *when* and *why* aspects of using the cognitive strategies. RC involves five kinds of metacognitive strategies: planning, information management, monitoring, debugging, and evaluation. *Planning* involves goal setting, activating relevant background knowledge, and budgeting time; *information management* refers to strategy sequences used online to process information more efficiently (e.g., organizing, elaborating, summarizing, selective focusing); *monitoring* includes the self-testing skills necessary to control learning; *debugging* strategies are used to correct comprehension and performance error; and *evaluation* refers to appraising the products and regulatory processes of one's learning (Schraw and Dennison 1994). Learners (students and teachers) who rate high in using metacognitive strategies backed up with considerations about *how*, *when*, and *why* to use these strategies are learners who are able to be "self-aware, knowledgeable, and decisive" in their approach to their own learning and teaching (Schraw 1998). Thus, according to these theories, self-awareness either *conscious* (implicit) or explicit is a prerequisite to constructing metacognitive knowledge. According to Zimmerman (2008), metacognition is part of cyclical dynamic learning processes for proactive learners, across three phases of task performance: pre-action (planning), in-action (monitoring), and post-action (evaluation).

Research indicates that metacognition develops slowly and is quite poor in students and teachers (e.g., Veenman et al. 2006). Some researchers and theorists (Butler and Winne 1995) suggest that "metacognitive self-regulatory" processes (i.e., part of the metacognitive component that relates to regulation of cognition), including planning, monitoring, and evaluation, may not be *conscious or explicit* in many learning and teaching situations that might hinder the internalization of these processes (Kistner et al. 2010; Kramarski and Revach 2009). As such, a learner needs to be taught explicitly how to activate metacognitive processes and to be given "ample opportunity to practice" those processes (p. 17, Mevarech and Kramarski 2014), both for the teachers themselves (as learners) and as teachers for their students (Kramarski and Michalsky 2009; Perry et al. 2006; Randi 2004; Vrieling et al. 2012).

Building on that recommendations for explicit metacognitive practice (Kistner et al. 2010), previous researchers claim that mathematics education lacks a practical and theoretical language for communicating about teachers' activity (Hill et al. 2005), the current study suggests a *theoretical-practical* model for explicit integration of *pedagogical metacognition* in mathematics lessons. Pedagogical metacognition relates to understanding/knowing how to implement or integrate metacognition to reinforce teacher's knowledge construction (Kohen and Kramarski 2012a;

The Multi-Dimensional Cog/Meta _T Model				
Cognition/Metacognition			**Teaching instruction**	
Cognition	**Metacognition**	**Considerations**	**Explicit strategies**	**Engagement activations**
Simple strategies Information processing	Planning Information Management	Declarative (*What?*) Procedural (*How?*)	**Presenting** : Theories Strategies Concepts on cognition/ metacognition	**Knowledge construction:** Process oriented/ Student-centrum instruction
Problem solving	Monitoring Debugging	Conditional (*Why?*)	**Naming:** Strategies Concepts Consideration : *What, How, When, Why*	**Teacher as regulator** : External Intermediate Internal
Critical thinking	Evaluation	Along the **three phases** of the lesson	**Modeling** : Thinking aloud Explanations Questioning	**Directed instruction:** Whole class / Individual student
				Motivation Feedback
Web-based learning environment: Ready-made clips Prompts Forum discussions				

Fig. 13.1 The multidimensional Cog/Meta_T Model for promoting pedagogical metacognition in teaching

Kramarski and Kohen 2016; Kramarski and Michalsky 2010, 2015; Wilson and Bai 2010).

The model is built in three parts to promote teachers' knowledge:

(a) *Cognition/metacognition* theoretical framework with its justified considerations.
(b) *Teaching instruction* of *explicit strategies* oriented to student *engagement activation*.
(c) *Web-based learning environment*. As such, it is called a multidimensional *Cog/Meta_T* model. Figure 13.1 presents the three parts of the model.

13.2.1 The Three Parts of the Multidimensional Cog/Meta_T Model

13.2.1.1 Part A: Cognition/Metacognition and Consideration Dimension

The theoretical framework includes four cognitive elements, five metacognitive elements, and knowledge strategy considerations (*what, how, when, and why*) through the three phases of the lesson (pre-/in-/post-action).

13.2.1.2 Part B: Teaching Instruction Dimension

Two types of *teaching instruction* are suggested to raise *metacognition* with its knowledge considerations: explicit strategies and engagement activities.

Explicit Strategies Researchers argue that teachers' metacognitive self-regulation knowledge is mostly tacit and remains unconscious until teachers are challenged to use that knowledge explicitly, like explaining metacognitive self-regulation strategies to their students. The more teachers know about metacognition self-regulation, the better they can make it visible to their students (Perry et al. 2006; Randi 2004; Schön 1983). Similar conclusions have been presented by mathematics researchers (e.g., Borko et al. 2015; Schoenfeld 1992; Verschaffel et al. 2000), that students' problem-solving failures do not always result from lack of mathematical knowledge but rather because they are *unaware* of how to activate their knowledge.

In contrast to findings indicating that explicit metacognitive strategy instruction in mathematics lessons is associated with a gain in students' performance (Kistner et al. 2010; Kramarski and Revach 2009), explicit strategy instruction is still rare in classrooms. In our model we adopted three stages to make metacognitive self-regulation process explicit. The first stage requires *presentation of* the theories of metacognitive concepts and phases for raising explicit *awareness* among learners that metacognition exists and differs from cognition and increases academic success. The next step is *naming* concepts and teaching strategies and, more importantly, to help learners construct *explicit knowledge* considerations about *when*, *how*, and *why* to use strategies (Schraw 1998). The third step is to *apply* some recommended metacognitive explicit strategies. *Modeling*, *thinking aloud*, *explanations*, and questioning are techniques of externalizing one's thought processes. Teachers can think out loud to externalize their thought processes, serving as an "expert model," so students can hear effective ways of using metacognitive knowledge and skills (Veenman et al. 2006). *Explaining* might include the mental processes, not simply telling about them while performing a task such as solving a problem or answering a question (Gama 2005).

Questioning This is an effective way of prompting learners' (students' and teachers') metacognition in mathematics. Prompts are an external stimulus, with the objective of enhancing metacognition. Questions can guide the learner's performance through the three phases of the solution (pre-/in-/post-action); it can improve self-awareness and control over thinking and thereby improve teacher and students' mathematics performance (Kramarski and Mevarech 2003; Kramarski and Revach 2009; Mevarech and Kramarski 1997, 2014; Schoenfeld 1992; Zimmerman 2008). For example, the IMPROVE questions model[1] designed in mathematics (e.g., Kramarski and Revach 2009; Kramarski and Michalsky 2013; Mevarech and Kramarski 1997, 2014) helps students/teachers to understand the task's or problem's goals or main idea (e.g., *What is the problem/task? What is similar/different from that task and other tasks?*) and encourages learners to plan and select appropriate strategies and to monitor and control their effectiveness (e.g., *What is the strategy? and why?*). Questions also play an important role in helping learners to think backward and forward by evaluating their strategies and efforts in the solution phases (e.g., *Does the plan/solution make sense? Can I plan/solve the task in another way?*).

Engagement Activation Researchers claim that the way students are engaged in teaching instruction largely determines the quality of their learning (van Beek et al. 2014; Turner et al. 2014). In our study engagement activation relates to the mode of *knowledge construction,* the role of the *teacher as regulator* in instruction, *directing instruction* to whole class/individual students, and raising *motivation* and exchanging *feedback* in the context of metacognition in mathematics learning.

Knowledge Construction in learning demands a process-oriented teaching approach which consists of instruction that puts the student in the centrum of learning and the *teacher's role* in supporting and enhancing the student as a self-regulator (van Beek et al. 2014; Bolhuis 2003; Schraw 1998). A prerequisite for that instruction is to make explicit prior concepts and strategies which are relevant to the topic and process of learning (Bolhuis 2003). Teachers should *engage* students to use these strategies in learning by means of questions and methods of presenting an argument/explanation. Teachers have to stimulate students to try out new learning and metacognitive strategies. Teachers might create challenging environments and provide complex tasks that stimulate employment of explicit metacognitive strategies (van Beek et al. 2014).

Teacher as Regulator There are three aspects of the teaching model that *facilitate* and enhance metacognitive self-regulation by the teacher or student:

1. *External regulation*: In this type of regulation, the teacher regulates all learning actions. The teacher determines the students' learning processes by undertaking explicit educational activities himself/herself. Teachers' activities are instructing, telling, and specifying.
2. *Intermediate regulation*: Teacher and students divide the task regulation. The teacher stimulates students to learn actively through assignments, questions, and

[1] Elaboration of the IMPROVE model can be found in Mevarech and Kramarski (2014, p. 68). The model comprises five stages: introducing the topic, metacognitive questioning and practice, reviewing materials, obtaining mastery, and verifying skills, enrichment, and remedial activities.

study tasks. Teachers' activities are modeling, explicating, demonstrating, stimulating, supporting, questioning, probing, and discussing.

3. *Internal regulation*: In this type, the students choose their own learning activities and carry out the main component of the learning functions. Teachers' activities are to let students think, discuss, correct, and reflect on themselves (van Beek et al. 2014). The three approaches present a *continuum* from strong teacher control to students' control of learning and can be implemented by instructions *directed either to the whole class or individual students.*

Motivation and Feedback Teachers have to create an effective climate that enables the experience of interest associated with motivation for learning success and praising learners (feedback). Complex tasks afford opportunities for learners to address multiple goals and focus on meaningful content. Perry et al. (2006) found that complex tasks are highly correlated with increased opportunities to engage students in metacognitive self-regulation. Although feedback has a major influence on learning, the type of feedback and the way it is given can be differentially effective, including the timing when given and the level at which it works. It could be directed to task performance, the process needed to perform the task, metacognitive and self-regulation, or the learner himself (Hattie and Timperley 2007).

13.2.1.3 Part C: Web-Based Learning Environments as Tools for Metacognition

Web-based learning environments have been looked upon as tools that support cognitive/metacognitive processes (Azevedo 2005; Jonassen 2000). As a nonlinear environment, web-based learning provides new possibilities in synchronous, asynchronous, autonomous, and collaborative modes for preservice teachers in learning and teaching by giving access to open-ended activities, moving beyond theoretical declarative knowledge into complex learning and teaching. For example, analyzing dynamic and simulated mathematics videotaped teaching scenarios of preservice teachers or of their colleagues, through their ability to record interactions with users, in asynchronic (i.e., forums) environments can become powerful reflection tools (Jonassen 2000; Kohen and Kramarski 2012a; Kramarski and Michalsky 2010, 2015; Wegerif 2004) which help to manage productive mathematics discourse in a social context that supports conceptual development (Cobb et al. 1990). Results generally indicate that preservice teachers' use of metacognitive learning strategies increases significantly in web-based learning environments with increased metacognitive opportunities (Kohen and Kramarski 2012a; Kramarski and Kohen 2016; Kramarski and Michalsky 2010, 2015; Vrieling et al. 2012). As such, we suggest that productive mathematics preservice teachers' interaction in web-based learning needs to be encouraged by using explicit question prompts to practice metacognition in teachers' feedback exchanges, while analyzing teaching scenarios in forum interactions (see Sect. 3 "Method"). Embedding question prompts in the web activities enables learners to focus attention on their own thoughts, processes, and activities while interacting with online materials and peers. Overall, research supports prompting in web environments as a catalyst to evoke the use of metacognitive self-regulation strategies (Davis 2003; Kohen and Kramarski 2012b; Kramarski and Michalsky 2009; Schraw 1998).

The current study applied the proposed model in a mandatory blended microteaching course with web-based learning for training mathematics preservice teachers (see Sect. 3 "Method"). We expected that mathematics preservice teachers that employed the Cog/Meta_T model with the dual theoretical components (cognition/metacognition and teaching instruction) while analyzing teaching scenarios in the blended course would advance their pedagogical metacognition on both types of measures (1) cognition/metacognition with strategy knowledge considerations and (2) teaching instruction of explicit strategies oriented to metacognitive engagement activities.

As an initial step to learn about the effectiveness of the model, we present a case study analysis on two mathematics preservice teachers that were exposed to the Cog/Meta_T model in a blended microteaching course. According to Stake (2000), a case study incorporates observations and analyses of human activity in a certain place and time. In-depth analysis of each case and a comparison between the two cases can shed light on the benefits and pitfalls of the proposed model.

The study has two research questions:

1. How do the microteaching scenarios illustrate *the quantity, quality, and patterns* of teachers' pedagogical metacognition practice along the teaching phases (pre-/in-/post-action)?
2. Are the two preservice teachers similar/different on *both types* of their pedagogical metacognition gains: cognitive/metacognitive and teaching instruction?

13.3 Method

13.3.1 The Case Study: Background

The case study focuses on two preservice teachers Mia and Ella who participated in a 2-year university teacher training program in Israel, in parallel to their undergraduate studies in mathematics. The preservice teachers in the program were in their second year of teacher training and participated in a one-semester microteaching course that involves 14 meetings in which two rounds of teaching simulations, 5 min in length, are videotaped. Mia and Ella were chosen to participate as a case study, since in their first round of teaching episodes, they taught the same subject of parallel lines and demonstrated the same quality of teaching. Both lacked pedagogical metacognition capacities, for example, Mia's episode: "Let's observe examples of parallel lines… Parallel lines refer to… According to these examples, we can see that… Let's take another example…" and Ella's episode, "So, what have we said that parallel means?"

Furthermore, no emphasis was placed on students' self-construction knowledge in their learning engagement. The teaching process was mostly focused by the pre-service teacher and directed to the entire class, instead of activating and challenging the students to take part in the activity. For example, after giving a chance to a student to explain one of her examples of the parallel lines by saying "it's two lines that...," Ella stopped him in the middle of the sentence and continued the explanation by herself. When another student answered a wrong question, Mia said "someone thinks otherwise?" and with no time given for the class to respond, she continued her explanations. Finally, neither teacher interrupted their teaching to make sure that students understood or to give feedback oriented to the process.

In addition, they shared general common characteristics: (a) they were at an equal level in mathematics (an average grade of 85 in their mathematics undergraduate studies[2]); (b) during the time of the study, they had no experience in mathematics teaching; and (c) they appeared to be at ease in front of the camera in the first round of teaching episodes.

In order to compare their teaching episodes, they were asked to design and present to their peers a similar topic for their second teaching episode: an arithmetic series. Ella had prepared a lesson on understanding the nature of the series, while Mia chose to prepare a lesson on calculating the sum of the series and connected it to the Hanukkah festival.[3]

13.3.2 *The Cog/Meta_T Practical Program*

As described earlier the Cog/Meta_T model was embedded in a blended course with web-based learning. The microteaching involved teaching episodes that were designed and presented as a teaching simulation by the preservice teachers to their peers, who act as real students. The teaching topic had to address one of the topics required in the mathematics curriculum for high school students, as mandated by the Israeli Ministry of Education. The teaching episodes were videotaped and followed immediately by a reflective discussion with the participant's peers and the instructor.

The main exposure to the Cog/Meta_T model was carried out in web-based learning environment activities related to: (a) exposure to metacognitive theories and concepts and explicit strategies to enhance students' engagement in knowledge construction as presented in the introduction; (b) practice with pedagogical tasks of varying complexity, particularly by analyzing ready-made clips of teaching episodes; (c) practice with prompts that stimulate usage of metacognitive elements and teaching instruction; and (d) forum discussion and sharing knowledge between the

---

[2] At the time they were assigned to the teaching program

[3] Hanukkah is an 8-day Jewish festival commemorating the rededication of the Holy Temple in Jerusalem, by kindling one additional light on each night of the holiday. Therefore, it is suitable for demonstrating the arithmetic series topic.

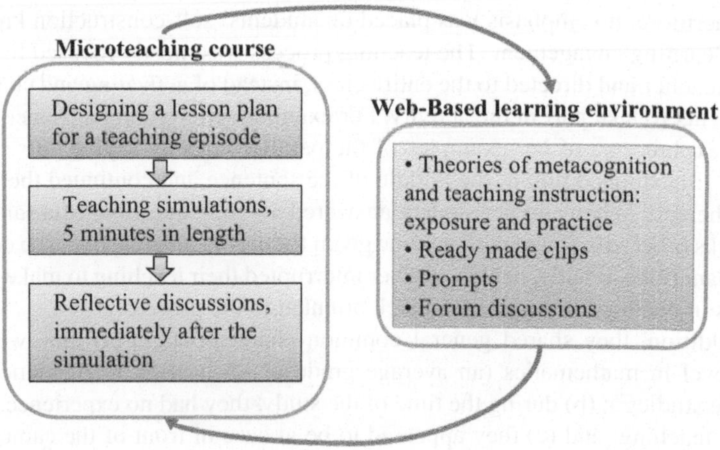

Fig. 13.2 The process of implementing the Cog/Meta_T model in a blended microteaching course with web-based learning

preservice teachers on planning teaching episodes and for further discussion of the teaching episodes that were conducted in the classroom by the preservice teachers (see Fig. 13.2). The preservice students were guided to log into the web-based learning environment once a week before each meeting in order to perform a task or to study a specific learning unit, but were encouraged to log in at any time they felt a need to discuss or consult their peers or the instructor. Accessing the web-based learning environment has been weighted into the final course grade. Appendix presents an example of a screenshot. Prompts with question stimuli (based on the models of Kramarski and Revach 2009; Kramarski and Michalsky 2015; Santagata and Guarino 2011) popped up consecutively on the bottom right of the screen for analyzing clips and encouraging discussions and feedback in the forums by explicitly thinking back and forward regarding metacognitive elements and instruction:

- *What* do I **notice** about Cog/Meta_T elements in the teaching scenario? That question can encourage descriptive abilities of noteworthy events, actions, and decisions.
- *How* can I **explain** it? That question can cultivate reasoning abilities and use of evidence.
- *When* and *how* can I **improve** metacognitive instruction in another way? That question can help in generalizing and proposing alternatives (prediction).
- *Why?* This can support justifications of decisions/considerations in teaching.

Concurrently, the preservice teachers were stimulated to base their microteaching simulations and reflective discussions using the Cog/Meta_T model. They were exposed to flashcards with the same printed question prompts as presented in the Web. Attention was paid to implementing the model components across the three phases of a teaching episode: *pre-action, in-action, and post-action*.

These types of prompts help teachers become more *self-aware* in their metacognitive approach to teaching (Kramarski and Michalsky 2009; Kramarski and Revach 2009) and promote pedagogical content knowledge integration processes (Kramarski and Michalsky 2015; Santagata and Guarino 2011).

13.3.3 Methodology

In the current study we analyzed Mia's and Ella's teaching episodes, based on the two types of Cog/Meta_T model (cognition/metacognition and teaching instruction). The teaching episodes' data were transcribed and viewed multiple times line by line to identify events in the data, categories, and conceptual connections between a category and its subcategories. Categories were discussed to define and refine the concepts and subconcepts to elicit their interpretations, explanations, and meanings, until we reached full agreement by two experts on the elements/categories (Strauss and Corbin 1990). The cognition/metacognition categories were easily identified according to the metacognitive theories of Schraw (1998) and Zimmerman (2008) as described in the introduction and as a consequence of previous experience in similar analyses (Kohen and Kramarski 2012a, b). The distinction between the cognition and metacognition categories was based on Schraw's framework (Schraw 1998), by which this distinction is dependent on the way of carrying out the learning task. If it involves *the procedure* of what and how to perform a task, then it is attributed as cognitive. However, if it involves *considering* the understanding of when and why the task is performed, then it is attributed as metacognitive.

Similarly, the teaching instruction types categories of the model (explicit strategies and engagement activities) were identified (Strauss and Corbin 1990).

We found five kinds of explicit strategies, rehearsal, questioning, examples, summary, and thinking, that could be presented in a cognitive level (what and how considerations) or in a metacognitive level (*when* and *why* consideration) demonstrated by naming and modeling conceptual concepts (see examples on p. 11). The four engagement activities categories were knowledge construction, directed instruction, motivation, and feedback, as described in the introduction.

In our analysis knowledge construction is the main engagement category that reflected the process/centrum learning, teacher's role, and teaching activities as presented in the introduction and in Fig. 13.1.

13.3.4 Data Analysis

Mixed methods (quantitative and qualitative) for analyzing data were performed. The cognition/metacognition dimension of the model was assessed through the *incidence* (i.e., frequency) and *quality* of elements (1–3), as assessed by the three level types of cognitive/metacognitive considerations: *what* (level 1), *how* or *when* (level

2), and *why* (level 3). In addition, the sequential pattern of the incidences along the three phases (pre-/in-/post-action) is presented. The teaching instruction dimension was assessed through frequencies of the explicit strategy usages and engagement activities categories. In addition, a qualitative analysis was performed relating to an elaborated description with examples of explicit strategies demonstrated by metacognitive strategies and engagement activities.

13.4 Findings

The findings section is based on analyzing scenarios excerpted from Mia's and Ella's actual teaching in an arithmetic series (5 min) with reference to the Cog/Meta_T components. These excerpts include the *cognitive/metacognitive* component with their *considerations* scores (1–3, respectively, for what, how or when, and why) related to *teaching instruction* of explicit strategies and engagement activities oriented to metacognition over the three lesson phases (pre-/in-/post-action).

13.4.1 Case Study Analysis

13.4.1.1 Cognitive/Metacognitive Components and Their Considerations

Table 13.1 (part 1 and part 2) presents Mia's and Ella's *incidence*, *quality*, and *sequential pattern* of cognitive and metacognitive components, i.e., their implementation considerations (part 1) and the element pattern in sequence on pre-/in-/post-action phases (part 2) during one actual 5-min scenario.

Comparison of the two cases in part 1 revealed that overall, Mia and Ella didn't differ in their total incidence of cognitive/metacognitive elements, demonstrating 38 and 40 teachers' events, respectively. But compared to the beginning of the course, where they used only cognitive elements in their teaching episodes, this time both demonstrated the use of metacognitive elements. Although both experienced the Cog/Meta_T program, we found differences between these two preservice teachers. Mia revealed more metacognitive elements (67.6%) than Ella (42.5%). Also, as seen in part 1, Mia alternated between the different aspects of metacognitive elements, focusing mainly on the planning (19.2%), monitoring (34.6%), and information management elements (30.8%) and less on the debugging (3.8%) and evaluation element (11.6%). Ella outperformed Mia only in usage of planning (35.3%). She showed similar usage of monitoring elements (35.3%), but used the information management elements (23.5%), and evaluation elements (5.6%; *mean* = 2) less. Also, she ignored the debugging element.

Beyond Mia's higher incidence of metacognitive elements usage, these elements differed in their quality, as Mia's scores for most of the metacognition elements were skewed more to the highest score (3) than Ella's scores, indicating well-justified

Table 13.1 Mia's and Ella's cognition and metacognition actual teaching of one 5-min scenario: incidence, quality, and sequential pattern on pre-/in-/post-action phases

COG/META_T elements identified in the teaching scenario		Mia		Ella	
		Incidence	Quality	Incidence	Quality
Part 1					
Cognition metacognition total		12 (32.4%)	1.50	23 (57.5%)	1.43
		26 (67.6%)	2.20	17 (42.5%)	1.37
		38 (100%)		40 (100%)	
Metacognition					
P	Planning	5 (19.2%)	2.20	6 (35.3%)	1.33
IM	Information management	8 (30.8%)	2.13	4 (23.5%)	1.75
M	Monitoring	9 (34.6%)	2.33	6 (35.3%)	1.75
D	Debugging	1 (3.8%)	2	–	–
E	Evaluation	3 (11.6%)	2.33	1 (5.9%)	2
Total		26 (100%)	2.20	18 (100%)	1.34
Part 2	Sequential pattern (over 5-min scenario)				
Pre-action		P, IM, COG, COG, COG, P, IM, M, P, P, IM, IM		P, COG, COG, COG, M, COG, COG, M, COG, P, M	
		[9 META elements – 75%[a]; mean = 2]		**[5 META elements – 46%[a]; mean = 1.20]**	
In-action		*M, IM, COG, M, E, M, COG, M, M, M, M, M, COG, COG. IM, IM*		*P, IM, COG, COG, IM, M, COG, COG, COG, COG, COG, P, P, COG, IM, COG, COG, P, IM, cog, M, COG, COG, COG*	
		[12 META elements – 75%[a]; mean = 2.42]		**[10 META elements – 38%; mean = 1.70]**	
Post-action		E, E, COG, COG, M, COG, D, COG, P, IM **[6 META elements – 60%[a]; mean = 2.3]**		E, M, COG, COG, **[2 META elements – 40%[a]; mean = 2.0]**	

Note. META_T = metacognition in teaching. Quality of META_T events was scored in a range of 1–3
[a]Percent = amount of the META elements divided by the overall elements in the phase

considerations for metacognition by Mia, compared to more technical considerations by Ella. Further analysis indicated that, respectively, 11 of Mia's 26 metacognitive elements were scored 3 (42.3%), whereas only 3 of Ella's 18 metacognitive elements were scored 3 (16.7%). This indicated consistently higher metacognition with *why* considerations on the whole for Mia's scenario ($M = 2.20$) as compared to Ella's scenario ($M = 1.37$), which was more technical, employing the *what* and *how* considerations.

Part 2 of Table 13.1 demonstrates these differences between the two preservice teachers, based on sequential patterns of cognition/metacognition elements over a 5-min scenario. The element sequence revealed Mia's notably flexible capacity to utilize various metacognitive elements across the *entire actual teaching phases*, with a stable ability to use *why* considerations.

As for the *pre-action*, most (75%) of Mia's elements were identified as metacognitive as compared to Ella, who demonstrated less than half (46%) metacognitive elements. However, for pre-action, Mia demonstrated a mean score of 2, revealing mixed considerations, while Ella's score was 1.20, revealing mostly considerations of *what* and sometimes considerations of *how* or *when*.

The following excerpts from Mia's lesson demonstrate part of the sequential pattern, for manifesting her mixed considerations for metacognitive elements. In the brackets, we indicate the metacognitive element and its considerations and scoring.

> *For this lesson I chose to teach you a topic connected with the festival of Hanukkah... I want to present something very interesting today that will make you think* [P, why, 3] ... *Now, I will remind you of the definition of an arithmetic series* [P, what, 1] *(while writing on the board, says): It is a series of members in which the difference between two adjacent numbers is fixed* [IM, what, 1]. *Clear?* [M, what, 1] *In today's lesson we will learn how to compute the sum of an arithmetic series in the spirit of the festival* [P, how, 2]...

The next excerpt of Ella's usage of metacognition in the pre-action phase of the lesson demonstrates less frequently systematic and less justified considerations:

> *Today we're going to talk about a very, very interesting topic in math called Series* [P, what, 1]... *In series we have order, like a television series, right?* [M, what, 1]... *We won't begin with the first episode, jump to the sixth episode, go back to the second episode, right? We have a fixed order* [M, how, 2]... *whatever is not understood, ask!* [M, what, 1].

Similar patterns appeared in the *in-action* phase, and only relatively better quality was revealed, indicating greater usage of justified considerations by Mia and greater usage of *how* and *when* considerations by Ella. As seen in Mia's sequential pattern, she repeatedly used the monitoring element. The next excerpt demonstrates that this usage was also accompanied by high-level considerations:

> *...what do we see? What is the total of each pair that we get each time (points to a pair example)?* [M, why, 3] ... *What does that mean that the sum of each pair is 101? Explain.* [M, why, 3] ... *Something constant. Nice. Perhaps someone has an idea why the sum is constant? Explain.* [M, why, 3] ... *Because it's an arithmetic series? (repeats student answer) Rachel, do you want to think more about the answer you gave?* [M, why, 3] *What do you think about what Rachel answered? Given the activity in the forum, one of you must certainly have a response...* [M, why, 3]

However, Ella demonstrated a sequence of mixed cognitive and metacognitive elements, most of which were not justified, as can be seen in the next excerpt:

> *What's special about this series? Every series has a certain uniqueness, according to the order it is arranged, according to rules that actually define the transition from one member to the next...* [M, what, 1] ... *In this series, for example, how can we move from the first member to the second? Who can tell me how we move from 0 to 1?* [COG, how,2] ... *We add 1 (repeats student's answer). Nice (writes on the board +1). How do we go from the second member to the third?* [COG, how,2] ... *Let's see another example (writes on the board: 1,2,4,8...)* [P, how,2].

Interesting differences in the metacognitive sequential pattern were found in the *post-action* phase. Mia presents a holistic-cyclical self-regulation perspective (Zimmerman 2008), starting with evaluation elements for summarizing the lesson's learning goals (thinking back):

> *So let's sum up the topic. What did we learn today? Can someone tell me?* [E, how, 2]

and concluded the lesson with planning activities (thinking ahead) based on what was learned:

> *For the next lesson, I ask each of you to think about another actual example that can be represented by an arithmetic series* [P, why, 3], *and to compute the sum of the series* [IM, how, 2].

Ella started her post-action phase with a short summary of the last calculated steps:

> *Take note, what we find in common between all three transitions is times 2, times 2, times, 2 [E, how]*

and finalized the lesson by providing a procedural conclusion:

> *...That way we can know how to continue the series* [COG, how, 2].

13.4.1.2 Teaching Instruction Dimension (Strategies and Engagement)

Table 13.2 presents Mia's and Ella's teaching instruction relating to implementation of explicit strategies and engagement activities, in the same 5-min scenario.

13.4.1.3 Explicit Strategies

By contrast with the beginning of the study, where in their initial episode both Mia and Ella focused on simple strategies like giving examples and memorizing the material, at the end of the program we found that both were more flexible in using new explicit strategies. We didn't divide these strategies into cognitive/metacognitive, because most of them were used infrequently. However, comparing their specific usage of the various strategies, Mia and Ella were pretty similar on the use of the *rehearsal* strategy (6.3% and 2.6%, respectively) and *summary* strategy (6.3% and 5.1%, respectively, for Mia and Ella). Both Ella and Mia modeled the rehearsal strategy by thinking aloud and giving explanations; for example, Ella said:

> *Let's see what the hidden word 'series' reminds you*

However, while Ella demonstrated considerations of *what* strategy should be implemented, Mia showed a different quality of modeling considerations, involving *why* considerations. It is clearly noticeable in the following example, where Mia explains the rational of choosing the topic that is connected with the festival of Hanukkah:

Table 13.2 Teaching instruction (strategies and engagement activation) from Mia's and Ella's one 5-min scenario

Teaching strategies[a]		
Explicit strategies[b]	**Mia**	**Ella**
Rehearsal	2 (6.3%)	1 (2.6%)
Examples	–	2 (5.1%)
Summary	2 (6.3%)	2 (5.1%)
Thinking	5 (15.6%)	–
Questioning	13 (34.2%)	5 (12.8%)
Engagement activation	**Mia**	**Ella**
Knowledge construction	10 (31.3%)	5 (12.8%)
Directed instruction		
Whole class	22 (68.8%)	39 (100%)
Individual student	10 (31.3%)	–
Motivation	6 (18.8%)	6 (15.4%)
Feedback	12 (37.5%)	8 (20.5%)

Notes:
[a]Frequencies were calculated by the number of incidence of each category divided by the total number of statements, $n = 32$ for Mia, $n = 39$ for Ella
[b]Frequencies of the rehearsal, examples, and thinking strategies present both the cognitive and metacognitive level; the questioning strategy presents the metacognitive level

I will review very briefly what we learned in the last lesson... so we can move on to the topic of arithmetic series, that is connected with the festival of Hanukkah

Similarly, they modeled the summary strategy by thinking aloud, by explaining *why* the Gauss topic had been taught:

We discussed finding the rules of summing an arithmetic progression. We were assisted by Gauss' computation

Ella, however, modeled a lower level of consideration of *how* to sum up the lesson as presented earlier.

OK? That's how we can know how to continue the series...

A difference between Mia and Ella, in both quantity and quality terms, was revealed regarding their ability to explicitly employ the *questioning* strategy, in particular metacognitive questioning (34.2% for Mia and 12.8% for Ella). For example, Mia presented explicit metacognitive monitoring by questioning, which also involves *why* considerations:

What is the total of each pair that we get each time (points to a pair example)?... ***What does that mean*** *that the sum of each pair is 101?*

However, Ella's monitoring questions showed low modeling considerations, mostly for *what*. For example:

I've written three dots, because this series might continue... Now, ***what's special about this series?***

Finally, the *thinking* strategy was used only by Mia (15.6%), who demonstrated this strategy explicitly by also modeling high considerations of *why* and even more so presenting and naming metacognition concepts:

So I wanted to present something very interesting today that will make you think... ***the purpose is*** *to exercise thinking, the* ***metacognition****... and not throw out anything that comes to mind...*

Mia also modeled the *thinking* strategy as she thought aloud about her own and her students' actions during the three phases of the lesson:

I will remind you of the definition of an arithmetic series (pre action phase)

Perhaps someone has an idea why the sum is constant? Explain (in action phase).*There's an error here* (post action phase).

Also, the *example* strategy was demonstrated only in Ella's lesson (5.1%,), e.g., her first sentence, for starting the *in-action* phase of the class:

Let's begin with the first example... Now – we begin from left to right, like in math. The far left member is our first member...

13.4.1.4 Engagement Activation

Similar to the outcomes of the previous components of the model, we found that at the end of the program, Mia and Ella demonstrated attempts to engage students in learning activities. However, the two preservice students mostly differed in the extent to which they *directed instructions* to the whole class or to individual students. Mia alternated between the two options; she directed questions to the whole class (68.8%), e.g.:

What did we learn today? Can someone tell me?

but also to individual students (31.3%), asking for explanations and personal opinion:

Dan, how did you reach the solution? Explain

In contrast, all (100%) of Ella's explanations and questions were directed to the whole class and ignored personal opinion and students' explanations.

These two types of directed instruction used by the two preservice students made an attempt to engage students in *knowledge construction*. However, they differed not just in the frequency (31.3% for Mia vs. 12.8% for Ella) but also in the quality of activities that demand students to be active in their learning for constructing their knowledge.

Mia demonstrated a very stable process-oriented approach directed by student-centrum instruction in her teaching. Her *teacher's role as a regulator* was to stimulate students' *internal regulation* ability, which was manifested by probing students to think and giving them time to do so, e.g.:

Do you want to think more about the answer you gave?

Mia also challenged students to try new tasks:

For the next lesson, I ask each of you to think about another actual example that can be represented by an arithmetic series.

or to connect new knowledge with previous knowledge:

Given the activity in the forum, one of you must certainly have a response...

She also focused a lot on regulating her students by explicating and questioning and challenging them to discuss and reflect on classmates' problem-solving activities and mistakes. Overall, Mia's lesson included 21 exchanges that were used for teacher/student interactions, e.g.:

What do you think about what Rachel answered?

or student/student interactions, e.g.:

Miri (direct her answer to Rachel): "Ah, you computed it without the Shamash."
Dan (explains to Rachel): "Because of the Shamash, on the first day we light two candles, and on the last day, 9 candles"

However, Ella demonstrated a stable teacher-centrum instruction approach with an *external teacher's regulation role*. She focused on instructing the subject matter in detail, step by step, and noting central concepts for the whole class, e.g.:

...in fact this series is defined by the rule of adding 1.That's how we will be able to know how to continue the series
...OK, or another 1. Come let's think about another way. Just think about another way to move from 1 to 2?... times 2. Nice (repeats student's answer...

She does not give time for thinking and sharing solutions with classmates and sometimes even ignored students' answers). Overall, only 12 teacher/students' exchanges took place in Ella's lesson with no student/student exchanges.

Almost similarly, Mia and Ella adopted an authentic approach to stimulate students' *interest and motivation* in teaching the arithmetic series lesson (18.8% for Mia and 15.4% for Ella). For example, Mia used the Gauss task for teaching how to sum an arithmetic series, as she said:

I've prepared a kind of slide for you, an interesting slide. I will show it in a moment. Actually, this slide describes Gauss' method of computation...

Further, she also gave the Hanukkah festivals as an example of the Gauss task:

Now, looking at this video-clip [showing an animation of eight candles, lit one by one] ... first of all note that there is an arithmetic series here...

Also, Ella used the "television series" example that the students suggested as a demonstration of an arithmetic series:

Taking a television series as an example – what do we actually have?... We have a first episode, second episode, third episode, right? If it's also a good series, we reach the final episode...

Furthermore, both raised their awareness in providing *feedback* to students' replies. Again, it is notable that Mia used more feedbacks in her teaching (37.5% for Mia vs. 20.5% for Ella), but we could also have noticed that most of Ella's feedback was directed to students' final results. However, Mia demonstrated feedback, directed to student process performance and regulation efforts (monitoring and evaluation):

Good. And then... Rachel, do you want to think more about the answer you gave?... (Turning to the students) What do you think about what Rachel answered? Given the activity in the forum, one of you must certainly have a response...

To sum up, the case analysis of the two preservice teachers revealed that generally, both Mia and Ella benefited from the Cog/Meta_T model in their teaching capacities as compared to the beginning of the study, when they focused mainly on simple cognitive elements with no emphasis on explicit strategies and students' engagement activities. However, Mia and Ella demonstrated different levels of development on the dual pedagogical metacognitive dimensions backed up with justified considerations. Regarding the cognition/metacognition dimension of the Cog/Meta_T model, Mia appeared to be more successful (incidence, quality, and sequential pattern) than Ella at demonstrating metacognitive elements with a high level of *why* considerations across the three phases. In contrast, Ella was more successful than Mia in using essential metacognitive elements (e.g., planning and monitoring) and still had difficulties in the ability to back up her lesson choices with a high level of *why* considerations across the three phases.

Regarding the teaching instruction dimension of the Cog/Meta_T model, at the end of the program, both teachers were flexible in using explicit strategies (e.g., rehearsal, summary, metacognitive questioning). They were oriented to students' knowledge construction, and both adopted an authentic teaching approach to stimulate students' interest and motivation, when teaching the arithmetic series lesson. Yet, Mia appeared to be more successful than Ella in her tendency to activate process-oriented learning and to stimulate students' inter-regulation ability, by explicating, questioning, and challenging them to discuss and reflect on classmates' problem-solving activities and mistakes. Unlike Mia, Ella demonstrated a teacher-centrum instruction approach with an external teacher's regulation role and low justified considerations ability.

13.5 Discussion

The current study suggests a multidimensional, theoretical, and practical Cog/Meta_T model as a springboard to enhance mathematics teachers' pedagogical metacognition (cognition/metacognition and teaching instruction) as a part of their pedagogical content knowledge, while analyzing teaching scenarios in a blended web-based learning course. The case study analysis on two mathematics preservice teachers that were exposed to the Cog/Meta_T model provides initial insights about the model's effectiveness.

According to Spruce and Bol (2014) and Zimmerman (2000), the ability of the two preservice teachers to implement metacognitive elements along the entire scenario (pre-/in-/post-action) is an indicator of high capacity in implementing metacognition. Spruce and Bol (2014) found that mathematics teachers most frequently encouraged student metacognition during the in-action phase of learning in their classroom, while ignoring them in the pre-/post-action phases. It seems that the multidimensional Cog/Meta_T model with its embedded question prompts practice in the web-based learning environment helped Mia and Ella to integrate metacognition into their teaching (Krauskopf et al. 2012).

Our findings corroborate other studies where metacognitive support (i.e., prompts in a web-based learning environment) was provided to mathematics and science in-/preservice teachers to use for reexamining learning goals and processes, which may help them shift their attention from technical actions to a higher level of metacognitive processing, whereby they consider goals, monitor strategies, and evaluate performance effectiveness (e.g., Davis 2003; Kramarski and Revach 2009; Kohen and Kramarski 2012a; Kramarski and Michalsky 2009, 2010, 2015).

Despite these interesting findings, two questions should be discussed. The findings show that both Mia and Ella increased their pedagogical metacognitive knowledge as was manifested in the types of questions they raised throughout the entire scenario (pre-/in-/post-action) (Table 13.1). Thus, how sure can we be that their increased pedagogical metacognitive knowledge is in fact a result of the training? Second, why did differences emerge between the two in their outcomes, despite their exposure to the same practice?

As noted, both preservice teachers started their teacher training program with minimal knowledge of metacognition. Also, according to the curriculum of the other courses in the teaching program, they were not exposed explicitly to the metacognitive topic. Thus, we can assume that their metacognitive knowledge development was affected by the three dimensions of the Cog/Meta_T theoretical-practical model (see example of practice in Appendix).

Furthermore, the source of the differences in the outcomes between Mia and Ella is hard to identify, because we have no additional data on the two teachers that could shed light on the outcomes and the process. Moreover, it is possible that despite the Cog/Meta_T model offered the two preservice teachers a rich theoretical and practical pedagogical metacognitive training embedded in a technological environment, the various components of the model may have created cognitive load for Ella, who had difficulty coordinating between the different representations of content and using effective strategies to monitor her learning (Kramarski 2012).

According to researchers (e.g., Azevedo 2005), this difficulty is common when learners are exposed to technological environment, as it requires more time for adaptation and learning. Finally, we should remember that the Cog/Meta_T model was implemented as part of only one-semester microteaching course, involving relatively short time of practice. The difference could also be a consequence of personal characteristics (different paces of learning) and/or initial features that were not assessed in the study.

The differences between the two students support recommendations about the need to pay more attention to the complexity of metacognition and teaching instruction, to foster individual novice teachers with different intellectual, self-confidence, and pedagogical needs in the same class (Kramarski and Michalsky 2013; Tomlinson 2005). We recommend to further investigate individual effects in future studies with a control group and followed by interviews.

13.5.1 Practical Implications, Future Research, and Limitations

This study contributes three main unique perspectives on mathematics teachers' professional development, regarding theory, methodology, and practice. First, this study contributes to a conceptual understanding of the crucial role of integrating metacognition with teaching. The multidimensional Cog/Meta_T framework is innovative in its blend of two complementary theories of teacher's professionalism.

The combination of Cog/Meta_T components appears to provide a *theoretical* effective lens for preservice teachers to understand how to integrate metacognition into teaching instruction (i.e., pedagogical content knowledge, in Shulman 1986). This model explicitly encourages teachers' self-awareness and focuses attention onto ways of intentionally applying metacognition in their teaching practice by justifying their decisions and actions (see Fig. 13.3).

The study's *methodological* contributions lie in its detailed two lesson analyses, assessing and illustrating cognition/metacognition and teaching instruction, regarding the three teaching lesson phases.

The development of the dual pedagogical metacognitive dimensions at the end of the program by the two preservice teachers, on the one hand, and the variance between the two preservice teachers' outcomes, on the other hand, contribute to the validity of the entire model that was achieved by pedagogical experts' analysis of the content categories in each dimension and by interjudge reliability (see Sect. 3 "Method" section). The resulting variance between the two preservice teachers indicates the models' sensitivity in assessing differences in teachers' ability on the dual dimensions (e.g., Kohen and Kramarski 2012a), thus responding to Avargil, Lavi, and Dori's (Chap. 3) claim that the largest gap in metacognition empirical research is the development of assessment tools and their validation. The validity of the metacognitive dimension with justified considerations in a web-based learning course was partially tested in our previous studies (e.g., Kohen and Kramarski 2012a; Kramarski and Michalsky 2010, 2015). However, the entire model with the teaching instruction dimension (explicit strategies and engagement activities) is a new combination. Future research should test the validity of the entire model on a large sample with experimental and control groups.

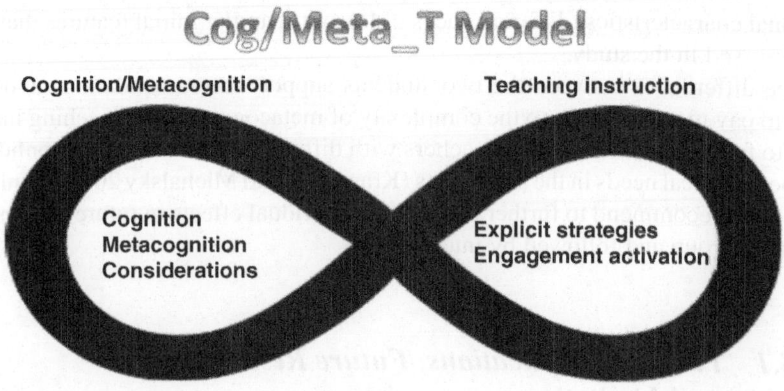

Fig. 13.3 The multidimensional Cog/Meta_T framework

Practically, the current study's unique Cog/Meta_T model for preservice teachers, comprising a web-based learning environment and metacognitive self-regulatory prompts, bears important implications for teachers' metacognitive practice.

Our previous studies focused on implementing a metacognitive self-regulation model in a pedagogical context and web-learning environment that reflects the macro level of preservice teachers' professional development (e.g., Kohen and Kramarski 2012a, b). The current multidimensional Cog/Meta_T model adds an explicit pedagogical microlevel that extends the practical tools for teachers' professional development, thus corresponding with Mevarech and Fan's claim (Chap. 12), according to which students need explicit exposure to metacognitive skills, in order to implement them in practice for solving mathematics problems. Moreover, the teaching, practice, and internalization of metacognition could be generalized as part of preservice and in-service teacher education in diverse science technology environment mathematics (STEM) learning domains and in traditional class programs without technology usage (Kramarski et al. 2013). As Zeichner and Liston (1987, p. 25) argue, "reflective teaching seeks to help student teachers become more aware of themselves and their environments in a way that changes their perceptions of what is possible."

Despite this study's potential contributions, several limitations deserve consideration. Our analysis was based on the work of two teachers who were exposed to the same Cog/Meta_T model. Whether it is applicable in the context of other teachers' work with the same model still remains to be investigated. For example, the generalizability of the instructional model to the training of elementary teachers who have limited mathematics backgrounds, to teacher training programs outside of Israel, to the teaching of different areas of mathematics such as algebra versus calculus, and to minority students preparing to be mathematic teachers. Furthermore, the preservice teachers were teaching to their peers. How well the teachers would be able to sustain the cognition/metacognition teaching strategies with reluctant students or students who were unable to answer their cognitive metacognitive questions has not yet been explored.

We suggest investigating the effect of this model among large samples of mathematics teachers from different cultures, as recommended by Dori, Mevarech, and Baker in the introduction to this book (Chap. 1). We also suggest to take into account participants' characteristics and beliefs in metacognition and different pedagogies. This investigation should be followed by mixed quantitative and qualitative methodological methods (interviews, videotaping, and questionnaires) to understand in depth the learning process of teachers' professional development, among pre-/in-service teachers' practice during their training and follow-up in their classes in real-time teaching. Furthermore, future analysis would corporate analyses of the online interactions between the participants and facilitators through the Web. Testing the model in a variety of contexts, with a broader range of students and tools, in real classrooms will provide mathematic educators with the knowledge they need to improve the training of future teachers of mathematics.

Finally, the Cog/Meta_T model is based on the combination of "explicit strategies" and "engagement activation." This combination raises the question of the extent to which the two components are needed in the technology environment. Future research should compare in an intervention study the effectiveness of the combination of the two components to each component alone. It will help in understanding the possible additional contribution of the "explicit strategies" for constructing knowledge beyond the "engagement activation" of the teacher and learner's role, on the one hand, and to the possible contribution of the "engagement activities" in the technology environment for constructing explicit knowledge of metacognition, on the other hand.

To conclude, this study contributes an explicit training program of a pedagogical metacognitive theoretical-practical model. The message of the model is the importance of the interaction between the two fields (metacognition and pedagogy) and the need for flexibility and adaptation to *different paces of learning*, as we can see in the case study of the mathematics preservice teachers.

13.5.2 Recommendations

We suggest a list of recommendations, targeted mostly to teachers' educators who wish to promote metacognition among mathematics pre-/in-teachers:

- Metacognition is essential in mathematics education. Teachers need to be *explicitly* taught how to activate metacognitive processes and to have ample opportunities to practice.
- Investigating the Cog/Meta_T model in other *STEM domains*, and among broader populations of preservice teachers (e.g., elementary teachers), will provide teachers' educators extensive knowledge for preparing future teachers.
- The Cog/Meta_T model can be *generalized to other contexts*, besides PD programs of preservice teachers, e.g., students in real classroom settings.
- There is a need for flexibility and adaptation of the model to *different paces of learning* in class by teachers (as found in the case study analysis).
- *Future research* should investigate the effect of the model among large mathematics teachers' samples, as compared to a control group by mixed methodological methods (interviews, videotaping, and questionnaires) and follow-up of teachers in their classes in real-time teaching.

Appendix: Screenshot of a Cog/Meta_T Task for Analyzing a Ready-Made Clip of a Teaching Episode(Fig. 13.4)

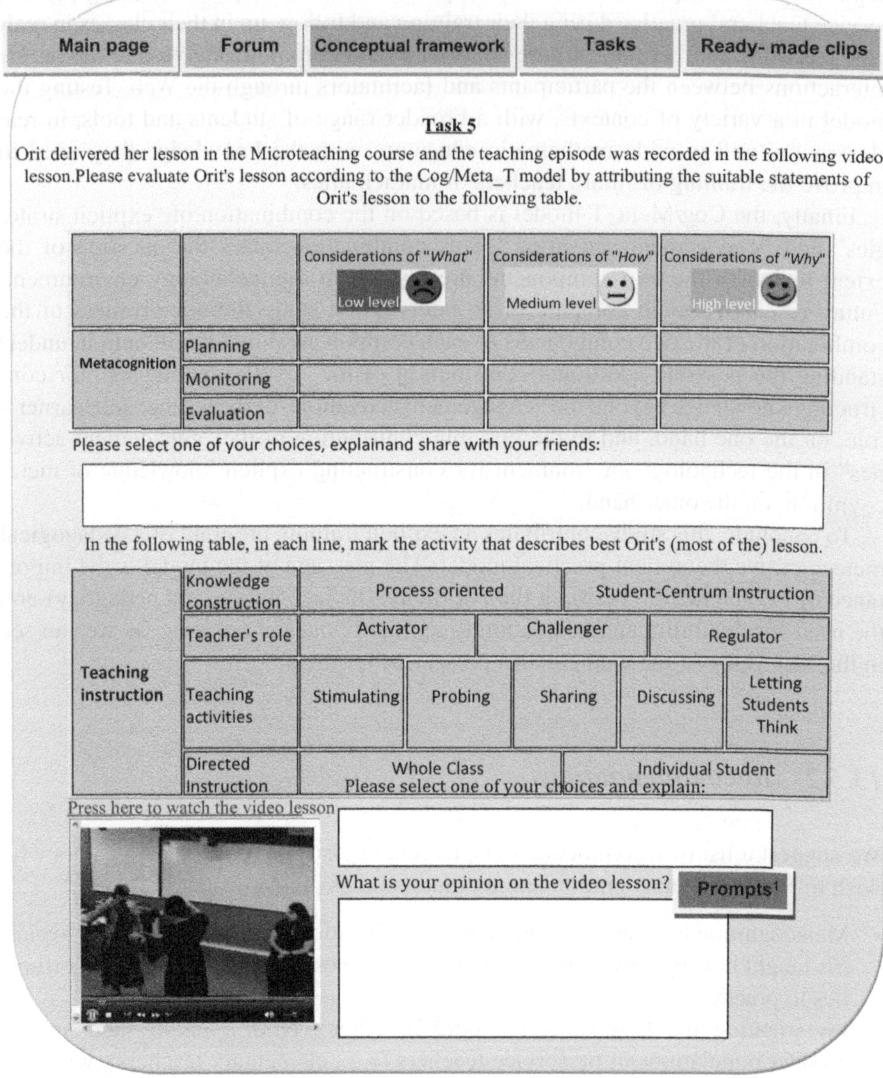

Fig. 13.4 Screenshot of a Cog/Meta_T task for analyzing a ready-made clip of a teaching episode

Note[1]: *what* do I **notice** on Cog00/Meta_T elements? "*How* can I **explain** it? *When* and *how* can I **improve** metacognitive instruction in another way? and *why?*

References

Artzt, A. F., & Armour-Thomas, E. (1998). Mathematics teaching as problem solving: A framework for studying teacher metacognition underlying instructional practice in mathematics. *Instructional Science, 26*(1), 5–25.

Avargil, S., Lavi, R., & Dori, Y. J. (2018). Students' metacognition and metacognitive strategies in science education. In Y. J. Dori, Z. Mevareach, & D. Bake (Eds.), *Cognition, metacognition and culture in STEM education* (pp. 33–64). Springer.

Azevedo, R. (2005). Using hypermedia as a metacognitive tool for enhancing student learning? The role of self-regulated learning. *Educational Psychologist, 40*(4), 199–209.

Bolhuis, S. (2003). Towards process-oriented teaching for self directed lifelong learning: A multidimensional perspective. *Learning and Instruction, 13*(3), 327–347.

Borko, H., Jacobs, J., Koellner, K., & Swackhamer, L. E. (2015). *Mathematics professional development: Improving teaching using the problem-solving cycle and leadership preparation models*. New York: Teachers College Press.

Brown, A. L. (1987). Metacognition, executive control, self-regulation and other more mysterious mechanisms. In F. E. Weinert & R. H. Kluwe (Eds.), *Metacognition, motivation and understanding* (pp. 65–116). Hillsdale: Erlbaum.

Butler, D. L., & Winne, P. H. (1995). Feedback and self-regulated learning: A theoretical synthesis. *Review of Educational Research, 65*, 245–281.

Cobb, P., Wood, T., & Yackel, E. (1990). Classrooms as learning environments for teachers and researchers. *Journal for Research in Mathematics Education Monographs, 4*, 125–146.

Davis, E. A. (2003). Prompting middle school science students for productive reflection: Generic and directed prompts. *The Journal of the Learning Sciences, 12*(1), 91–142.

Dignath-van Ewijk, C., Dickhäuser, O., & Büttner, G. (2013). Assessing how teachers enhance self-regulated learning: A multiperspective approach. *Journal of Cognitive Education and Psychology, 12*(3), 338–358.

Dori, Y. J., Mevarech, Z., & Baker, D. (2018). Introduction. In Y. J. Dori, Z. Mevarech, & D. Baker (Eds.), *Cognition, metacognition, and culture in STEM education* (pp. 1–8). Springer.

Flavell, J. H. (1979). Metacognition and cognitive monitoring: A new area of cognitive developmental inquiry. *American Psychologist, 34*(10), 906–911.

Gama, C.A. (2005). Integrating metacognition instruction in interactive learning environments. Doctoral dissertation, University of Sussex.

Hattie, J., & Timperley, H. (2007). The power of feedback. *Review of Educational Research, 77*(1), 81–112.

Hill, H. C., Rowan, B., & Ball, D. L. (2005). Effects of teachers' mathematical knowledge for teaching on student achievement. *American Educational Research Journal, 42*(2), 371–406.

Jonassen, D. H. (2000). *Computers as mindtools for schools: Engaging critical thinking* (2nd ed.). Upper Saddle River: Prentice-Hall.

Kistner, S., Rakoczy, K., Otto, B., Dignath-van Ewijk, C., Büttner, G., & Klieme, E. (2010). Promotion of self-regulated learning in classrooms: Investigating frequency, quality, and consequences for student performance. *Metacognition and Learning, 5*(2), 157–171.

Kohen, Z., & Kramarski, B. (2012a). Developing a TPCK-SRL assessment scheme for conceptually advancing technology in education. *Studies in Educational Evaluation, 38*(1), 1–8.

Kohen, Z., & Kramarski, B. (2012b). Developing self-regulation by using reflective support in a video-digital microteaching environment. *Journal for Education Research International, 2012*, 10. doi:10.1155/2012/105246.

Kramarski, B. (2012). Stimulating self-regulated learning in hypermedia to support mathematical literacy of lower-achieving students. In *Technology as a support for literacy achievements for children at risk* (pp. 157–169). Dordrecht: Springer.

Kramarski, B., Desoete, A., Bannert, M., Narciss, S., & Perry, N. (2013). New perspectives on integrating self-regulated learning at school (special issue). *Education Research International*, Editorial paper, Article ID 498214, 1–4.

Kramarski, B., & Kohen, Z. (2016). Promoting preservice teachers' dual self-regulation roles as learners and as teachers: Effects of generic vs. specific prompts. *Metacognition and Learning*, 1–35. doi:10.1007/s11409-016-9164-8.

Kramarski, B., & Mevarech, Z. R. (2003). Enhancing mathematical reasoning in the classroom: The effect of cooperative learning and metacognitive training. *American Educational Research Journal, 40*, 281–310.

Kramarski, B., & Michalsky, T. (2009). Investigating preservice teachers' professional growth in self-regulated learning environments. *Journal of Educational Psychology, 101*(1), 161–175.

Kramarski, B., & Michalsky, T. (2010). Preparing preservice teachers for self-regulated learning in the context of technological pedagogical content knowledge. *Learning and Instruction, 20*(5), 434–447.

Kramarski, B., & Michalsky, T. (2013). Student and teacher perspectives on IMPROVE self-regulation prompts in web-based learning. In R. Azevedo & V. Aleven (Eds.), *International handbook of metacognition and learning technologies* (pp. 35–51). New York: Springer Publisher.

Kramarski, B., & Michalsky, T. (2015). Effects of a TPCK-SRL model on teachers' pedagogical beliefs, self-efficacy and technology-based lesson design. In C. Angelie, & N. Valanides (Eds.), *Technological Pedagogical Content Knowledge (TPCK): Exploring, developing, and assessing TPCK* (pp. 89–112). New York: Springer Publisher.

Kramarski, B., & Revach, T. (2009). The challenge of self-regulated learning in mathematics teachers' professional training. *Educational Studies in Mathematics, 72*(3), 379–399.

Krauskopf, K., Zahn, C., & Hesse, F. W. (2012). Leveraging the affordances of YouTube: The role of pedagogical knowledge and mental models of technology functions for lesson planning with technology. *Computers & Education, 58*(4), 1194–1206.

Mevarech, Z., & Fan, L. (2018). Cognition, metacognition and mathematics literacy. In Y. J. Dori, Z. Mevareach, & D. Bake (Eds.), *Cognition, metacognition and culture in STEM education* (pp. 261–278). Springer.

Mevarech, Z. R., & Kramarski, B. (1997). IMPROVE: A multidimensional method for teaching mathematics in heterogeneous classroom. *American Educational Research Journal, 34*(2), 365–395.

Mevarech, Z. R., & Kramarski, B. (2014). *Critical maths for innovative societies: The role of metacognitive pedagogies*. Paris. (196 pages): OECD publisher. doi:10.1787/9789264223561-en.

National Council of Teachers of Mathematics – NCTM. (2000). *Principles and standards for school mathematics*. Reston: National Council of Teachers of Mathematics – NCTM.

Peeters, E., Backer, F.D., Reina, V.R., Kindekens, A., & Buffel, T. (2013). *The role of teachers' self-regulatory capacities in the implementation of self-regulated learning practices*. Procedia – Social and Behavioral Sciences. www.elsevier.com/locate/procedia.

Perry, N. E., Phillips, L., & Hutchinson, L. (2006). Mentoring student teachers to support self-regulated learning. *The Elementary School Journal, 106*(3), 237–254.

Pintrich, P. R. (2000). The role of goal orientation in self-regulated learning. In M. Boekaerts, P. R. Pintrich, & M. Zeidner (Eds.), *Handbook of self-regulation* (pp. 451–502). San Diego: Academic.

Program for International Student Assessment—PISA. (2003). *Literacy skills for the world of tomorrow: Further results from PISA 2000*. Paris: Program for International Student Assessment—PISA.

Randi, J. (2004). Teachers as self-regulated learners. *Teachers College Record, 106*, 1825–1853.

Santagata, R., & Guarino, J. (2011). Using video to teach future teachers to learn from teaching. *ZDM, 43*(1), 133–145.

Schoenfeld, A. H. (1992). Learning to think mathematically: Problem solving, metacognition, and sense making in mathematics. In D. A. Grouws (Ed.), *Handbook of research on mathematics teaching and learning* (pp. 165–197). New York: MacMillan.

Schoenfeld, A. H. (2011). Toward professional development for teachers grounded in a theory of decision making. *ZDM, 43*(4), 457–469.

Schön, D. A. (1983). *The reflective practitioner*. New York: Basic Books.

Schraw, G. (1998). Promoting general metacognitive awareness. *Instructional Science, 26*(1), 113–125.

Schraw, G., & Dennison, R. S. (1994). Assessing metacognitive awareness. *Contemporary Educational Psychology, 19*, 460–475.

Shulman, L. S. (1986). Those who understand: Knowledge growth in teaching. *Educational Researcher, 15*(2), 4–14.

Spruce, R., & Bol, L. (2014). Teacher belief, knowledge, and practice of self-regulated learning. *Metacognition and Learning*. doi:10.1007/s11409-014-9124-0.

Stake, R. E. (2000). Case studies. In N. K. Denzin & Y. S. Lincoln (Eds.), *Handbook of qualitative research* (pp. 435–454). Thousand Oaks: Sage.

Strauss, A., & Corbin, J. (1990). *Basics of qualitative research: Grounded theory procedures and techniques*. Newbury Park: Sage.

Tomlinson, C. A. (2005). Quality curriculum and instruction for highly able students. *Theory Into Practice, 44*(2), 160–166.

Turner, J. C., Christensen, A., Kackar-Cam, H. Z., Trucano, M., & Fulmer, S. M. (2014). Enhancing students' engagement: Report of a 3-year intervention with middle school teachers. *American Educational Research Journal, 51*(6), 1195–1226.

van Beek, J. A., de Jong, F. P. C. M., Minnaer, A. E. M. G., & Wubbels, T. (2014). Teacher practice in secondary vocational education: Between teacher-regulated activities of student learning and student self-regulation. *Teaching and Teacher, 40*, 1–9.

Veenman, M. V. J., Bernadette, H. A. M., Hout-Wolters, V., & Afflerbach, P. (2006). Metacognition and learning: Conceptual and methodological considerations. *Metacognition and Learning, 1*(1), 3–14.

Verschaffel, L., Greer, B., & De Corte, E. (2000). *Making sense of word problems*. Lisse: Swets and Zeitlinger.

Vrieling, E., Bastiaens, T., & Stijnen, S. (2012). Effects of increased self-regulated learning opportunities on student teachers' motivation and use of metacognitive skills. *Australian Journal of Teacher Education, 37*(6), 102–117.

Wegerif, R. (2004). The role of educational software as a support for teaching and learning conversations. Computers and Education (Special edition of selected papers from the CAL'03 conference).

Wilson, N. S., & Bai, H. (2010). The relationships and impact of teachers' metacognitive knowledge and pedagogical understandings of metacognition. *Metacognition and Learning, 5*(3), 269–288.

Zeichner, K. M., & Liston, D. P. (1987). Teaching student teachers to reflect. *Harvard Educational Review, 57*(1), 23–24.

Zimmerman, B. J. (2000). Self-efficacy: An essential motive to learn. *Contemporary Educational Psychology, 25*(1), 82–91.

Zimmerman, B. J. (2008). Investigating self-regulation and motivation: Historical background, methodological developments, and future prospects. *American Educational Research Journal, 45*(1), 166–183.

Chapter 14
Mathematical Modeling and Culturally Relevant Pedagogy

Cynthia O. Anhalt, Susan Staats, Ricardo Cortez, and Marta Civil

14.1 Introduction

The encounter with persons, one by one, rather than categories and generalities, is still the best way to cross lines of strangeness. (Bateson 2000, p. 81)

In this chapter, we propose a new pedagogical approach that brings together two domains that rely on students' knowledge of everyday situations, mathematical modeling and culturally relevant pedagogy (CRP). Culturally relevant teaching (Gay 2000; Ladson-Billings 1995) utilizes the students' backgrounds, knowledge, and experiences to inform the teacher's lessons and methodology, which requires teachers to create bridges between students' home cultures and the school. Through knowledge of family practices, teachers have the opportunity to connect the curriculum in mathematics and adapt various ways to learn about the everyday, lived experiences of students and their families.

Mathematical modeling is a process in which students use their knowledge of an everyday situation to engage in cycles of mathematical inquiry. Students' cultural backgrounds can play a central role within rich mathematical modeling activities, which ask students to create problem-solving methods for nonroutine tasks in everyday contexts. These opportunities have the potential for teachers to leverage diverse students' everyday lived experiences for meaningful engagement with challenging mathematics through modeling tasks.

C.O. Anhalt (✉) • M. Civil
Department of Mathematics, University of Arizona, Tucson, AZ, USA
e-mail: canhalt@math.arizona.edu

S. Staats
Department of Curriculum and Instruction, University of Minnesota, Minneapolis, MN, USA

R. Cortez
Mathematics Department, Tulane University, New Orleans, LA, USA

Y.J. Dori et al. (eds.), *Cognition, Metacognition, and Culture in STEM Education*, Innovations in Science Education and Technology 24,
https://doi.org/10.1007/978-3-319-66659-4_14

In this chapter we provide ideas that mathematics educators and teachers can use to consider contexts that are relevant to students' lives for creating mathematical modeling tasks. We initially discuss culture from an anthropological perspective and its influence on mathematics teaching and learning, followed by the tenets of CRP, with a focus on Funds of Knowledge (Greenberg 1989; Moll et al. 1992) as the approach we used in the project we describe. In this project, we engaged a group of secondary preservice teachers (PTs) in a mathematical modeling module that brings together the tenets of CRP, culture, and local community contexts. The experiences gained by the PTs throughout the module provide a glimpse into the possibilities that teacher preparation programs can offer in the context of CRP in mathematics classrooms. We conclude the chapter with implications for teaching focusing on balancing the rigor of the mathematics, cultural connections, and helping students develop a critical analysis of the social implications.

14.2 Culture and Its Influence on Mathematics Education

One of the earliest definitions of culture captures commonplace understandings of culture today, that culture is the "knowledge, belief, art, morals, law, custom, and any other capabilities and habits acquired by man as a member of society" (Tylor 1920, 1871, p. 1). In this view, culture involves the relatively consistent, visible, unchanging aspects of life—beliefs and belongings—that serve as markers of social difference among people. While this way of thinking of culture is embedded in everyday life, it proved to be insufficient for researchers and educators whose work responds to the complexities of culture.

14.2.1 Changing Concepts of Culture

González (2008) traces over a hundred years of the theoretical twists and turns of the anthropological culture concept subsequent to Tylor's definition. In Tylor's period, anthropologists believed that cultures evolved and improved through specific stages. The development of direct observation through fieldwork reduced this scientific racism but also strengthened the position that cultures determine human behavior. By the 1970s and 1980s, two general approaches to culture were prominent: culture as symbols and culture as activity (Henze and Hauser 1999; Sewell 1999). The first of these positions holds that culture refers to knowledge—a system for making meaning of the world. A second position focuses on people's means of taking action: "Culture is not a coherent system of symbols and meanings but a diverse collection of 'tools' that, as the metaphor indicates, are to be understood as a means for the performance of action" (Sewell 1999, p. 46).

Viewing culture as action, and as interaction, however, has not reduced the complexity of the concept. Many communities are culturally varied and foster multiple identifications; even when an individual identifies with a particular cultural group, the person may not know about or practice the elements associated with this culture (Henze and Hauser 1999). González (2008), for example, comments that "an Irish Catholic teacher can see that the Haitian family that lives next door differs in some crucial ways from a Haitian family that lives across town…the Haitian family that lives across town may be in some respects more like her own family than the Irish Catholic family that lives across the street" (p. 96). Ultimately, no authoritative framework for understanding cultural change, variation, and identification has emerged in the discipline of anthropology. As anthropologist James Clifford put it, "culture is a deeply compromised concept that I cannot yet do without" (Clifford 1988; p. 10 in Sewell 1999, p. 38). Researchers involved in studies of culture must define an interest within one of many dimensions of complexity.

14.2.2 Culture in Mathematics Education

During the 1980s, educational researchers began to incorporate social perspectives, shifting from a psychological or cognitive model of knowledge to the idea that thinking and learning are grounded in social interaction (Lerman 2000), and this social interaction serves as mediation for cognitive development from socially guided learning (Vygotsky 1978). This "social turn" (Lerman 2000) was grounded in a concern with acknowledging and addressing social inequality in research and in classrooms. Despite the intractability of the definition of culture, many educators regard the culture concept as vitally important for improving equity in schooling. In some respects, this shift recalls the debate of culture as knowledge versus action. Mathematical knowledge was viewed as the product of action, discussion, and construction, rather than simply as an intergenerational transfer of knowledge.

Bishop (1988), for example, argues that mathematics is a cultural practice. He suggests that several types of cultural activities can lead to culturally based mathematical ideas: counting, locating, measuring, designing, playing, and explaining. He proposes that a "culturally fair" curriculum could be designed from the standpoint of this structure, which would allow local mathematical concepts to enter the classroom, along with widely shared forms of academic mathematics. "Is it indeed possible by this means to create a culturally-fair mathematics curriculum—a curriculum that would allow all cultural groups to involve their own mathematical ideas whilst also permitting the 'international' mathematical ideas to be developed?" (Bishop 1988, p. 189). The field of ethnomathematics, too, addressed issues of cultural fairness through ethnographic inquiries into mathematical practices embedded within cultural activities (Ascher 1991; d'Ambrosio 1985, 2006). Ethnomathematics faces the conundrum that activities are most clearly recognized as mathematics when they are translated into traditional mathematical forms (Civil 2016; Wagner and Lunney Borden 2012). Though this issue is unresolved, several scholars have recommended the general approach of asking community members to identify activities that they consider mathematical and to develop curriculum from this starting point (Borba 1997; Wagner and Lunney Borden 2012).

As culture is embedded in issues of fairness and equity, the unruly nature of the concept creates tensions in basic questions such as what activities or representations of activities count as mathematics and how educators can incorporate mathematical community knowledge into classrooms as a bridge to widely recognized mathematical practices.

14.3 Culturally Relevant Pedagogy

Culturally relevant pedagogy has become one of the most influential responses to incorporation of cultural perspectives in education. By structuring curriculum and classroom interactions around students' cultures, CRP seeks to ensure that students are academically successful and that they develop a sense of social critique (Ladson-Billings 1995). CRP emphasizes the development of a collective rather than individualized identity (Ladson-Billings 1995; Tate 1995). The idea is that through a "pedagogy of opposition" (Tate 1995, p. 169), students resist assimilation into the cultural norms of the majority and use classroom learning to take action in their communities.

CRP calls for developing pedagogical approaches in which students (a) experience academic success, (b) develop and/or maintain cultural competence, and (c) develop a critical consciousness through which they challenge the status quo of the current social order. These three tenets constitute the basis for using students' strengths to promote academic success. Several researchers have used CRP in mathematics education, including Greer et al. (2009), Gutstein et al. (1997), Lipka et al. (2009), Moses and Cobb (2001), Tate (1995), and Turner and Font Strawhun (2007).

14.3.1 Dilemmas Posed by Culturally Relevant Pedagogy

Although CRP has become one of education's "best practices," its complexity means that it is often implemented in ways that diverge from its original principles. Of Ladson-Billings' three goals, cultural competence has been attended to more strongly than the other two principles (Young 2010). However, a limited perspective on culture, similar to Tylor's 1871 definition, underlies some of the problem. CRP misses the point when it merely involves "acknowledging ethnic holidays, including popular culture in the curriculum, or adopting colloquial speech" (Irvine 2010, p. 58). This can have the effect of emphasizing "the sense of otherness commonly felt by minority students" (Young 2010, p. 252; referring to Troyna 1987). Further, teaching practices for CRP are often developed in reference to homogeneous classrooms (Morrison et al. 2008). Teachers may assume that all students identify with one version of one culture. More broadly, a focus on cultural difference may reproduce a system of exclusion if teachers assume that different children require different pedagogies or if the target of academic development is to achieve a standard defined as the behaviors and level of achievement of the dominant group of students (Schmeichel 2012).

Aguirre and Zavala (2013) have addressed this dilemma through a lesson analysis tool that uses all three tenets of CRP to help teachers integrate mathematical thinking with components of CRP such as language, culture, and social justice. These authors refer to culturally responsive mathematics teaching (CRMT) as "a set of specific pedagogical knowledge, dispositions, and practices that privilege mathematical thinking, cultural and linguistic funds of knowledge, and issues of power and social justice in mathematics education (Aguirre and Zavala 2013, p. 1)." It remains a challenge to reach consensus on the content of such CRMT tools and how to prepare teachers for integrating CRP principles into mathematics instruction.

In general, critical consciousness is the component of Ladson-Billings' model that is less fully realized in classroom teaching (Young 2010). Teachers may feel uncomfortable with political analysis—many people in the United States prefer discussing culture instead of structural inequity or racism (Sleeter 2011). Reflection on personal identity is a necessary step for teachers from dominant social classes before they can implement classroom activities that support development of critical consciousness among diverse students.

14.3.2 Addressing Dilemmas Through Funds of Knowledge

In recent years, educational researchers have begun to acknowledge the difficulty of implementing each of the three elements of CRP. Nuanced understandings of culture, teaching across cultural differences, and integrating cultural and mathematical understanding are significant dilemmas for this pedagogical approach; debates over the meaning of culture are at the heart of all of these issues.

The Funds of Knowledge approach (González et al. 2005; Greenberg 1989; Moll et al. 1992; Tapia 1991) can address some of the difficulties in implementing the three tenets of CRP. Through ethnographic visits to some of their students' homes, teachers learn about their students' and their families' knowledge and experience their funds of knowledge. This process places families as the knowledge experts and the teacher as a learner.

Following the Funds of Knowledge approach, we adopt González' dynamic view of culture as "lived experience. The focus is on 'practice,' that is, what it is that people do and what they say about what they do. The processes of everyday life, in the forms of daily activities, emerge as important" (González 2008, p. 96). This perspective asks teachers, researchers, and students to actively investigate the forms that culture takes in a particular community.

The Funds of Knowledge approach overturns deficit concepts of students. By identifying reservoirs of community expertise, and creating projects and classroom activities around them, teachers can engage students' knowledge more deeply.

Because it involves teachers' active learning in households and communities, this approach may be able to uncover cultural complexity of communities better. It avoids the pitfalls of homogeneity, and it accounts for cultural change and cultural borrowing. It can help teachers develop critical consciousness within themselves and prepare them to help students find their political voice. The active search for community knowledge represented in the Funds of Knowledge approach can avoid essentializing assumptions about students' cultures.

Examples of mathematics education work within the Funds of Knowledge approach include the use of occupational interviews to uncover the mathematics behind some practices (e.g., a mechanic, a carpenter, or a seamstress) (Civil 2016; Civil and Andrade 2002; González et al. 2001). Civil (2007) describes two mathematically rich classroom experiences based on funds of knowledge work, one centered on construction with a class of second graders (see also Sandoval-Taylor 2005) that involved geometric thinking and measurement; the second one was a garden unit with a class of fourth and fifth graders, also exploring ideas of measurement and optimization (maximizing area of a garden plot given a fixed perimeter) (see also Civil and Kahn 2001). This chapter draws on the general concepts at the heart of Funds of Knowledge to propose an approach that brings together mathematical modeling and CRP. We next turn to a discussion on mathematical modeling.

14.4 Mathematical Modeling

14.4.1 Mathematical Modeling: Its History and Background

The mathematics literature has long discussed ancient cultures that used modeling to improve their everyday life, starting around 2000 BC (Schichl 2004). In this context, modeling meant the application of mathematics to solve problems arising in sciences (e.g., astronomy) and other aspects of everyday life. For centuries, mathematical modeling has been driven by the desire to describe nature's principles. More recently, the motivation for developing mathematical models comes from an increasing number of disciplines including the sciences, technology, engineering, economics, health care, politics, and more. Today, modeling is an area of mathematical research, and it is typically taught in universities as part of an applied mathematics curriculum.

By the mid-1980s, mathematical modeling was emerging in the UK and Europe as a pedagogical approach in secondary and early undergraduate mathematics curriculum (Berry et al. 1984). The approach emphasized an active and creative way of learning mathematics—"learning modeling"—rather than memorizing established approaches to solving formulaic problems—"learning models" (Burkhardt 1984). In the United States the National Council of Teachers of Mathematics (NCTM) underscores the use of representations to interpret physical, social, and mathematical phenomena in mathematical modeling (2000).

14.4.2 Perspectives and Definition of Mathematical Modeling

Drawing from Lesh and Zawojewski (2007), English and Sriraman (2010) write that "modeling problems are realistically complex situations where the problem solver engages in mathematical thinking beyond the usual school experience and where the products to be generated often include complex artifacts or conceptual tools that are needed for some purpose, or to accomplish some goal" (p. 273). In the Common Core State Standards, model with mathematics is one of the Standards for Mathematical Practice and is defined as "the process of choosing and using appropriate mathematics and statistics to analyze empirical situations, to understand them better, and to improve decisions" (CCSSI 2010, p. 72). The expectation is that all elementary and secondary school students will develop modeling proficiency, which includes applying the mathematics they know to solve problems not originally posed as mathematics problems and making simplifications and choices that must be validated and possibly revised. Several authors have written about modeling implications and issues connected to the Common Core (e.g., Anhalt and Cortez 2015; Felton et al. 2015; Tam 2011). In the Guidelines for Assessment and Instruction in Mathematical Modeling Education (GAIMME) report, Garfunkel and Montgomery (2016) define mathematical modeling as a process that uses mathematics to represent, analyze, and make predictions or otherwise provide insight into real-world phenomena.

It is important to recognize that mathematical modeling is not only a part of K-16 education but also an active area of research among professional mathematicians. For this reason, there are multiple sources that define mathematical modeling. The definitions have some variations but are essentially very similar: mathematical modeling is an iterative process whereby we use mathematics to understand or analyze some situation that often comes from outside mathematics. To illustrate the definition, consider the following situation:

> *The weather forecast calls for heavy rain for several hours. It is expected that the water level of a river that goes through town will rise above its banks in one particular section and cause major flooding, so the residents want to protect themselves from the flood by elevating the riverbank using sandbags. How long will this take?*

To estimate the answer, we can use mathematics. If we can find out the dimensions of the bags filled with sand, the proper way to stack them, the desired height of the sandbag wall, and the length of the river section that needs to be protected, we could develop a mathematical formula that tells us how many sandbags we might need. We can then estimate the time it will take to fill the bags and build the protection wall depending on the number of helpers and additional assumptions. The formulas themselves constitute the model in this case. The entire process is mathematical

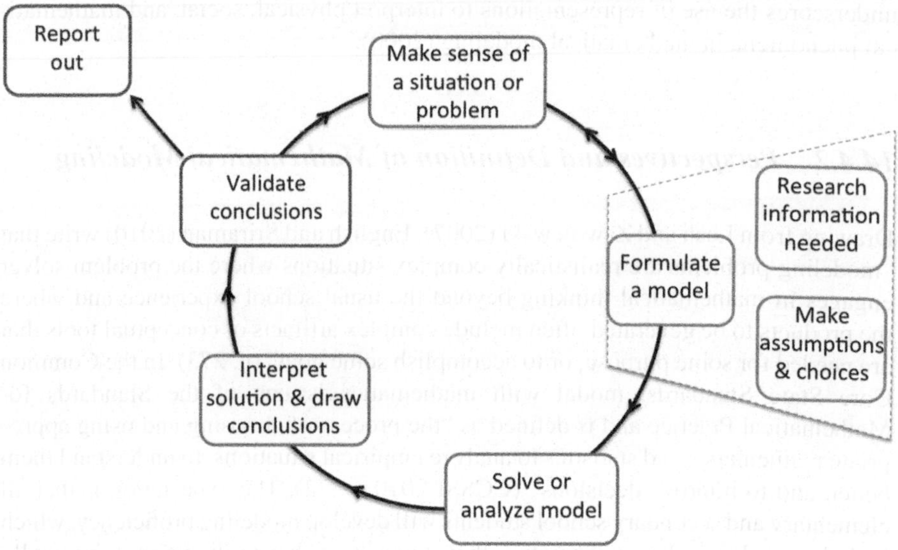

Fig. 14.1 A representation of the mathematical modeling process

modeling. Even after doing this, we may find that additional variables or parameters need to be taken into account and the formulas will have to be adjusted. For example, the sandbag thickness near the bottom of the wall may be smaller due to the weight of the sandbags on top; or the time between filling the bags and placing them on the wall may get longer as the wall grows. Such adjustments to the model can be made iteratively.

14.4.3 Elements of Mathematical Modeling

One of the components of the modeling process is the formulation of a model. A mathematical model is a "simplification of reality that is phrased in the symbolic language of mathematics [that] can take the form of equations, algorithms, graphical relations, and sometimes even paragraphs" (SIAM 2012, p. 11). In education, a definition of *models* is given by Doerr and English (2003) as "systems of elements, operations, relationships, and rules that can be used to describe, explain, or predict the behavior of some other familiar system" (p. 112). The modeling process, however, has additional elements that are usually represented as stages of a cycle like those shown in Fig. 14.1. Similar representations have been depicted in many sources, including textbooks (Mooney and Swift 1999) and mathematics and mathematics education journals (e.g., Blum and Leiss 2005; CCSSI 2010; Felton et al. 2015; Meier 2009; Yoon et al. 2010).

The figure emphasizes the iterative nature of the modeling process and identifies salient elements, all of which could be expanded further. Starting with a situation to be analyzed, the first stage is to make sense of it and understand the questions that need answers. The next major stage is the formulation of a model, which involves formulating a mathematical problem that represents a simplified or distilled version of the original situation. The model formulation step may involve substages such as determining essential variables, making assumptions about any missing information, and choosing appropriate mathematics (e.g., statistics or linear functions) for the model. Typically, sense-making continues during this part of the process, and some research is necessary to make reasonable assumptions.

Once the model is formulated as a set of equations, a graph, or a table of values, the problem-solving step leads to a mathematical solution that needs to be interpreted in the original context. Conclusions about the original situation are drawn from this interpretation, and the conclusions must be evaluated in a validation stage in order to determine if they make sense in terms of the original situation. Since the mathematical answer is influenced by the assumptions and choices made earlier, the conclusions may not be satisfactory based on the needed accuracy, the applicability of the solution, or some other factor. If this is the case, a new iteration is entered where assumptions and choices are revised with an eye on overcoming the shortcomings of the first model. The cycle may be repeated once or multiple times until satisfactory conclusions are reached and can be reported.

14.5 Mathematical Modeling and Culturally Relevant Pedagogy

Culturally relevant pedagogy is based on the assumption that when academic knowledge and skills are situated within the lived experiences and frames of reference of students from various cultural backgrounds, they are more personally meaningful, have higher interest appeal, and are learned more easily and thoroughly (Gay 2000; Ladson-Billings 1995). Although CRP and mathematical modeling are both significant and well-respected contemporary pedagogies, there has been relatively little explicit scholarship on ways to integrate their strengths. The cycle of mathematical modeling is inherently challenging and reflective, and it depends on contextual knowledge of everyday situations. For these reasons, we suggest that mathematical modeling corresponds naturally to the tenets of CRP. In particular, mathematical modeling pedagogy can address some of the weakness of implementation of CRP that has been observed of the past few decades. Mathematical modeling activities are as follows: (a) motivate mathematics content, (b) promote discussion between students, and (c) integrate contexts relevant to students. In what follows we describe how these three characteristics of modeling relate to CRP.

Motivating Mathematical Content A modeling task can address specific content and build upon content previously learned. At the same time, because the models developed by students are limited by their mathematical knowledge and experience in recognizing essential variables and their relationships, the task also serves as a springboard for discussing content that is new to students. By grounding models in students' lived experiences, the cultural context can motivate students—to offer them a reason to conduct mathematical activities. Ladson-Billings' intention was that CRP would be implemented holistically and that each of the three tenets would support each other. Conceptualizing modeling in this way helps strengthen this dimension of CRP, that students must succeed in a rigorous academic environment and that personal knowledge is a factor that leads to this success.

Promoting Discussion Between Students The modeling process promotes mathematical discourse as it requires justifying the choices and assumptions made along the way, the selection of variables and mathematical concepts for the model, and the choices of representations. Substantial communication is also needed to report the solution and critique others'. Appropriately designed modeling problems provide opportunities for students to actively use mathematical language to communicate meaning about and negotiate meaning for mathematical situations. Culturally based modeling activities touch on students' cultures and demand that students communicate the connections between the context and the mathematics they have used in their models. This promotes understanding of students' cultures and brings the significance of students' cultural background to the foreground. Teachers who intentionally plan modeling discussions can include topics that assist students' integration of cultural knowledge as it is realized in community and household activities.

Integrating Relevant Contexts The entire modeling activity may be motivated by activities that are familiar to students. That is, modeling allows us to draw on students' funds of knowledge and design activities that are culturally relevant. The modeling process contains opportune moments to draw cultural knowledge into mathematical problem-solving. In the initial stages of the cycle, students usually establish simplifying assumptions that allow them to create their mathematical models. Some of these assumptions are motivated by mathematical needs—students will choose to use mathematical structures that they understand—but also, they will be based on students' knowledge of the context of the task. Explaining why an assumption is reasonable will rely partly on students' lived experiences.

To complete the first cycle of the modeling process, students reflect on the strengths and weaknesses of their model, they discuss whether their solutions are reasonable, and they plan a subsequent cycle of improvements to the model. These interpretation and validation stages of modeling are moments in the modeling cycle in which students' contextual knowledge is important. Typically, validation involves critical reflection on the mathematical scope and accuracy of the model, but this critical reflection could extend to questions of equity, access, and fairness when these concepts are relevant to the model's context. We suggest that the critical reflections of the final stages of the first modeling cycle are appropriate times for the teacher to engage students in discussions to explore and strengthen critical consciousness.

When exploring ideas taken from students' background cultural knowledge, a natural yet sometimes uncomfortable next step is to ask students how they view the idea in the context of the world views, such as taking into consideration the political, social, and/or economic perspectives that are associated with the problem situation. Exploring these aspects of the problem situation is important, yet it is essential to focus on how the mathematics helps explain the situation. Integrating CRP into the modeling cycle can show students how to use mathematics to interact more powerfully in the social world, so that mathematical modeling activities can promote taking action.

14.6 Strengthening Implementation of CRP Through Mathematical Modeling

CRP is a significant advancement for connecting lived experience to mathematical explorations, but as we have seen, there are challenges in the implementation of each of its three tenets. Teachers must be able to implement all three tenets in an integrated way so that they inform one another. Incorporating a strategy for culturally relevant teaching into the mathematical modeling cycle can address some of these dilemmas. Modeling improves the rigor of the curriculum, and the modeling cycle allows teachers to plan discussions that address cultural competence and critical consciousness at specific stages. Using this approach, teachers can attempt to implement Ladson-Billings' construct in the manner in which it was intended. In the following section, we describe a mathematical modeling activity involving mathematical functions in a community cultural context, along with topics that have the potential to raise critical questions about the social world.

14.6.1 Community Contexts for Mathematical Modeling

This modeling activity, presented in Fig. 14.2, was created as an illustration of a problem that allows students to experience the mathematical modeling process as they work through the problem.

There are several reasons why this problem is a good choice as a modeling activity. First, there is no particular correct answer that students must find. In fact, the problem does not ask to find a number or a specific expression or formula. Instead, the students are asked to propose functions that have qualitative features, which allows students to use creativity and prior knowledge to suggest functions. Since there are multiple correct possibilities, there is opportunity for students to reveal their knowledge and personal choices.

Second, the mathematical modeling requirements are the same regardless of the specific pictures of fences that the students bring. However, the variety of options that can result from different fence shapes and different gate purposes provides multiple directions for students to explore and opportunities to justify specific choices to formulate their models. It is known that "different purposes may result in

Neighborhood fences and gates: Design using mathematics

As you walk around your neighborhood, you will see lots of fences and gates in the front yards of houses. The design of the fences can be described by mathematical functions.

1) Walk around your neighborhood and take pictures or draw sketches of yard fences or gates of different shapes. If your house has a fence be sure to include it.

2) Find mathematical functions that can be used to design the fences in the pictures or sketches. Include the domain of the functions. Since the pictures don't have coordinate axes, you will need to make choices about the height and width of your functions and possibly about other parameters. Be sure to list the choices you make and your reasons for making those choices.

3) Fences and gates can have different purposes. Use your imagination to sketch or describe a new fence shape that you find interesting and that has a unique shape. Think of where your fence might be used and what purpose your fence might have. Then find a function that describes your fence and explain the choices you made and how those choices are connected to the purpose for your fence.

4) Provide a set of instructions that you can give to a picket fence builder. Your instructions should include the number of pickets and their width, the height of each picket, the separation between pickets and the order in which they should be installed.

5) Look for public places or private homes in your neighborhood that do not have fences. Propose a fence or a gate for one of these places based on its purpose and provide a mathematical function for it.

Fig. 14.2 The "neighborhood fences and gates" problem

different mathematical models of the 'same' reality" (Jablonka 2007, p. 193). For instance, a decorative fence may be relatively low and have more spacing between pickets compared to a security fence or a fence to keep a pet from running out of the yard. These considerations affect the choice of parameters needed for the model. As an example, Fig. 14.3 shows photographs of fences and a gate with different heights and shapes. Based on observations of the photos, some students may choose to create a single function for the entire fence, while other students may reason that a fence is made of repeating segments and choose to provide a function for the segment only. Additionally, the parabolic-looking fence on the right photograph can be modeled by a polynomial, a trigonometric function, or some other curve. These choices are part of model assumptions.

Third, the functions that students produce constitute an initial model. The interpretation of the graph of their functions as the shapes of fences or gates can conclude a first pass of the iterative modeling process. There are several options for revising the model, some based on the shapes (*Are the functions high enough? Do they dip too low in some places? Are they aesthetically pleasing or should they be modified?*) and some based on the representation of the functions. For instance, a bar graph of a function may give a better visual idea of what the fence will look like. Figure 14.4 shows two representations of the same function for values of x in the

Fig. 14.3 Examples of neighborhood fences and gates

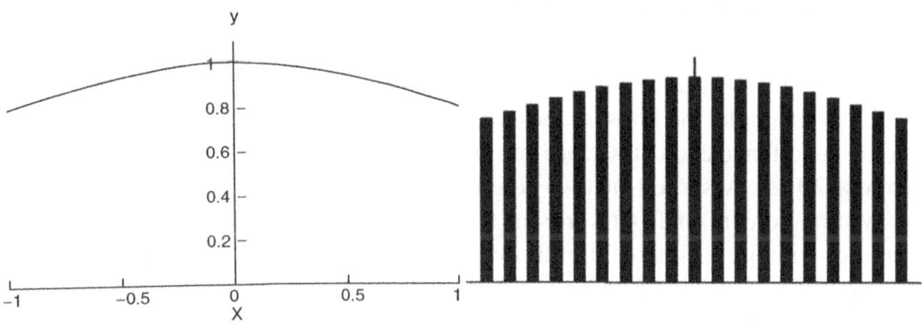

Fig. 14.4 Example of two ways of representing the function $y(x)=\frac{1}{5}\cos\left(\frac{\pi x}{2}\right)+\frac{1}{4}$

interval [−1,1]. The bar graph on the right gives a better sense of the fence. Refinement in the functions themselves or their representations can be attempted as iterations within the modeling process.

14.6.2 Mathematical Content of the Task

The Fences task involves substantial content related to functions, some of which is mentioned explicitly and some that is intended to surface as the students work on the task. The domain of the function is explicitly requested, and it is directly connected to the mathematical modeling assumptions. For instance, if the bottom of the fence is the x-axis and the top of the fence is given by the graph of $y = f(x)$, the domain of the function $f(x)$ determines how wide the fence will be as well as the minimum and maximum heights of the fence. If the fence is assumed to be at least 5 ft high, the range of the function must satisfy this assumption.

A traditional question might provide a specific function $f(x)$ and ask the students to determine the domain and range. The Fences task is perhaps more challenging since it asks the students to produce a function whose range has particular features like "it cannot include values of y less than 5." The students also have to make assumptions regarding the height and width of the fence segments and translate those assumptions into the range and domain of their functions.

Part #3 of the activity is wide open for students to explore new functions and provides the opportunity to discuss concepts like even functions, functions that are neither even nor odd, periodic functions, piecewise functions, etc. Part #4 addresses the more practical side of the activity and expects students to generate a table of values for each of the fence pickets to be created.

14.6.3 Connection to Culture and Community

Fences are part of residential landscapes that contribute to a community's cultural space. For example, the housescapes, including house colors, religious images, decorations, and fence enclosures, are common features of many Latino neighborhoods (Arreola 2012). A survey of two neighborhoods in central Phoenix revealed that "Sixty-nine percent of front yards in Garfield [mostly Hispanic] were completely enclosed, with only 18 percent of those in Coronado [mostly non-Hispanic] fenced" (Manger 2000, p. 6), but residents in both neighborhoods perceived the purpose of fences to keep pets or children in the yard, to keep trespassers out, or to demarcate boundaries (Manger 2000).

The Fences task asks students to look through their neighborhoods for examples of fences or gates and to consider their purpose. In this way, the students will bring a piece of their neighborhood to the classroom and share it as part of the activity while simultaneously considering cultural implications of the purposes for fences. Throughout the problem, the focus of the activity is the mathematical functions that represent the tops of fences and the connection between mathematics to aspects of the students' lives.

14.6.4 Connection to Critical Consciousness

Part #3 of the problem alludes to the fact that some fences may be purely decorative or have purposes related to security or privacy. Without explicitly mentioning these purposes, the problem lets the students suggest possibilities and opens the door for a discussion on perceptions of crime in neighborhoods and social implications of such perceptions. Importantly, this part of the problem is not divorced from the mathematics as it asks students to think about and justify how the purpose of the fence/gate affects the mathematical choices they make in their design. The last part of the problem (item #5) makes a connection between the students' findings and suggestions to a concrete action that may improve or otherwise effect a change in their neighborhoods. The purpose that students cite may be aesthetics, security, or something else. Throughout the process, the problem emphasizes the mathematical knowledge required of the students.

14.6.5 Connection to Academic Success

As aforementioned, the problem addresses functions and function representation in a nontraditional way. In contrast to traditional textbook problems which typically provide a function to be graphed or provide sufficient information to determine a unique function, this task requires students to suggest functions that have certain general features, which may be met by several functions. Students must understand how to produce functions with the given features and, further, provide new features of their choice and produce functions that meet them. This kind of task requires a high level of understanding of functions.

14.7 Implications for Teaching: Balancing the Tenets of Culturally Relevant Pedagogy

Practicing CRP in the context of mathematical modeling may seem like a daunting task to many teachers since both are demanding in terms of time and knowledge. Nevertheless, we have made a case for the natural integration of CRP and mathematical modeling because teaching mathematics with the expectation that all students succeed academically is at the heart of both. Since mathematical modeling draws on students' mathematical knowledge while offering opportunities for new mathematical content to be developed, teachers can support students in critical thinking about their approach to mathematical modeling. For this reason, modeling tasks have the potential for teachers to leverage diverse students' everyday lived experiences for meaningful engagement with challenging mathematics. The way the students maneuver around the modeling process is informed by their culture and "ways of thinking" which are formed by their everyday lived experiences.

As students show evidence of logical reasoning, especially for improving their initial models by reevaluating their assumptions, teachers can use this opportunity to extend student thinking and ask for justification, motivation, and explanation of the improvements. Given that any mathematical model can be improved in some way, classroom discussions can develop both critical consciousness and mathematical strategies once students have completed the first cycle of modeling. The following questions are designed to help teachers chart a discussion pathway from the mathematics that students use to self-awareness of how culture influences their decision-making to social consciousness and critical views of the world.

- What mathematics did you use to create your initial model of the problem?
- What other mathematics could you have used?
- How is your model similar or different to other models created by peers?
- What information did you need to research to make assumptions for your model?
- What influenced you to choose the assumptions you came up with?

- What "ways of thinking" from your background knowledge and culture impact your decisions in the modeling process?
- What aspects of your model do you think can be revised and improved?
- What aspects of your model help you think about social issues that impact people in various places in the world? These social issues could include issues related to economic, social equity, fairness, safety and protection, and political influences in people's lives.

As a concrete example, the following are possible questions about the Fences task presented earlier as it relates to middle or high school students:

- What did you notice about the type of fences and gates that you found? From what materials are the fences made? What is the purpose for the fences that you found?
- What do you think is the cost of these different kinds of fences?
- What is the relationship between the cost of the fences and the design of the fences? What is the relationship between the cost of the fences and the purpose for the fences?
- If your family wanted to put a fence in your front or backyard, what would you choose for materials or design? How could you determine the cost of the materials?
- How much artistic or aesthetic value would you like your fence to have? Is this important to you or your family?
- When families settle in a new country, are there costs involved for people who want to maintain aspects of their culture? How could this affect fence choices?
- Which of our class fence designs would cost the most? The least?
- If you wanted to make your fence more culturally aesthetic, and only increase the cost by a little, how would you do it?

These discussion questions attempt to tie choices about cultural conservation and aesthetics to household financial decision-making. In many case studies of culturally relevant teaching, the discussion begins with students identifying problems in their community (e.g., Ladson-Billings 1995; Tate 1995; Turner and Font Strawhun 2007; Turner et al. 2009). In the mathematical modeling context, discussions of critical consciousness can occur between the formulation of an initial model and making decisions for possible improvements. This teaching trajectory allows the teacher to observe the type of mathematics that the students use in the initial model and then guide them to increase the level of mathematics, specifically when the teacher knows the kinds of connections that students could make to improve the model. This is a useful strategy when teachers feel pressure to align student mathematical work with curriculum standards. Conducting a critical consciousness discussion between modeling iterations also helps achieve the original intention of creating a unified sense of purpose for mathematics and critical consciousness.

14.8 Implications for Teacher Education

Achieving a balance of rigorous mathematics content, cultural competence, and critical consciousness through mathematical modeling is a complex endeavor yet an attainable goal that needs much attention. It is necessary for teacher education to provide experiences with mathematical modeling that can prepare teachers to engage their students in the mathematical modeling process. Two critical aspects of this teacher preparation are (1) becoming comfortable posing modeling problems that are open-ended and different from traditional textbook problems and (2) understanding the concept of making assumptions, as this is something that they may not have experienced explicitly before in mathematics.

For effective teacher preparation, teacher educators must become fluent with the nature of the mathematical modeling cycle as an approach to solving open-ended problems in familiar contexts. In order to promote creativity, teacher educators should resist steering teachers toward predetermined modeling approaches but rather support their own thinking to develop their models. Time should be taken to uncover how teachers' backgrounds influence their modeling approaches and to have open discussions with teachers about their cultural influences on learning, especially in decision-making during mathematical modeling. For teacher professional development, it is important to include projects in which teachers of various grade levels in K-12 collaborate to experience diverse mathematical modeling tasks to develop understanding of the modeling process while simultaneously work toward inclusion of the CRP tenets.

For prospective and in-service teachers, understanding the various aspects of CRP can take place through readings, discussion, and engagement in a problem-based project to bring together the various elements. Building on these components, teachers can strategize and build a progression for ways of teaching mathematical modeling with the relevant cultural aspects that help shape critical consciousness for students. Further deepening of this aspect would require implementation, analysis of student work, continuation of collaboration through discussions, building more context-rich and relevant modeling tasks, and continuous reflection for improving the teaching of mathematical modeling.

14.9 A Brief Look at the Mathematical Modeling Module with Cultural Aspects

We close this chapter with a brief look at the module we implemented with a group of PTs, in a sophomore-level mathematics pedagogy class in a department of mathematics. Because none of the PTs in the class had taken a mathematics course in mathematical modeling nor had they had course work relating mathematics and culture, we assumed that most of the ideas would be new to most of them. We first introduced the construct of culture by assigning a reading, "What is Culture?" by

González (2008), to discuss the role of culture on the learning and teaching of mathematics. This centrality of culture was followed by the introduction of mathematical modeling problems that touched on cultural aspects.

Consistent with the Funds of Knowledge approach, our definition of culture was that of lived experiences. We had several activities to engage students with discussions around culture, including having students build individual "identity maps" to share their individuality, readings (Gay 2002; González 2008), a video (Teaching Tolerance 2010), discussions around major points pertaining to culture and CRP, and guest speakers (Norma González on culture and Funds of Knowledge; Marta Civil on topics of culture and Funds of Knowledge in the mathematics classroom). The PTs' immediate questions and concerns were about the teaching of secondary mathematics concepts and how to include culturally relevant aspects.

We followed this with a mathematical modeling problem that tied in with culturally relevant aspects, "'Cuts & Styles' is a hair salon that claims to serve over a million customers per year. Is this reasonable? Under what conditions could this be true? Create a mathematical model for this situation." This simple and open-ended problem required the PTs to consider many assumptions drawn from their knowledge and experiences in hair cutting including knowing particular owners of local salons. These experiences were shared in small group discussions before formulating a model. This problem allowed for PTs to analyze and compare traditional textbook problems with aspects of this particular problem. The assumptions were based on their personal experiences of getting haircuts and considering the variables involved, such as the location (to determine how busy the place could be) and the number of minutes for haircuts and styles for short and long hair. Economics became part of the discussion on cost of haircuts and styles; some students claimed that they did not cut their hair often because of the expense, which also led to discussions about various places and the cost associated.

Another assignment included reading and discussing the pertinent pieces of the Common Core State Standards in Mathematics (CCSSI 2010), specifically the mathematical modeling cycle as described in the high school conceptual category (pp. 72–73) and the K-12 mathematical practice, model with mathematics (p. 7). In addition, the PTs engaged in transforming traditional textbook problems into modeling problems. This proved to be challenging because there was a sense of uneasiness with leaving out parameters and posing open-ended problems. One example of a traditional textbook problem (Jacobs 1982, p. 140) that was given to the PTs was the following:

A person's shoe is a function of the length of his or her foot. Formulas for this function for men's and women's shoes are given below: x represents the length of a person's foot in inches and y represents the corresponding shoe size.

Men's shoe size is $y = 3x - 25$.
Women's shoe size is $y = 3x - 22$.

(a) Graph both functions on one pair of axes. What do you notice about their graphs?

(b) If a man and a woman have feet of the same length, who has the larger shoe size?
(c) If a man and a woman have the same shoe size, who has the longer foot?

Following this activity, the PTs engaged in the Fence mathematical modeling activity shown in earlier in this chapter.

The PTs shared that this task solidified their understanding of the community context and its relevance to mathematics learning. Each PT found images of fences and gates around their own community (mostly around the university campus) and created functions related to the designs of the fences and gates in their images. This activity promoted much discussion on the elements of modeling, mainly around the notion of creating function models of the top edge of fences and within their designs. Additional discussion was around the purpose of these fences in their communities including the safety of their neighborhoods and possibly how the taller and less aesthetic fences correlated to some kind of safety factor. We recognize that PTs are not traditionally asked to consider culture in preparing mathematics problems, so this example proved to be fruitful in underscoring the tenets of CRP and the elements of mathematical modeling.

In the end, it was evident that one module in one course with several mathematical modeling example problems may not provide enough experiences for PTs to feel fully comfortable or confident in incorporating the components of CRP. Several PTs indicated that they understood how the cultural backgrounds of students can have a role in the learning process of modeling problems but that the critical consciousness connections with mathematics were less developed for them. This general reflection from the PTs lead us to believe that more mathematical modeling problems that incorporate the CRP tenets are necessary to help PTs develop their understanding of the mathematical modeling process for teaching it through the lens of academic rigor in mathematics while integrating students' cultural backgrounds and helping students develop critical consciousness of meaningful social issues.

There are some limitations to integrating the practice of CRP with mathematical modeling. First, while modeling is prominent in mathematics applied to everyday situations, it is more difficult to connect abstract mathematical concepts to cultural knowledge and attempting to do so may oversimplify either the mathematics or the cultural understanding. The PTs who engaged in the module made this observation. Second, the modeling process requires students to translate back and forth between the situation context and a mathematical model. This translation is informed by students' lived experiences, but once a mathematical model is constructed, the students enter a problem-solving realm in order to compute a solution of the equations—or other mathematical constructs—in the model. This stage can be unrelated to the context of the problem (some algorithms used to solve problems can have cultural connections). Consequently, the link between culture and mathematics is temporarily interrupted, which can cause a loss of continuity in the CRP process. Similarly, a disproportionate emphasis on the social issues that a situation evokes can relegate mathematics to a mere tool rather than a discipline whose understanding must be solidified and expanded.

14.10 Conclusion

Through rich mathematical modeling problems, students are able to work within the tenets of CRP: achieving through mathematics, building cultural self-awareness, and developing critical consciousness. Various elements of the mathematical modeling process require students to make decisions, for example, formulating a model requires that assumptions be made, or in interpreting and validating the model after analyzing the results. We argue that students' background knowledge including cultural backgrounds, lived experiences, and mathematical knowledge inform the modeling process. Because mathematical modeling requires students to consider relevant information they may know about the problem situation, decision-making, formulating a model, and finally interpreting and validating the outcomes, we argue that the process requires ownership of the mathematics and navigation through the modeling cycle. Students bring in their "ways of thinking" about the mathematics and the social contexts and implications that the problem situation presents to them.

By incorporating rich mathematical modeling problems that involve students' researching of their own communities, we can provide opportunities for learning school curriculum mathematics in a way that is most relevant to students. Having specific knowledge of the cultural background of the students in a class makes it possible for a teacher to present modeling tasks that connect to the students' lives and promote discussions about issues that students care about. Once the students take ownership of a problem, they can engage in more meaningful discussions about the mathematics and other social issues that may be important to them.

Mathematical modeling can be thought of as a way to bring together a set of mathematical concepts, selected by the students, and apply them strategically to address a situation that comes from any part of life. This freedom to use mathematical reasoning to address issues in contexts outside typical school mathematics is precisely why mathematical modeling lends itself nicely to CRP. Aspects of students' cultures related to school regulations, social inequities, truth in advertisement, hobbies, health, etc. can be investigated and discussed with the use of mathematical models. Learning to use modeling as a framework for accessing students' funds of knowledge, as Bateson (2000) would have it, helps teachers lead students across lines of strangeness into a world of socially aware mathematical exploration.

Teachers may encounter some tension between incorporating authentic cultural knowledge into the modeling process while staying true to the goals and modes of analysis of the discipline of mathematics. Other chapters in this book allude to this tension—or aspiration—in the contexts of science and engineering education. In Chap. 4 of this book, for example, Sjöström and Eilks (2018) provide a nuanced discussion of the dimensions of critical-reflexive *Bildung* in science education, the knowledge of self, society, and capacity for action. The principle of *Bildung* resonates with culturally responsive pedagogy through the valuation of increased awareness of a cultural self and the understanding that STEM disciplines can create a better and more just world. In Chap. 8 of this book, Purzer, Moore, and Dringenberg (2018) describe engineering design as an iterative process that alternates between acquiring and applying knowledge (Fig. 8.3). The knowledge acquisition stage recognizes that the initial problem statement will be ambiguous and partial, so that students need to learn to question and communicate deeply with the client. In Chap. 10 of this book, Carberry and Baker (2018) recognize that engineers need to engage users more deeply than the discipline sometimes values, to become sensitive to cultural, economic, and power-laden fault lines that can sink an engineering project. Awareness of the Funds of Knowledge approach with its direct and deep engagement in communities could contribute to culturally sensitive design processes in many STEM fields.

While we stress that there is no consensus on modeling pedagogies in any of the STEM fields, we also note that those fields that use an iterative design or pedagogical process may be able to incorporate perspectives from our chapter. Our proposed pedagogical model asserts that the stages of mathematical modeling provide valuable moments to access students' culturally based knowledge and to use this knowledge as a resource for learning. We offer this approach as a step forward in the development of culturally relevant modeling pedagogy.

References

Aguirre, J. M., & Zavala, M. (2013). Making culturally responsive mathematics teaching explicit: A lesson analysis tool. *Pedagogies: An International Journal, 8*(2), 163–190. doi:10.1080/1554480X.2013.768518.

Anhalt, C., & Cortez, R. (2015). Mathematical modeling: A structured process. *Mathematics Teacher, Focus Issue, February, 108*(6), 446–452.

Arreola, D. D. (2012). Placemaking and Latino urbanism in a Phoenix Mexican immigrant community. *Journal of Urbanism: International Research on Placemaking and Urban Sustainability, 5*(2–3), 157–170. doi:10.1080/17549175.2012.693749.

Ascher, M. (1991). *Ethnomathematics: A multicultural view of mathematical ideas*. Pacific Grove: Brooks/Cole.

Bateson, M. (2000). *Full circles, overlapping lives: Culture and generation in transition*. New York: Random House.

Berry, J., Burghes, D., Huntley, I., James, D., & Moscardini, A. O. (Eds.). (1984). *Teaching and applying mathematical modelling*. Chichester: Ellis Horwood.

Bishop, A. (1988). Mathematics education in its cultural context. *Educational Studies in Mathematics, 19*, 179–191.

Blum, W., & Leiss, D. (2005). "Filling up"—the problem of independence-preserving teacher interventions in lessons with demanding modeling tasks. In *Proceedings of the fourth Congress of the European Society for Research in Mathematics Education* (pp. 1623–1633). Presented at the European Research in Mathematics Education IV, Sant Feliu de Guíxols, Spain.

Borba, M. (1997). Ethnomathematics and education. In A. Powell & M. Frankenstein (Eds.), *Ethnomathematics: Challenging eurocentrism in mathematics education* (pp. 261–272). Albany: SUNY.

Burkhardt, H. (1984). Modelling in the classroom: How can we get it to happen? In J. S. Berry, D. N. Burghes, I. D. Huntley, D. J. G. James, & A. O. Moscardini (Eds.), *Teaching and applying mathematical modelling* (pp. 39–47). Chichester: Ellis Horwood.

Carberry, A. R., & Baker, D. R. (2018f). The impact of culture on engineering and engineering education. In Y. J. Dori, Z. Mevarech, & D. Baker (Eds.), *Cognition, metacognition, and culture in STEM education* (pp. 217–239). Springer.

Civil, M. (2007). Building on community knowledge: An avenue to equity in mathematics education. In N. Nassir & P. Cobb (Eds.), *Improving access to mathematics: Diversity and equity in the classroom* (pp. 105–117). New York: Teachers College Press.
Civil, M. (2016). STEM learning research through a funds of knowledge lens. *Cultural Studies of Science Education, 11*(1), 41–59. doi:10.1007/s11422-014-9648-2.
Civil, M., & Andrade, R. (2002). Transitions between home and school mathematics: Rays of hope amidst the passing clouds. In G. de Abreu, A. J. Bishop, & N. C. Presmeg (Eds.), *Transitions between contexts of mathematical practices* (pp. 149–169). Boston: Kluwer.
Civil, M., & Kahn, L. (2001). Mathematics instruction developed from a garden theme. *Teaching Children Mathematics, 7*, 400–405.
Clifford, J. (1988). *The predicament of culture: Twentieth-century ethnography, literature, and art*. Cambridge, MA: Harvard University.
Common Core State Standards Initiative. (2010). *Common core state standards for mathematics*. Washington, DC: National Governors Association Center for Best Practices and Council of Chief State School Officers. Retrieved from http://www.corestandards.org/assets/CCSSI_Math%20Standards.pdf
d'Ambrosio, U. (1985). Ethnomathematics and its place in the history and pedagogy of mathematics. *For the Learning of Mathematics, 5*(1), 44–48.
d'Ambrosio, U. (2006). *Ethnomathematics: Link between traditions and modernity*. Rotterdam: Sense Publishers.
Doerr, H. M., & English, L. D. (2003). A modeling perspective on students' mathematical reasoning about data. *Journal for Research in Mathematics Education, 34*(2), 110–137.
English, L., & Sriraman, B. (2010). Problem solving for the 21st century. In B. Sriraman & L. English (Eds.), *Theories of mathematics education: Seeking new frontiers* (pp. 263–290). Berlin: Springer.
Felton, M., Anhalt, C., & Cortez, R. (2015). Going with the flow: Challenging students to make assumptions. *Mathematics Teaching in the Middle School.*, Focus Issue on Mathematical Modeling, *20*(6), 342–349.
Garfunkel, S., & Montgomery, M. (Eds.). (2016). *Guidelines for assessment and instruction in mathematical modeling education (GAIMME) report*. Boston: Consortium for Mathematics and Its Applications (COMAP)/Society for Industrial and Applied Mathematics (SIAM).
Gay, G. (2000). *Culturally responsive teaching: Theory, research, and practice*. New York: Teachers College Press.
Gay, G. (2002). Preparing for culturally responsive teaching. *Journal of Teacher Education, 53*(2), 106–116.
González, N. (2008). What is culture? In A. Roseberry & B. Warren (Eds.), *Teaching science to English language learners: Building on students' strengths* (pp. 89–97). Arlington: National Science Teachers Association.
González, N., Andrade, R., Civil, M., & Moll, L. (2001). Bridging funds of distributed knowledge: Creating zones of practices in mathematics. *Journal of Education for Students Placed at Risk (JESPAR), 6*(1–2), 115–132. doi:10.1207/S15327671ESPR0601-2_7.
González, N., Moll, L., & Amanti, C. (Eds.). (2005). *Funds of knowledge: Theorizing practice in households, communities, and classrooms*. Mahwah: Lawrence Erlbaum.
Greenberg, J. B. (1989). *Funds of knowledge: Historical constitution, social distribution, and transmission*. Paper presented at the annual meetings of the Society for Applied Anthropology, Santa Fe, NM.
Greer, B., Mukhopadhyay, S., Powell, A. B., & Nelson Barber, S. (Eds.). (2009). *Culturally responsive mathematics education*. New York: Routledge.
Gutstein, E., Lipman, P., Hernandez, P., & De los Reyes, R. (1997). Equity, mathematics reform, and research: Crossing boundaries in search of understanding. *Journal for Research in Mathematics Education, 28*(6), 709–737.
Henze, R., & Hauser, M. (1999). *Personalizing culture through anthropological and educational perspectives* (Educational Practitioner Report #4). Santa Cruz: Center for Research on

Education (CREDE), Diversity and Excellence. Retrieved from www.crede.ucsc.edu/products/print/eprs/epr4.html

Irvine, J. (2010). Culturally relevant pedagogy. *Education Digest, 75*(8), 57–61.

Jablonka, E. (2007). The relevance of modelling and applications: Relevant to whom and for what purpose? In W. Blum, P. Galbraith, H. Henn, & M. Niss (Eds.), *Modelling and applications in mathematics education, The 14th ICMI Study* (pp. 193–200). New York: Springer.

Jacobs, H. (1982). *Mathematics: A human endeavor* (2nd ed.). New York: Freeman Publishers.

Ladson-Billings, G. (1995). But that's just good teaching! The case for culturally relevant pedagogy. *Theory Into Practice, 34*(3), 159–165.

Lerman, S. (2000). The social turn in mathematics education research. In J. Boaler (Ed.), *Multiple perspectives on mathematics teaching and learning* (pp. 14–44). Westport: Ablex.

Lesh, R., & Zawojewski, J. S. (2007). Problem solving and modeling. In F. Lester (Ed.), *The second handbook of research on mathematics teaching and learning* (pp. 763–804). Charlotte: Information Age Publishing.

Lipka, J., Yanez, E., Andrew-Ihrke, D., & Adam, S. (2009). A two-way process for developing effective culturally based math: Examples from math in a cultural context. In B. Greer, S. Mukhopadhyay, S. Nelson-Barber, & A. Powell (Eds.), *Culturally responsive mathematics education* (pp. 257–280). New York: Routledge.

Manger, W. F. (2000). The "idealized" Mexican American housescape. *Material Culture, 32*(1), 1–36.

Meier, S. (2009). Mathematical modelling in a European context – a European network-project. In M. Blomhøj, & S. Carreira (Eds.), *Mathematical applications and modelling in the teaching and learning of mathematics*. Proceedings from Topic Study Group 21 at the 11th International Congress on Mathematical Education (pp. 207–216). Monterrey, Mexico, July 6–13, 2008.

Moll, L. C., Amanti, C., Neff, D., & Gonzalez, N. (1992). Funds of knowledge for teaching: Using a qualitative approach to connect homes and classrooms. *Theory Into Practice, 31*(2), 132–141.

Mooney, D., & Swift, R. (1999). *A course in mathematical modeling. Mathematical Association of America.*

Morrison, K., Robinson, H., & Rose, D. (2008). Operationalizing culturally relevant pedagogy: A synthesis of classroom-based research. *Equity and Excellence in Education, 41*(4), 433–452.

Moses, R. P., & Cobb, C. E. (2001). *Radical equations: Math literacy and civil rights*. Boston: Beacon Press.

National Council of Teachers of Mathematics. (2000). In National Council of Teachers of Mathematics (Ed.), *Principles and standards for school mathematics*. Reston.

Purzer, Ş., Moore, T. J., & Dringenberg, E. (2018). Engineering cognition: A process of knowledge acquisition and application. In Y. J. Dori, Z. Mevarech, & D. Baker (Eds.), *Cognition, metacognition, and culture in STEM education* (pp. 167–190). Springer.

Sandoval-Taylor, P. (2005). Home is where the heart is: A funds of knowledge-based curriculum module. In N. González, L. C. Moll, & C. Amanti (Eds.), *Funds of knowledge: Theorizing practices in households, communities, and classrooms* (pp. 153–165). Mahwah: Lawrence Erlbaum.

Schichl, H. (2004). Models and the history of modeling. In J. Kallrath (Ed.), *Modeling languages in mathematical optimization* (pp. 25–36). Boston: Kluwer.

Schmeichel, M. (2012). Good teaching? An examination of culturally relevant pedagogy as an equity practice. *Journal of Curriculum Studies, 44*(2), 211–231.

Sewell, W. (1999). The concept(s) of culture. In V. Bonnell & L. Hunt (Eds.), *Beyond the cultural turn* (pp. 35–61). Berkeley: University of California Press.

Sjöström, J., & Eilks, I. (2018). Reconsidering different visions of scientific literacy and science education based on the concept of *Bildung*. In Y. J. Dori, Z. Mevarech, & D. Baker (Eds.), *Cognition, metacognition, and culture in STEM education* (pp. 65–88). Springer.

Sleeter, C. (2011). An agenda to strengthen culturally responsive pedagogy. *English Teaching: Practice and Critique, 10*(2), 7–23.

Society for Industrial and Applied Mathematics. (2012). Modeling across the curriculum. Report on a SIAM-NSF Workshop. Arlington, Virginia. August 30–31, 2012. http://connect.siam.org/siam-reports-on-modeling-across-the-curriculum

Tam, K. (2011). Modeling in the common Core state standards. *Journal of Mathematics Education at Teachers College, 2*, 28–33.

Tapia J. (1991). *Cultural reproduction: Funds of knowledge as survival strategies in the Mexican American community*. Unpublished doctoral dissertation, University of Arizona, Tucson. Taylor, D., & Dorsey-Gaines, C. (1988).

Tate, W. (1995). Getting to the root: A culturally relevant approach to mathematics pedagogy. *Theory Into Practice, 34*(3), 166–173.

Teaching Tolerance. [screen name]. (2010). Culturally relevant pedagogy. [video file]. Retrieved from https://www.youtube.com/watch?v=nGTVjJuRaZ8&feature=related

Troyna, B. (1987). Beyond multiculturalism: Towards the enactment of ant-racist education in policy, provision and pedagogy. *Oxford Review of Education, 13*(3), 307–320.

Turner, E., & Font Strawhun, B. (2007). Posing problems that matter: Investigating school over-crowding. *Teaching Children Mathematics, 13*, 457–463.

Turner, E., Varley Gutiérrez, M., Simic-Muller, K., & Díez Palomar, J. (2009). "Everything is math in the whole world": Integrating critical and community knowledge in authentic mathematical investigations with elementary Latina/o students. *Mathematical Thinking and Learning, 11*, 136–157. http://dx.doi.org/10.1080/10986060903013382

Tylor, E. (1920, 1871). *Primitive culture*. New York: J. P. Putnam's Sons.

Vygotsky, L. S. (1978). *Mind in society: The development of higher psychological processes* (M. Cole, V. John-Steiner, S. Scribner, & E. Souberman., Eds.) (A. R. Luria, M. Lopez-Morillas, & M. Cole [with J. V. Wertsch], Trans.) Cambridge, MA: Harvard University Press. (Original manuscripts [ca. 1930–1934]).

Wagner, D., & Lunney Borden, L. (2012). Aiming for equity in (ethno)mathematics research. In B. Herbel-Eisenmann, J. Choppin, D. Pimm, & D. Wagner (Eds.), *Equity in discourse for mathematics education: Theories, practices, and policies* (Mathematics Education Library, pp. 69–88). New York: Springer.

Yoon, C., Dreyfus, T., & Thomas, M. (2010). How high is the tramping track? Mathematising and applying in a calculus model-eliciting activity. *Mathematics Education Research Journal, 22*(2), 141–157.

Young, E. (2010). Challenges to conceptualizing and actualizing culturally relevant pedagogy: How viable is the theory in classroom practice? *Journal of Teacher Education, 61*(3), 248–260.

Chapter 15
Discussion

Yehudit Judy Dori, Zemira R. Mevarech, and Dale R. Baker

The primary motivation for inviting authors to participate in writing this book was the need to include at least one representative of each of the four STEM domains, i.e., science, engineering, technology, and mathematics, in conjuction with each one of the three educational themes comprising the title of the book: cognition, metacognition, and culture. These two dimensions—the domain with its four values and the theme with its three values—gave rise to a matrix with 12 cells, each of which is a chapter in its own right. We ended up with 15 chapters: an introduction and a discussion, two chapters on technology and culture, and two chapters on engineering culture. Chapter 5 concerns both cognition and metacognition in cyberlearning. Table 15.1 presents the chapters according to their domains and the theme dimensions.

The cross product of the four STEM domains and the three themes—cognition, metacognition, and culture—has provided a rich, encompassing framework for this edited book's authors to express the extensive research and different views on STEM education from a variety of vantage points. The views expressed in the book chapters are indeed quite diverse. Thus, one might claim that the book is not focused enough. Yet, we maintain that the different contexts of the chapters and multiple foci of studies highlight the differences in the various domains of study covered in the book, notably, differences between science education and engineering education.

Y.J. Dori (✉)
Faculty of Education in Science and Technology, Technion, Haifa, Israel

The Samuel Neaman Institute for National Policy Research, Haifa, Israel
e-mail: yjdori@technion.ac.il

Z.R. Mevarech
David Yellin Academic College of Education, Bar-Ilan University, Jerusalem, Israel

D.R. Baker
MaryLou Fulton Teachers College, Educational Leadership and Innovation, Arizona State University, Tempe, AZ, USA

Y.J. Dori et al. (eds.), *Cognition, Metacognition, and Culture in STEM Education*, Innovations in Science Education and Technology 24,
https://doi.org/10.1007/978-3-319-66659-4_15

Table 15.1 Book chapters according to domain and theme

ThemeDomain	Cognition	Metacognition	Culture
Science education	Chapter 2: Teacher Cognition of Engaging Children in Scientific Practices; Crawford and Capps	Chapter 3: Students' Metacognition and Metacognitive Strategies in Science Education; Avargil, Lavi, and Dori	Chapter 4: Reconsidering Different Visions of Scientific Literacy and Science Education based on the Concpet of Bildung; Sjöström and Eilks
Technology education	Chapter 5: Designing for Collaborative Problem-Solving in STEM Cyberlearning; Crippen and Antonenko		Chapter 6: Technology, Culture, and Young Science Teachers: A Promise Unfulfilled and Proposals for Change; Yerrick, Radosta, and Greene
			Chapter 7: Technology, Culture, and Values: Implications for Enactment of Technological Tools in Precollege Science Classrooms; Waight and Abd-El-Khalick
Engineering education	Chapter 8: Engineering Design Cognition: a Process of Knowledge Acquisition and Application; Purzer, Moore, and Dringenberg	Chapter 9: Metacognition and Meta-assessment in Engineering Education; Wengrowicz, Dori, and Dori	Chapter 10: The Impact of Culture on Engineering and Engineering Education; Carberry and Baker
			Chapter 11: Engineering Education in Higher Education in Europe; Corlu, Svidt, Gnaur, Lavi, Borat, and Çorlu
Mathematics education	Chapter 12: Cognition, Metacognition, and Mathematics Literacy; Mevarech and Fan		Chapter 14: Mathematical Modeling and Culturally Relevant Pedagogy; Anhalt, Staats, Cortez, and Civil
		Chapter 13: Promoting Mathematics Teachers' Pedagogical Metacognition – a Theoretical-practical Model and Case Study; Kohen and Kramarski	

These differences stem from the diversity of cultures, including norms, values, and worldviews, as well as the different approaches to STEM education, which influence how cognition and metacognition are viewed.

For example, what teachers need to know in order to teach science to children (Chap. 2 in this book) is very different from what teachers need to know for teaching engineering design to children (Chap. 8 in this book). Both require complex cognitive and metacognitive processes, but despite having the end goal of educating children, they employ very different ways of thinking and reflecting on the educational practices. These differences are the result of the cultural contexts from which science education and engineering education derive their goals and the differences in the knowledge and skills required to teach science on one hand and engineering design on the other hand.

Teaching engineering design to undergraduate students (Chap. 10 in this book) is another cognitively complex process, which is influenced by the culture of engineering as practiced by professionals. It requires university instructors to have a set of knowledge that is different than the set that classroom teachers teaching engineering design need to have, reflecting the different goals of educating professional engineers rather than school children.

It is also evident that what constitutes literacy as a cognitive and metacognitive activity varies as a function of domains and the cultures that are characteristic of those domains. For example, there are major differences between mathematical literacy, discussed in Chap. 12, and science literacy, which Chap. 4 elaborates on. Thus, very different stances on literacy surface, with Mevarech and Fan (2018, in this book) on one side, while Sjöström and Eilks (2018, in this book), are on the other side. The former authors argue that mathematical literacy is the comprehension and use of mathematics as measured by the PISA assessment, while the latter claim that literacy in science consists of global citizenry.

Culture, as explored by our authors, depends upon grain size. Some authors have chosen to look at country-level influences, while others have looked at specific social influences, domain-specific cultures, and even the cultures associated with levels of schooling. Culture, especially at the national level, influences also the ways European and North American scholars address cognition, metacognition, and culture. However, what all of the authors addressing culture have in common is that culture influences what we think is important to know and be able to do. Culture also influences decisions about pedagogy. In particular, culture can both impede and support the adoption of technologies that have the potential to advance both learning and metacognition, as discussed in Chaps. 6, 7, and 8.

Across the book chapters, the discussion on metacognition has focused on the essence of metacognition and on how it can support teaching and learning. Metacognition in science, engineering, and mathematics education is the focus of Chaps. 3, 9, and 13, respectively. While authors agree on a core set of metacognitive strategies, primarily planning and monitoring, the metacognition they examine is different depending upon who performs the monitoring and the reflecting processes. The authors differentiate, for example, between teachers' pedagogical metacognition and students' metacognitive strategies while solving problem. Authors also grapple with the difficulty of fostering and supporting metacognition and determining whether, and to what extent, metacognition is taking place. Determining if metacognition is actually happening is often inferential, as it is measured indirectly by observations such as how well students solve problems. Wengrowicz, Dori, and Dori (Chap. 9 in this book) tackle the task of assessing metacognition by adding a layer of meta-assessment in the realm of pedagogy, making assessment a student-centered

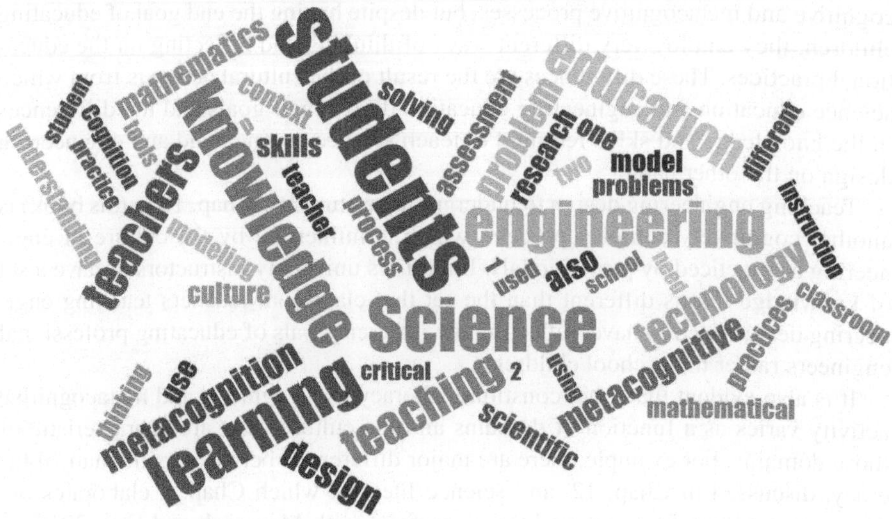

Fig. 15.1 Word cloud of the 50 most frequent words in the book

activity and meta-assessment as a complementary activity of the course staff, who assesses the individual students' quality of assessing peers' team projects.

Given the diversity of views about cognition, metacognition, and culture, this book presents the breadth of perspectives and the richness of research into these STEM domains and themes with their variety of approaches and definitions.

The word clouds in Figs. 15.1 and 15.2 are a graphic, fun way to determine where our authors focused their attention and to demonstrate the diversity of perspectives on cognition, metacognition, and culture under the umbrella of STEM domains.

15.1 Major Book Concepts and Keywords

Figures 15.1 and 15.2 are word clouds of the approximately 117,000 words in the book,[1] created by software developed by Jason Davis.[2] These figures represent the 50 and 100 most frequent words in the book, respectively. Figure 15.1 highlights where the authors were focusing their attention by their frequency as indicated by the size of the word. Words that appear with high frequency include students, science, knowledge, engineering, learning, teachers, education, teaching, technology, design, problem, metacognition, metacognitive, mathematics, scientific, research, model, mathematical, solving, understanding, thinking, modeling, problems,

[1] Excluding this chapter, the chapters' references, and the words "et" and "al".

[2] https://www.jasondavies.com/wordcloud/

Fig. 15.2 Word cloud of the 100 most frequent words in the book

cognition, practice, course, skills, research, culture, teacher, assessment, practices, process, instruction, literacy, cognitive, social, processes, information, tools, classroom, and school.

Doubling the number of words in Fig. 15.2 from 50 to 100 allowed some new relevant words to show up. These were STEM, development, studies, inquiry, content, tasks, society, activities, strategies, environment, Bildung, study, approach, and individual.

This collection of words provides a faithful mental image of the major topics dealt within this book. Interestingly, while "students" is the most frequent word, the singular "student" is over four times smaller. Similarly, "teachers," one of the most frequent words, is twice as frequent as "teacher." This might be an indication that the authors' reviews, studies, and discussions in the book tend to generalize for students and teachers as groups of stakeholders, rather than focusing on the individual student and teacher. The words STEM and culture are not as frequent as one might expect from a book on these subjects, but science, engineering, and education do appear amongst the most frequent words followed by technology and mathematics. Indeed, there is tension in the literature between unifying and separating the teaching of all or some of the STEM domains. However, this may be an artifact, because we called for papers that addressed science, technology, engineering, and mathematics as separate domains.

If we combine "metacognition" with "metacognitive," we get a word count that is about as frequent as science or knowledge, and if we combine "science" with "scientific," this becomes at least as big as most frequent word, "students."

While the word cloud examines frequency of single words, many of the keywords that the chapter authors defined are phrases that comprise two or more words. By switching from words that authors actually used to keywords they defined in their respective chapters, we emphasize how cognition, metacognition, and culture relate to the various STEM domains (see Table 15.2).

A more parsimonious way of looking at key words and phrases is to say that our authors are interested in students' learning and acquiring knowledge and skills in various contexts with support from teachers and technology.

In what follows, we integrate the keywords in Table 15.2 to elicit main ideas and insights by STEM domains and themes. We walk through the major ideas, as indicated by the keywords, using the underlying logic of tracing the path from knowledge to practices, such as nature of science, higher-order thinking, and scientific literacy. We then discuss major constructs of metacognition and the relation of both cognition and metacognition to culture.

Table 15.2 Keywords by STEM domain, theme, and chapter

Theme Domain	Cognition	Metacognition	Culture
Science Education	Chapter 2	Chapter 3	Chapter 4
	Higher order thinking	*Knowledge of cognition*	*Bildung*
	Inquiry-oriented pedagogy	*Metacognition* *Metacognitive strategies*	*Cultural perspectives*
	Nature of science	*Metacognitive science learning*	*Education for sustainability*
	Scientific inquiry	*Regulation of cognition*	*Global citizenship education*
	Scientific practices	*Metacognition assessment tools*	*Scientific literacy*
	Teacher cognition		*Transformative learning*
Technology Education	Chapter 5		Chapter 6
	Authentic practices		*Culture*
	Collaborative problem solving		*Culture and values*
	Cyberlearning		*Digital natives*
	Reflection		*Flipped classroom*
	Scaffolding		*Innovative teaching/learning*
			Technology
			Chapter 7
			Technology
			Culture and values *Culture of technology*
			Nature of technology

(continued)

Table 15.2 (continued)

Theme Domain	Cognition	Metacognition	Culture
Engineering Education	Chapter 8	Chapter 9	Chapter 10
	Knowledge acquisition	*Assessment*	*Engineering culture*
	Knowledge application	*Meta-assessment*	*Engineering education*
	Knowledge production	*Metacognition*	*Engineering education culture*
	Engineering design cognition	*Peer assessment*	
		Project-based learning	*Enculturation as an engineer*
	Design/inquiry design	*Student-oriented meta-assessment*	Chapter 11
			Anglo-American tradition
			Continental European tradition Engineering education
			Innovation
			Problem-based learning
Mathematics Education	Chapter 12		Chapter 14
	Complex, unfamiliar, and non-routine tasks		*Culturally relevant pedagogy*
	Mathematics literacy		
	Metacognitive self-directed questioning		*Funds of Knowledge*
			Mathematical modeling
	IMPROVE		
		Chapter 13	
		Metacognitive self-regulation	
		Pedagogical metacognition	
		Reflection	
		Web-based environment	

15.1.1 Cognition

We start with the first theme, *cognition*. In this book, authors refer to cognition of students and of teachers as a process of knowing, understanding, and thinking.

They frame *knowledge acquisition* as the process of building background knowledge and developing competencies for solving increasingly complex problems. Well-defined problems help scaffold knowledge acquisition in classroom contexts, often with focus on acquisition of theoretical scientific concepts. *Knowledge application*, as viewed by our authors, may include the process of knowledge transfer as students engage in complex design projects or mathematics situations. It requires determining and applying relevant knowledge from the students' prior knowledge, assuming that these students retain that knowledge from prior studies.

Knowledge production results from the ongoing iteration between knowledge acquisition and application, inducing deeper understanding of the problem context and consequently better design with a variety of features (Chap. 8 in this book).

Teacher cognition is a knowledge base for teaching, which includes formal propositional knowledge, practical knowledge, and beliefs, as discussed in Chap. 2. Cognition is thus an active learner-centered process. It is not supported by traditional educational practices such as lecture- or textbook-based instruction, challenging notions of efficiency and most forms of standardized paper-and-pencil assessments.

Scientific practices, also referred to as *inquiry teaching*, are consistent with recent education reforms in the United States. They align with how scientists carry out scientific investigations and inquiry. Scientific practices include multiple aspects of scientific work that are important for children to engage in and learn as students in the classrooms. They include human aspects of science, such as the recognition that science is empirically based, creative, and tentative (Chap. 2 in this book).

Scientific inquiry is a term encompassing the various ways in which scientists study the natural world. It involves a process of investigation, framed by asking testable questions and collecting and interpreting data to develop explanations about the natural world (Chap. 2 in this book).

Authentic practices, discussed in Chap. 5, encompass the instruments, strategies, and heuristics of a professional, including how to communicate in a range of technical forms to a variety of audiences, how to work with others in teams, and how to mentor and apprentice others who are less experienced. Engaging in what scientists do as authentic, scientific practice and inquiry have wide-ranging implications for curriculum development, teacher preparation, and assessment. Changes in curriculum development, teacher preparation, and assessment are currently underway, but they will not succeed unless there is a change in what policy makers and the public think of as teaching and learning. Beyond educating students, STEM educators will have to engage in educating the public through persuasive communication. This persuasion includes hard data that engaging in authentic scientific practices and inquiry results in positive academic outcomes for all children, as measured by a variety of assessment tools.

At the core of mathematics literacy are *complex, unfamiliar, and nonroutine (CUN) tasks* (Chap. 12 in this book). *Funds of knowledge*, defined in Chap. 14, are bodies of knowledge and skills that are essential for household or individual functioning and well-being and which have culturally developed and historically accumulated. This view of mathematics, advocated by mathematics educators, runs counter to the views of mathematicians who refer to mathematics as a logical closed system, worthy of study in and of itself, with no need to reference or map to the real world. These differences have contributed to the debate over whether STEM fields can be integrated or should remain distinct. In the United States, attempts to integrate mathematics with other areas of science, technology, and engineering in the K-12 curriculum have been met with resistance. The recently developed US Common Core Standards for mathematics present mathematics in a third way, addressing mathematics as both content and mathematics practices. This third way of viewing mathematics is also debated by 'pure' mathematicians.

Global citizenship education is, according to UNESCO, education that aims to empower learners to play active roles in facing and resolving global challenges and to facilitate learners' transition to being proactive contributors to a more peaceful, tolerant, inclusive, and secure world. Closely related to global citizenship education is *education for sustainability*: education, public awareness, and training that are presumed to be keys for achieving sustainability. In Agenda 21, the UN suggested that there is a need for education for sustainable development and provided a corresponding definition. In recent years, the idea of education for sustainable development is under constant debate, and similar terms like education for sustainability or sustainability education are used interchangeably. A critical view on these two concepts is presented in Chap. 4.

In the past, notions related to global citizenship and sustainability have been part of STEM standards and documents, but despite their presence, classroom instruction has focused heavily on core content. The Next Generation Science Standards, developed by the National Academies in the United States, addresses skills needed for global citizenship and places greater emphasis on sustainability than previous standards. However, it remains to be seen whether these standards will be implemented rather than ignored, as similar standards were in the past. Much depends upon whether assessments contain a substantial number of items that look into student understanding in these areas. Whether these standards are addressed in classrooms will also depend upon the political climate and views about the purpose of education in countries around the world. Those who deny climate change and view globalization negatively could exert influence to prevent global citizenship and sustainability from being addressed in the curriculum and classroom.

Underlying a host of cultural perspectives is the *nature of science* (NoS, see Chap. 2 in this book)—the idea that things in the universe exist and events in it occur in consistent patterns that are comprehensible by systematic gathering of information through various forms of direct and indirect observations and testing this information by methods including, but not limited to, experimentation. The principal product of science is knowledge in the form of naturalistic concepts and the laws and theories related to those concepts. Building on the notion of scientific inquiry, Chap. 2 elaborates on the related *inquiry-oriented pedagogy*—a teaching method that fosters students' deep understanding of NoS by engaging students in the practices of science, including posing questions, developing evidence-based explanations, building and using models, analyzing data, creating and defending arguments, using critical thinking, and communicating conclusions.

An almost synonymous term, very often used to refer to these skills collectively, is *higher-order thinking*, which indicates the kinds of students' cognitive activities that are beyond the lower level kinds of thinking, such as recall of memorized bits of information and following a well-defined algorithmic procedure to solve a quantitative problem by plugging numbers in formulae. Higher-order thinking, in contrast, means applying, analyzing, evaluating, transferring, and creating. Examples of higher-order thinking in science include constructing arguments, asking research questions, making comparisons, considering and evaluating controversial issues, establishing causal relationships similar to how scientists think, and incorporating moral issues into scientific debates (Chap. 2 in this book).

A major concept in science education is *scientific literacy*. In 2015, the Organisation for Economic Co-operation and Development (OECD) Programme for International Student Assessment (PISA) defined scientific literacy as "the ability to engage with science-related issues, and with the ideas of science, as a reflective citizen" (PISA 2015, p. 7). There are different visions of what scientific literacy should encompass. These range from learning science for further training, via knowledge and understanding, to making the natural world and technological ramifications of science meaningful and accessible to the learner, toward critical visions of science learning for societal participation and action (Chap. 4 in this book). The latter vision emphasizes *transformative learning* and critical global citizenship. Indeed, transformative learning is a process in which the individual transforms and extends prior knowledge, beliefs, and attitudes. Transformative learning also includes active participation in critical discourse, in which the individual is exposed to the experiences and views of others in order to extend her own views on the world and society. *Mathematics literacy*, discussed in Chap. 12, which is essential for modern citizens, refers to the application of mathematical knowledge and skills in various situations. It includes reasoning mathematically and using mathematical concepts, procedures, facts, and tools to describe, explain, and predict phenomena.

Understanding the nature of science, fostering the practices of science, scientific and mathematical literacy, and higher-order thinking are more important than ever. In the new digital age, where anyone can post statements online, tweet, or blog, and the existence of facts is being debated, students need the skills and knowledge to navigate through this world. The ability to distinguish between a claim and evidence, to evaluate the evidence to support a claim, and to relate a claim to a larger conceptual framework is strongly related to global citizenship and sustainability. So too is mathematical literacy, where arguments are based on numbers and numbers are used to make predictions resulting from mathematical models.

The authors of Chap. 8 argue that *engineering design cognition* is a reciprocal interplay between knowledge acquisition and knowledge application that the designer engages in throughout the problem-solving process. While teaching practices often emphasize acquisition more than application or vice versa, the problems engineers tackle are often novel, typically due to their context, such as location or primary users. The authors present diverse definitions of *design* or *design inquiry* to emphasize the multifaceted views of design similar to the argument about the nature of science. Design or design inquiry has (a) multiple solutions, among which the designer selects the best one viable for the given context; (b) a set of strategies used by designers, which starts with building deep understanding of the problem and its context; and (c) a cognitive activity that involves reasoning (Chap. 8 in this book).

Behind the greater emphasis on engineering, design, design cognition, and design inquiry is the recognition that many of the problems the world faces today are engineering problems with engineering solutions. Exposing students to engineering before university serves to increase the number of students who will go on to study engineering. Exploring design cognition and design inquiry is a response to stakeholders and employers who are calling for better prepared engineers and serves to improve university instruction in engineering programs.

Innovation, defined as the generation, utilization, and circulation of new knowledge, is widely acknowledged among policy makers in Europe as being critical for countries to stay competitive in the twenty-first century, and this is true for all nations (Chap. 11 in this book).

15.1.2 Metacognition

As should be expected from this book's title, a major theme of almost all the chapters in this book concerns cognition- and metacognition-related issues.

Metacognition, first defined by Flavell (1979) as higher-level cognition or cognition about cognition, or thinking about thinking, encompasses a set of skills that enable learners to understand and monitor their cognitive processes. Discussed in Chap. 9, metacognition is concerned with knowledge of cognition—what students know about their knowledge and regulation of cognition and what students can do with this knowledge to better control their learning.

Metacognition-based pedagogical intervention is pedagogical intervention that aims at enhancing specific scientific and metacognitive skills. Such interventions are often accompanied by assessing their effect on the students (Chap. 3 in this book). It also refers to teaching learners how to learn and solve problems by guiding them to activate and implement metacognitive processes, such as planning, monitoring, control, debugging errors, and reflecting. As argued in Chap. 13, *pedagogical metacognition* is metacognition that relates to understanding and knowing how, when, and why to implement or integrate metacognition in teaching and learning. Meta-cognitively oriented teachers are aware of their students' learning processes and apply instructional methods that help students to be also aware of their learning.

Metacognitive self-regulation, discussed in Chap. 13, is an important aspect of the self-regulation cycle that relates to regulation of cognition and involves five kinds of strategies: planning, information management, monitoring, debugging, and evaluation.

Knowledge of cognition refers to what individuals know about their own cognition or about cognition in general, as discussed in Chap. 3. It includes at least three different metacognitive types of awareness: declarative knowledge ("about"), procedural knowledge ("how to"), and conditional knowledge ("why" and "when"). Knowledge of cognition is relatively stable but is age dependent. While it can be imperfect, one can often state it explicitly.

Regulation of cognition refers to regulatory skills and involves several *metacognitive strategies*, such as planning, evaluating, and monitoring. Some researchers refer to regulation of cognition as information management, or "debugging," which sometimes replaces the "planning" element. Regulation of cognition is relatively unstable and age independent, and several studies suggest that it can be learned (Chap. 3 in this book).

Metacognitive science learning is a learning process that employs one or more metacognitive skills in some science education domains and settings, as discussed in Chap. 3. More specifically, *metacognitive self-directed questioning*, suggested by the authors of Chap. 12, calls for using four kinds of metacognitive questions for inducing comprehension, connection, strategies, and reflection.

The metacognitive teaching method *IMPROVE* is an acronym of all the teaching steps: introducing the new materials to the whole class by modeling the metacognitive questioning, metacognitive questioning in small groups, practicing by using the metacognitive questioning, reviewing by using the metacognitive questioning, obtaining mastery on lower and higher cognitive processes, verification, and enrichment and remediation (Chap. 12 in this book).

Assessment, in the context of this book, is the process of collecting and processing data or evidence about the impact of education. Assessment of students' learning outcomes specializes into two major types, formative and summative, as elaborated in Chap. 9. In this book, assessment is a major focus in the metacognition theme but less so in the cognition theme. Researchers who study students' cognitive learning outcomes use a variety of assessment tools. Based on Chaps. 3 and 9, assessing metacognitive skills is important, but researchers are still struggling to measure these skills.

Metacognition assessment tools are quantitative and qualitative tools for assessing students' metacognition. Examples, presented in Chap. 3, include Physics Metacognition Inventory and Self-Efficacy Metacognition Learning Inventory—Science. The need to develop new approaches for engineering education has been a trigger for the development of *meta-assessment*, or assessment of assessment, as a technique for systematic evaluation of the assessment process itself, defined and discussed in Chap. 9. *Student-oriented meta-assessment* relates to assessment of how students assess their peers' outcomes. This *peer assessment*, namely, an arrangement in which individuals consider the amount, level, value, worth, quality, or success of the products or outcomes of learning of peers of similar status, is discussed in Chap. 9.

Research and development in the area of metacognition are driven by the recognition that learning is a process that teachers and technology can support, but it ultimately takes place within an individual by that individual. Moving learning from rote fact recall to deep processing is complex, and the various chapters addressing metacognition are attempts to determine how best to foster metacognition. Metacognition does not occur automatically; teachers must teach students the concept and language of metacognition explicitly and embed metacognitive activities in the content they teach. Metacognition is not altogether generic, and each of the book chapters that addresses metacognition makes this clear. Thus, although metacognition is thinking about one's own thinking, and aspects of one's thinking such as monitoring can be identified, it is thinking about one's own thinking in specific contexts.

15.1.3 Culture

The third theme, along with cognition and metacognition, is *culture*, defined in Chap. 6 as sociohistorical knowledge and practices embedded within communal, linguistic, economic, and technological contexts. Culture relates to values, and the concept *culture and values* reflects the context of the users, their associated beliefs about progress, and the patterns of organization. Interactions with culture and values impact technologies' realizations and diversity. Understandings of culture and values thus emerge from concurrent understandings of technology and ways people practice it, as argued in Chap. 7.

A key concept of the central and northern European culture of education is *Bildung*, a cornerstone of Chap. 4, which traces back to Wilhelm von Humboldt's works in the late eighteenth century. Bildung, according to von Humboldt, is a process of forming the personality toward individuality. The contemporary interpretation of Bildung, which started to develop in the 1950s, focuses on allowing the individual to develop capabilities and attitudes for self-determination, participation in, and solidarity with society. Bildung is one of the *cultural perspectives* of a particular group of individuals in a particular time that aim at understanding the complex sociocultural context. Other key cultural perspectives are language, knowledge, worldviews, beliefs, morals, customs, habits, and activities. Chapter 4 elaborates on how applying cultural perspectives to science education helps direct our attention to interactions and diversity across different cultures, curriculum models, disciplines, and students' academic classes or ages.

In order to facilitate connecting what learners learn to the worlds in which they live, pedagogy in STEM in general, and in mathematics in particular, has to be culturally relevant. *Culturally relevant pedagogy*, discussed in Chap. 14, is an approach to developing curricula and classroom practices that enable students to resist assimilation into the cultural norms of the majority and to use classroom learning to take action in their communities. Culturally relevant pedagogy supports students as they experience academic success, develop and maintain cultural competence, and develop a critical consciousness, through which they challenge the status quo of the current social order. Culturally relevant pedagogy can be instrumental in *mathematical modeling*—a process by which a real-world situation is analyzed, described, or understood using mathematics. The process, described in Chap. 14, is typically iterative. It involves stating assumptions, translating a description of the situation into mathematical equations, drawing conclusions from the mathematical solution, and revising the choices made along the way.

Culture is an aspect of cognition, metacognition, curriculum development, and standards that is often neglected. Addressing culture in instruction raises the question of whose culture: national culture or that of minority groups or all? Culture in instruction also runs the risk of trivializing or stereotyping some cultures and criticism of cultural appropriation. However, examining the role of culture in STEM has enriched our understanding of why some students disengage from the curriculum, why some students feel that they do not belong in a STEM career, and why some engineering projects have not been successful.

Technology is defined in Chap. 6 in the educational context as networked information and computing devices, methods, and technologies, such as iPads, Web 2.0 tools and resources, digital video recording and editing, scientific models and simulations, and digital scientific measurement via probeware.

Chapter 7 introduces the term *culture of technology*—the web of human interactions and activities that are mediated through the use, status, supply, and organization of technology, along with the human skills and knowledge associated with it. Culture of technology is deeply wrapped into lifestyles, expectations, values, and beliefs. For precollege science classrooms, it is affected by the context of use, science classroom users, and expectations of science teaching and learning. This is especially true for *digital natives*, discussed in Chap. 6, who are the generation of children, typically born after 1980, and who have grown up with considerable exposure to personal computers, video games, and the Internet. An example of the effect of a specific culture is the *Anglo-American tradition*. This engineering education tradition, discussed in Chap. 11, began in Anglo-American countries during the first Industrial Revolution (1760–1840s) and was characterized by high levels of self-government, where engineers were mainly entrepreneurs or freelance professionals who had learned advanced science and mathematics on the job. A parallel example is the *Continental European tradition*, which started in continental highly bureaucratic European countries, such as France, where engineers were mainly public servants who had learned advanced science and mathematics in school.

Analogous in some sense to the nature of science, the *nature of technology* (NoT, see Chap. 7 in this book) serves as an explanatory basis for how technologies evolve and interact with individuals, society, culture, institutions, and the economy. The NoT interrogates the artifact—the technology, its interactions with humans, and the role that culture and context play in these interactions. NoT engages the full system, product, or service life cycle, from conception, design, and development to enactment, usage, and discard. The NoT framework addresses five core dimensions that help to explain realizations of technology in precollege science classrooms. These dimensions address the role of *culture and values*, notions of technological progression, technology as part of systems, technology diffusion, technology as a fix, and technological expertise. The work of philosophers of technology, specified and discussed in Chap. 7, is a prime source of information needed for understandings of NoT.

One recent pedagogical approach enabled by technology is the *flipped classroom*. This approach uses online Web-based video resources to explain and explicate content that is traditionally provided through lecture, freeing up class time for more interactive learning activities, as elaborated on in Chap. 6. The flipped classroom is an example of an *innovative teaching and learning* approach that is being increasingly studied through an iterative process. This process includes creating classroom activities and using technology to drive inquiry and socially interactive learning activities and is applicable to teaching and learning of all STEM domains. More specifically, a *Web-based environment*, discussed in Chap. 13, is a learning and teaching environment that provides possibilities for learning and teaching in synchronous, asynchronous, autonomous, and collaborative modes by giving access to open-ended activities that move beyond theoretical declarative knowledge into complex learning and teaching, supporting cognitive and metacognitive processes. Such technologies free the time of both students and teachers for *reflection*—a systematic and socially situated practice of observation, evaluation, and modification of one's knowledge and social activity (Chap. 13 in this book). Web-based environments also facilitate *cyberlearning* through the use of networked learning technologies, discussed in Chap. 5, which have the potential to expand and transform learning opportunities, interests, and outcomes for all learners.

Philosophers of technology explain that technology is best understood in light of cultural structures and expectations, jointly termed *scaffolding*. As explained in Chap. 5, similar to how construction scaffolds support construction workers and extend their reach, instructional scaffolds support learners and allow them to perform tasks that they would not be able to do without this support. Examples include sentence starter prompts, explicitly defined roles and responsibilities, and partially completed examples. Addressing technology in a book about STEM is difficult. Technology changes so rapidly that research quickly becomes outdated. Technology also has the problem of overpromising. In the past, the technologies of television and language labs were touted as the tools that were going to transform education. Yet, the impact of these technologies was minimal. A variety of interesting current technology-driven experiments, such as cyberlearning in general and the flipped classroom in particular, have great potential while facing some serious challenges. Teachers need better training in the pedagogies of technology. Students, especially in lower socioeconomic classes, need easier access to the Internet, and schools need improved infrastructure to support these rapidly changing technologies. Educators should learn a lesson from the work in adaptive technologies for individuals with disabilities. In that work, one of the foci is on whether the individual can use an adaptive technology in school and not whether the adaptive technology helps the individual learn. Students and teachers can use many forms of technology, and our emphasis should be on whether those technologies support learning and under what circumstances.

Engineering education, another issue addressed in this book, is defined in Chap. 10 as the actions taken to educate and train novices to become practicing engineers. Engineering education includes the structure, curricula, and pedagogical approaches used to prepare students for careers as valued members of the engineering workforce.

Engineering culture is defined in Chap. 10 as a set of behaviors or beliefs that characterize the field of engineering. It encompasses the norms established by an engineering group, primarily industry, which are continually passed from one member to another. More specifically, *engineering education culture* is the set of behaviors, norms, and beliefs that characterize the field of engineering education. The authors argued that different educational units or environments, such as an engineering college, school, department, or program, may have a different engineering education culture and therefore typically exhibit diverse and different everyday actions when applying engineering education. In particular, this can apply to the process of *enculturation as an engineer*—understanding and adopting the traits of a professional engineer for becoming part of the engineering discipline. As explained in Chap. 10, as part of this process, individuals discover their identity as engineers within the greater field of engineering.

Many new approaches in engineering education include a component of *project-based learning* (PBL). This teaching and learning method, discussed in Chap. 9, focuses on developing a project in order to engage students in sustained, cooperative investigation, combining academic knowledge with real-world applications. Applying PBL in appropriate socioeconomic contexts, students learn to manage increasingly complex systems of scientific knowledge while gaining practical skills and contextual awareness in an organic, integrated fashion. PBL is confusingly also an abbreviation of *problem-based learning* (this abbreviation is more commonly used in Europe), and as argued in Chap. 11, a critical component of this kind of PBL is the scaffolding of a network of problems with increasing complexity, spanning from reproductive learning in the form of routine to complex problem-solving in contexts. It also involves the complexity of real-life settings, which promotes creative learning. Indeed, *collaborative problem-solving*, discussed in Chap. 5, is a core STEM practice and a critical and necessary twenty-first-century skill, as it relates to individuals working together to set shared goals, negotiate, and ultimately solve problems, particularly those with potentially multiple solutions and solution paths.

As mentioned earlier, engineering educators are concerned about preparing engineers for the world of work. In response to criticism about the lack of diversity in the engineering workforce and the call for better-prepared engineers, engineering educators have taken a hard look at the culture of engineering and changes it should undergo. Being practical minded, engineering educators have borrowed heavily from psychology, education, and the learning sciences to improve the preparation of engineers. Chapters 9, 10, and 11 reflect the concerns about diversifying the engineering workforce and preparing a more skilled engineering workforce. Many innovations in engineering education have been introduced in response to the 2007 book *Rising Above the Gathering Storm: Energizing and Employing American for a Brighter Future (National Academies of Engineering and Institute of Medicine,* 2007).[3]

The concern for educating a better-prepared engineer through better instruction is not just a US phenomenon. Aalborg University in Denmark has long been a leader in project-based engineering education, and engineering programs around the world have adopted many of these approaches to instruction.

[3] https://www.nap.edu/read/11463/chapter/1

15.2 Summary and Future Research

Reviewing research in all four STEM domains and referring to the three book themes enable us to draw generalizations and go beyond the state of the art of cognition, metacognition, and culture in STEM education.

- Cognition and metacognition play essential roles in enhancing STEM literacy and students' learning in all four STEM domains and at various age levels.
- Several authors have shown that metacognition can be enhanced in typical classrooms and under regular school conditions.
- Many of the current effective innovative teaching methods include one or more metacognitive components. Examples of these methods include problem- and project-based learning (PBL), inquiry-oriented pedagogy, IMPROVE, engineering design inquiry, flipped classroom, Web-based environment, cyberlearning, and collaborative problem-solving.
- Assessment tools that are based on metacognitive inventory have been proven to be effective for both formative and summative evaluation in small and large courses.
- Beside cognition and metacognition, culture provides another important perspective on STEM learning. For example, in many Western countries, girls tend to avoid majoring in STEM domains, but this is not the case in Eastern countries. Furthermore, while in some countries, such as Singapore, metacognitive skills are part of the mandatory mathematics curriculum; in other countries the teacher can decide whether she or he would like to include these higher-order thinking skills in STEM instruction.
- Educational systems are as good as the teachers within them. The role of the teachers becomes even more crucial in STEM education. Examples of teachers' professional development to improve pre- and in-service pedagogical content knowledge are training for inquiry-oriented pedagogy, reflecting on the teachers' own videos, and practicing mathematical modeling.

The authors of these book chapters have suggested various ways in which their research can be extended and refined. Before closing the book, in which work on three themes that are often studied separately is presented in an integrated way, we take this opportunity to stimulate the readers to investigate new research directions. We summarize the authors' suggestions and ours by chapters.

In Chap. 2, "Teacher Cognition of Engaging Children in Scientific Practices", Crawford and Capps (2018, in this book) focus on teachers' subject matter knowledge, pedagogical knowledge, pedagogical content knowledge of scientific practices, and the nature of science. They document the various types of knowledge of teachers prior to and following a professional development course that the authors conducted. Based on their study, it would be interesting to relate the qualitative findings concerning two teachers to the entire group of 30 teachers who participated in the professional development program overall. Two interesting questions in this context are: How did the group 30 teachers reflect on the model presented in this chapter? Can we identify initial categories of teachers' professional growth based on the model the authors suggested?

Chapter 3, "Students' Metacognition and Metacognitive Strategies in Science Education" by Avargil, Lavi, and Dori (2018, in this book), is a review that focused on students with respect to four study types: (1) theoretical papers, (2) papers focusing on assessment tools for metacognition, (3) metacognitive learning processes, and (4) metacognition-based pedagogical intervention. They found very little on the first two study types, so it would be important to deepen the theoretical research and develop more tools for assessing metacognition. While the focus of the review was students, the authors found out that the literature on teachers and metacognition is also scarce, so investigating metacognition from teachers' viewpoint is also in order, as is the case with younger students.

The three different visions of scientific literacy presented in Chap. 4, "Reconsidering Different Visions of Scientific Literacy and Science Education based on the Concept of *Bildung*" by Sjöström and Eilks (2018, in this book), are based on the concept of *Bildung* in the context of science education. Vision III, the most complex one, aims to foster a change in society, which would make it more sustainable. However, one could argue that implementing this complex approach in science education would require more resources, including money, time, and expertise, than doing so for either Vision I or II. Future work could use the theoretical basis of the complex vision presented in this chapter as a basis for designing curricula and action research to benefit the learner, science education, and the society.

In Chap. 5, "Designing for Collaborative Problem Solving in STEM Cyberlearning," Crippen and Antonenko (2018, in this book) provide a detailed description of their self-developed design framework for cyberlearning via collaborative solving of authentic problems. This framework contributes to STEM education research by providing an opportunity for studies into collaborative STEM problem-solving in online environments. The framework contributes also to STEM education in general by leveraging collaborative problem-solving to enhance meaningful learning. Future studies into the implementation of this framework in other STEM subjects would provide researchers with an account of how collaborative problem-solving enhances learning in the different STEM subjects. Such studies would also provide instructors with knowledge on how to best implement this framework in their classrooms. The authors discuss planning, monitoring, and reflection as components of metacognition relating to regulation of cognition within their framework. Future studies can help integrate into the framework also declarative, procedural, and condition components of knowledge of cognition.

Chapter 6, "Technology, Culture, and Young Science Teachers – A Promise Unfulfilled and Proposals for Change" by Yerrick, Radosta, and Greene (2018, in this book), is somewhat surprising in revealing that the preservice teachers studied were not quite the digital natives portrayed in the literature. While the preservice teachers did engage with technology for personal use, teaching with technology was "sparse." Moreover, the preservice teachers did not describe themselves as being creative in the use of technology in their teaching. The authors point out that not all technologies are equal and that some are more useful in supporting preservice teachers learning than others. The authors provide a cultural explanation for the low level of engagement in the example of an inquiry activity. However, one wonders if the preservice teachers' limited use of technology to teach inquiry has more to do with their misperceptions of what inquiry is than their proficiency in using technology for teaching and learning. Alternatively, the technology tasks that the students explored may not have been sufficiently engaging, resulting in their limited enthusiasm and effort and a disinclination to use technology in their own teaching. Furthermore, as the preservice teachers experienced, trouble with the technology can be disastrous in the science classroom in terms of management and completion of lessons. Yet, the preservice teachers considered the reflection videos important to improving their teaching. It is worth investigating whether the positive results are due to reflection, the type of technology used, or some combination thereof.

In Chap. 7, "Technology, Culture, and Values: Implications for Enactment of Technological Tools in Precollege Science Classrooms," Waight and Abd-El-Khalick (2018, in this book) provide a framework for integrating technology into the science classroom that takes into consideration the nature of technology and its place in culture and society. This framework enables instructors to better integrate technology into teaching and learning based on their specific teaching environment. It also helps researchers to conduct intervention studies concerning the use of technology in science education in various contexts. Future studies should strengthen the theoretical basis of the authors' framework by integrating insights from Vigotsky's seminal work on the place of technology in education and test the effectiveness of this framework in sciences other than biology and in other STEM subjects.

In Chap. 8, "Engineering Cognition: a Process of Knowledge Acquisition and Application," Purzer, Moore, and Dringenberg (2018, in this book) conceive engineering design cognition as a combination of knowledge acquisition, which is a goal of project-based learning, and knowledge application, which is a goal of problem-based learning, as key factors in knowledge production. Engineering design cognition can therefore serve as a potential framework for combining these two active learning approaches into a coherent and effective engineering design instruction approach. Intervention research with a combined project- and problem-based learning curriculum within the engineering design cognition framework could potentially provide fruitful learning and research outcomes.

Chapter 9, "Metacognition and Meta-assessment in Engineering Education" by Wengrowicz, Dori, and Dori (2018, in this book), combines theory and practice. It describes the assessment model, the course, rubric, and projects. The course and the pedagogy clearly require a great deal of upfront work on the part of instructors. The data that shows the effectiveness of the peer meta-assessment may convince others to try this approach. Students' excerpts indicate that they were asking metacognitive questions, showing that they were able to evaluate clarity and understandability, completeness, correctness, and documentation of their peers' projects. The Appendix in Chap. 9 shows that while the students in the large class found the task demanding, students in the small class noted that it had helped them learn the conceptual modeling languages. Given the characteristics of the students in the two courses and the difficulty of the task as expressed by students in the large class, the question that arises is whether this model of instruction is generalizable to students attending less selective institutions, especially those with larger classes. Future research on the effectiveness of the approach should compare and contrast the impact of the approach in a variety of instructional and cultural settings.

In Chap. 10, "The Impact of Culture on Engineering and Engineering Education," Carberry and Baker (2018, in this book) discuss the relationships between engineering, engineering education, and culture. They show how cultural perception of engineering and the resulting engineering education and engineering culture impact men and women's approach to and place in engineering. The authors provide educators, educational policy makers, and public communicators of engineering and engineering education with a basis for making engineering education more culturally accessible to women. The findings of the authors' survey could also be used in the development of tools for assessing students' attitudes toward engineering and engineering education. Future work suggested by the authors could focus on developing concise definitions of engineer, engineering, and engineering education, while empirical studies could explore students' reasons and arguments for choosing a career in engineering from a gender perspective. Findings from these studies would help inform both educational and public communication efforts.

In Chap. 11, "Engineering Education in Higher Education in Europe," Corlu, Svidt, Gnaur, Lavi, Borat, and Çorlu (2018, in this book) provide insight into how engineering education systems have developed across Europe. They link this development to the innovation score of the European Commission (EC) by comparing engineering education systems in three European countries. The chapter includes a detailed description of the systems in Denmark and Turkey and comparing them to the UK's engineering education system. Future studies could provide a more detailed characterization of engineering education systems and of innovation that go beyond the EC's definition. Additional comparisons of different European and other countries' engineering education systems can further elucidate the relations between these systems and innovation.

In Chap. 12, "Cognition, Metacognition, and Mathematics Literacy," Mevarech and Fan (2018, in this book) describe a pedagogical method for mathematics problem solving and its positive impact on students' mathematical literacy. This method is based on problem solving processes embedded within metacognitive scaffolding through question posing. The authors' method is implemented by the classroom teacher following an evaluation for its effectiveness. This method can be adopted by and adapted to other STEM subjects. The authors encourage teachers to place problem solving at the center of their students' mathematics learning process. This pedagogical method could be adapted to various collaborative authentic problem solving processes, thereby contributing not just to promoting students' mathematical literacy, but also to boosting their ability to solve authentic mathematics problems, both individually and collaboratively.

The cognition/metacognition and teaching instruction model in Chap. 13, "Promoting Mathematics Teachers' Pedagogical Metacognition – a Theoretical-practical Model and Case Study" by Kohen and Kramarski (2018, in this book), advocate moving the preparation of mathematics teachers from simply providing them with a series of instructional strategies to challenging the teachers to engage in student-centered teaching. This approach encourages knowledge construction through metacognition and self-regulation. As the authors point out, this is a sorely needed change in the approach to mathematics teacher preparation, since metacognition develops slowly and is quite poor among both students and teachers. Mathematics education lacks a vocabulary to communicate teachers' classroom activities, which this model provides. The case study of two female teachers has benefits and limitations as a test of the model. Further research is needed for the generalizability of the instructional model to the preparation of elementary teachers who have limited mathematics backgrounds, to teacher preparation programs outside Israel, and to minority students preparing to be mathematic teachers. Furthermore, the preservice teachers taught their peers. Testing the model in a variety of contexts with a broader range of students and in real classrooms will provide mathematics educators with the knowledge they need to improve the preparation of future mathematics teachers.

In Chap. 14, "Mathematical Modeling and Culturally Relevant Pedagogy" by Anhalt, Staats, Cortez, and Civil (2018), the authors state that two pedagogical approaches rely on students' knowledge of everyday situations: mathematical modeling and culturally relevant pedagogy. Yet, the question of how to combine these two approaches in the mathematics classroom has remained open. The chapter intends to remedy this situation by reviewing the relevant literature in two disparate disciplines and providing an in-depth exploration of a concrete example of how modeling and culturally relevant pedagogy can be combined. The rationale for combining the two approaches is that it will improve students' performance and be more motivating and relevant to them. The authors base this claim on the positive student outcomes found in the research literature. The authors provide a brief description of the impact of including a module of mathematical modeling and culturally relevant pedagogy in a mathematics pedagogy class. Given the short duration of the module, it is understandable that the impact was modest. However, the trial of the module did identify the difficulties preservice teachers, and most likely also in-service teachers, will encounter while trying to combine modeling with students' funds of knowledge. Another difficulty the preservice teachers encountered was the development of a critical consciousness through which they would challenge the status quo of the current social order. Future research should look into whether the combination of mathematical modeling and culturally relevant pedagogy is more effective than traditional mathematical pedagogies.

For STEM teachers to be able to perform these different roles, they must possess advanced cognitive and metacognitive skills, as well as cultural awareness. This will enable them to better monitor and improve students' learning and teach the students in ways that suit their cultural backgrounds. Effective monitoring of students necessitates assessment, making teacher assessment literacy another important competency for teachers to possess (Avargil et al. 2012; Dori and Avargil 2015; Xu and Brown 2016).

Anyone interested in implementing teaching focused on cognition, metacognition, and culture on a wide scale can draw encouragement from empirical demonstrations of improvements in cognition and metacognition. Examples include those described in Chaps. 2 and 3 for science students and teachers, respectively, in Chap. 9 for engineering students, and in Chaps. 12 and 13 for mathematics students and teachers, respectively. The impact of culture-aware teaching on learning outcomes, including cognitive and metacognitive skills, discussed in Chaps. 4, 7, 10, and 14, is therefore a fruitful direction for future research in STEM education.We conclude by noting that this book concerns three major topics of STEM education research—cognition, metacognition, and culture. The findings and conclusions presented throughout this book provide three overarching suggestions for STEM teachers: (1) nurturing cognitive skills in students to help them attain STEM knowledge in various domains, (2) developing students' metacognitive awareness to help them set learning goals and plan for achieving those goals, and (3) teaching students in culturally appropriate ways while helping them acquire cultural knowledge and values.

References

Avargil, S., Herscovitz, O., & Dori, Y. J. (2012). Teaching thinking skills in context-based learning: Teachers' challenges and assessment knowledge. *Journal of Science Education and Technology, 21*, 207–225.

Dori, Y.J., Avargil, S. (2015). Teachers' understanding of assessment. In R. Gunstone (Ed.), *Encyclopedia of science education*. SpringerReference. www.springerreference.com. Berlin: Springer. http://www.springerreference.com/docs/html/chapterdbid/303248.html. Accessed July 2016. doi: https://doi.org/10.1007/SpringerReference_303248 2013-08-21.

Flavell, J. H. (1979). Metacognition and cognitive monitoring. *American Psychologist, 34*, 906–911.

National Academies of Engineering and Institute of Medicine. (2007). *Rising Above the Gathering Storm: Energizing and Employing American for a Brighter Future*. Washington, DC: National Academies Press.

PISA (2015). PISA draft science framework. https://www.oecd.org/pisa/pisaproducts/Draft%20PISA%202015%20Science%20Framework%20.pdf/. Accessed July 2016.

Xu, Y., & Brown, G. T. (2016). Teacher assessment literacy in practice: A reconceptualization. *Teaching and Teacher Education, 58*, 149–162.

Correction to: Reconsidering Different Visions of Scientific Literacy and Science Education Based on the Concept of *Bildung*

Jesper Sjöström and Ingo Eilks

Correction to:
Chapter 4 in: Y.J. Dori et al. (eds.), *Cognition, Metacognition, and Culture in STEM Education*, Innovations in Science Education and Technology 24, https://doi.org/10.1007/978-3-319-66659-4_4

Chapter 4 was previously published non-open access. It has now been changed to open access under a CC BY 4.0 license and the copyright holder has been updated to "The Author(s)."

The updated version of this chapter can be found at https://doi.org/10.1007/978-3-319-66659-4_4

Y.J. Dori et al. (eds.), *Cognition, Metacognition, and Culture in STEM Education*, Innovations in Science Education and Technology 24,
https://doi.org/10.1007/978-3-319-66659-4_16

Editor Biographies

Yehudit Judy Dori is Dean of the Faculty of Education in Science and Technology at the Technion – Israel Institute of Technology and a Senior Researcher at the Samuel Neaman Institute for National Policy Research, Haifa, Israel. Professor Dori holds an MSc degree in life sciences and a PhD degree in science teaching, both received from the Weizmann Institute of Science. She has been intermittently Visiting Professor or Visiting Scholar at Massachusetts Institute of Technology (MIT), Cambridge, MA, since 2000. Between 2009 and 2013, she was Dean of Continuing Education and External Studies at the Technion. In 2013–2014, she was a Visiting Professor at Electrical Engineering and Computer Science, and in 2014–2015, she was Visiting Scientist at the Computer Science and Artificial Intelligence Laboratory, both at MIT. Her research interests encompass learning in technology-rich environments, educational assessment, scientific visualizations, and metacognition at high school and university levels. Professor Dori is also a pioneer in the integration of computerized molecular modeling and computerized laboratory in chemical education. From 2003 to 2008, she served as the Chairperson of the Israeli Professional Committee for Chemistry and since September 2016 is serving as the Chairperson of the Israeli Professional Committee of "Science and Technology for All." Prof. Dori is member of the editorial board of Journal of Science Education and Technology (JOST) and was twice member of the editorial board of Journal of Research in Science Teaching (JRST). She served on the NARST International committee and the NARST Membership and Elections committee. She was Guest Editor of two special issues on the educational reform at MIT in the Journal of Science Education and Technology (2007, 2008). Her book on metacognition in science education, which Prof. Dori coedited with Prof. Zohar, was published by Springer in 2012, and she is the first editor of this edited book on cognition, metacognition, and culture in STEM education, published by Springer in 2017.

Dale Baker is a Professor Emerita in MaryLou Fulton Teachers College at Arizona State University. She is the 2013 recipient of the Distinguished Contributions to

Y.J. Dori et al. (eds.), *Cognition, Metacognition, and Culture in STEM Education*, Innovations in Science Education and Technology 24,
https://doi.org/10.1007/978-3-319-66659-4

Science Education Through Research award given by NARST: a Worldwide Organization for Improving Science Teaching and Learning Through Research. She is also a fellow of the American Association for the Advancement of Science and the American Educational Research Association. Her article "Letting Girls Speak Out About Science" was selected as one of 12 most influential research papers published in the past 40 years in the 1995 anniversary issue of the *Journal of Research in Science Teaching*. She has also received the MaryLou Fulton Teachers College award for Research with Sustained Impact, the Outstanding Research Award in Classroom Applications from NARST, the Award of Merit for Outstanding Paper published in *Science Education*, and one of the Ten Best Papers Award from the Northern Rocky Mountain Educational Research Association. These awards were given for her research in gender equity in science and attitudes and reasoning in science. In addition to gender equity, Dr. Baker's current research interests include engineering education and science teacher professional development. She was Section 8: (Equity) Editor of the International Handbook of Research in Science Education and Coeditor, with Michael Piburn, of the *Journal of Research in Science Teaching* from 2001 to 2006. She currently serves on the Editorial Board of the *International Journal of Learning Technologies*. In addition, she reviews for 12 national and international journals. She has been principal investigator or coprincipal investigator for 29 funded projects with a total of $125,405,020 in awards from the national, state, and private foundations. She has been the author of 5 books, 9 book chapters, 40 articles, and 45 proceedings and has given 121 paper presentations.

Zemira Mevarech is a Professor (emerita) of education. She is currently the President of David Yellin Academic College of Education. Prof. Mevarech was the Chief Scientist of the Israeli Ministry of Education, Vice Rector at Bar-Ilan University, the Dean of the Faculty of Social Sciences, and the Head of the School of Education (two nonconsecutive terms). She was the coordinator of the Metacognitive SIG in the European Association for Research on Learning and Instruction (EARLI). Currently, she participates as an expert in the OECD project that focuses on promoting creativity and critical thinking in schools. Prof. Mevarech has a BSc degree from the Hebrew University in Mathematics, Physics, and Education; MA in education from Bar-Ilan University; and PhD from Chicago University in the area of Measurement, Evaluation, Statistics, and Assessment (MESA). Upon her return to Israel, she started to work at Bar-Ilan University, developing and studying innovative pedagogical methods, mainly in the area of mathematics education. Prof. Mevarech was one of the pioneers in studying metacognitive pedagogies in the area of mathematics education. She and her team developed the metacognitive pedagogy called IMPROVE that has been implemented in mathematics and science classrooms, at all levels of education, K–12 and higher education. The effects of IMPROVE was evident in a large number of studies and under different conditions. Tens of studies documented its effects on different schooling outcomes, including students' mathematics reasoning and achievement, social and emotional skills, science literacy, and teachers' pre- and in-service professional development. Studies reported also the effects of this metacognitive pedagogy in various ICT environments, including e-learning, a-synchronic learning networks, and domain-specific software. She has published several books and more than 80 research publications. The OECD has published her book "Critical Maths for Innovative Societies: The Role of Metacognitive Pedagogies." The book was translated into Japanese and Spanish.

Author Biographies

Fouad Abd-El-Khalick is Dean of the School of Education at the University of North Carolina at Chapel Hill. He holds a PhD in Science Education from Oregon State University and master's and BS degrees from the American University of Beirut. His research focuses on developing new classroom innovations and assessment techniques (e.g., alternative pedagogies, reflection, and grading systems), and identifying new ways to empirically understand how engineering students and educators learn (e.g., assessments of personal beliefs, expectations, values, risk-taking, and mindsets). He also maintains an active international program of educational research and development in Egypt, Lebanon, and Qatar. Professor Abd-El-Khalick served as an elected member (2004–2007) and currently is ex-officio member (2015–2019) of the Executive Board of the National Association for Research on Science Teaching (NARST). Currently, Abd-El-Khalick is Coeditor of the *Journal of Research in Science Teaching* (2015–2019) and Associate Editor of the *Journal of Science Teacher Education* (2013–2017). He has received the NARST Early Research Career Award (2003), Journal of Research in Science Teaching Award (2001), and Outstanding Doctoral Dissertation Award (1999). In 2011, Abd-El-Khalick was elected to the rank of Fellow of the Prestigious American Association for the Advancement of Science for, according to the Association, "his outstanding research on teachers' and students' conceptions of the nature of science that helped provide a foundation for this field of research."

Cynthia O. Anhalt is an Associate Research Professor and Director of the Secondary Mathematics Education Program in the Department of Mathematics at the University of Arizona. Her research focuses on mathematics teacher development and student reasoning. Specifically, she seeks to understand the work of teaching in working with culturally and linguistically diverse student populations. More recently, her research has focused on deepening and broadening teacher knowledge for teaching mathematical modeling at both elementary and secondary levels. Recently, she has led various externally funded initiatives, Mathematical Modeling in the Middle Grades (*M3*) and the Southern Arizona Mathematics Initiative

Y.J. Dori et al. (eds.), *Cognition, Metacognition, and Culture in STEM Education*, Innovations in Science Education and Technology 24,
https://doi.org/10.1007/978-3-319-66659-4

(*SAZMI*) designed to develop curriculum materials for implementation in professional development institutes with elementary and middle school teachers in mathematics content, in mathematical practices, and specifically in mathematical modeling. Currently, she is Principal Investigator of a Robert Noyce NSF Grant Project, *AZ Math Teach*, to increase the number of secondary mathematics teachers in her local context and across AZ and the nation. She has served as Coeditor of the *Teaching for Excellence and Equity* journal published by TODOS: Mathematics for ALL and serves as a reviewer for the *Mathematics Teaching in the Middle School* journal published by the National Council of Teachers of Mathematics. She has taught mathematics content and mathematics education courses for prospective and practicing teachers and graduate courses on research in mathematics education.

Pavlo (Pasha) Antonenko is an Associate Professor of Educational Technology in the School of Teaching and Learning at the University of Florida. Dr. Antonenko and his students investigate learning and teaching in complex and interactive learning environments, which integrate technology to organize learning activities into meaningful and relevant contexts. He has extensive expertise building curricula and technologies to support collaborative problem solving in face-to-face, blended, and online learning environments. Dr. Antonenko has served as PI for a number of federally funded research grants including the NASA-funded Engaging Native Americans in NASA-Centered STEM Cyberlearning and Career Awareness Activities. He works with STEM teachers and learners in Project iDigFossils: Engaging K–12 Students in Integrated STEM via 3D Digitization, Printing, and Exploration of Fossils. Antonenko is also the Director of the Neuroscience Applications for Learning (NeurAL) Lab at the University of Florida where he is studying attentional and cognitive dynamics during collaborative problem-solving and learning with educational multimedia.

Shirly Avargil PhD, is a Lecturer at Bar-Ilan University, Faculty of Social Science, School of Education, Ramat Gan, Israel. She graduated from the Technion, Israel Institute of Technology at 2011, and spent 2 years as a postdoctoral researcher at the University of Maine, USA. Her research includes students' metacognitive thinking in chemistry education, context-based science teaching, meaningful learning in chemistry that includes analogies and visualization, and science teachers' knowledge of assessment. She published her work at several peer review journals and presented her work at different conferences around the world. As a young scholar, she won the Jhumki Basu Scholars Program award of the Equity and Ethics Committee and the International Committee Scholarship, both sponsored by the National Association for Research in Science Teaching. She serves at the editorial board of the Journal of Research in Science Teaching and as a member of the national chemistry education committee in her country.

Oğuz Borat worked as academician and administrator at various universities; served as a member to the Environmental Research Group and the Scientist Development Group in the Scientific and Technological Research Council of Turkey (TUBİTAK); worked as a member of the board of directors in some well-known companies, working in the fields of engineering; was appointed as short-term expert and key-expert by the Council of Higher Education and the Ministry of National Education of Turkey in various VET and engineering projects funded by the World Bank and EU; served as director of the Project Coordination Center of Ministry of National Education, and was selected as the vice president of the Turkish Vocational Qualifications Authority; participated in Bologna implementation processes and in ABET program accreditation processes at universities; worked as the chair of Turkish Qualifications Framework Preparation Commission; has many books, articles and papers in the fields of engineering and VET. He is now serving in Istanbul Commerce University.

Daniel K. Capps is an Assistant Professor of Science Education in the Department of Mathematics and Science Education at The University of Georgia, USA. His research focuses on teacher knowledge and practice related to inquiry-based instruction, teacher professional development, and in designing instructional experiences to support teachers and students in learning about scientific practice. His work on scientific models has been funded by the National Science Foundation (DRL-1720996), and he has served on the editorial board of the *Journal of Science Teacher Education*.

Adam R. Carberry is an Associate Professor in The Polytechnic School within Arizona State University's Fulton Schools of Engineering. He earned a BSc in materials science engineering from Alfred University and received his MSc and PhD, both from Tufts University, in chemistry and engineering education, respectively. Carberry teaches use-inspired design courses utilizing a project-based approach. His research investigates the impact student characteristics (self-efficacy, epistemological beliefs, motivation, and associated value) and pedagogical approaches (standards-based grading, project-based learning, service learning, and reflection) can have on learning in engineering education. He was previously an employee of the Tufts' Center for Engineering Education and Outreach and manager of the Student Teacher Outreach Mentorship Program (STOMP) where he received the Tufts University Presidential Award for Citizenship and Public Service. He currently serves on the LEGO Education Advisory Panel as a member of the ambassador group, a distinction imparted on him for having previously facilitated over 20 LEGO Engineering workshops around the world.

Marta Civil is a Professor of Mathematics Education and the Roy F. Graesser Endowed Chair in the Department of Mathematics at the University of Arizona. Her research focuses on cultural, social, and language aspects in the teaching and learning of mathematics, linking in-school and out-of-school mathematics, and parental engagement in mathematics. Her work is located primarily in working class, Latina/Latino communities. She has led several externally funded initiatives that combine research and outreach and are aimed at developing culturally responsive learning environments. Some of these initiatives include project *Bridge* that extended the Funds of Knowledge work to mathematics education; a gender equity project that engaged low-income, culturally diverse children in hands-on mathematics and science explorations in informal and after-school settings; a parental involvement project that supported Latina/o parental engagement in mathematics education; and the Center for the Mathematics Education of Latinos/as (CEMELA), an interdisciplinary, multi-university consortium focused on research and practice on the connections between the teaching and learning of mathematics and the cultural, social, and linguistic contexts of Latina/o students. She teaches primarily mathematics and mathematics education courses for preservice and practicing teachers and graduate courses on research in mathematics education.

Ricardo Cortez is the Pendergraft William Larkin Duren Professor of mathematics in the Mathematics Department and Director of the Center for Computational Science at Tulane University in New Orleans. His research focuses on mathematical and computational models of fluid motion with scientific applications. Specifically, his work seeks to understand and predict the swimming motion of microorganisms and cells that play a role in reproduction or disease (bacteria). His research work extends to mathematical modeling in K–12, including the design of modeling problems and ways of improving teachers' and students' mathematical modeling proficiency. His work has been continuously funded for 20 years by federal and state agencies including a National Science Foundation CAREER award and a Woodrow Wilson Career Enhancement award. Prof. Cortez has also been recognized for his work to increase the representation of minority groups in the mathematical sciences.

Barbara A. Crawford is a Professor and recent Department Head of Mathematics and Science Education at The University of Georgia, USA. She currently serves as the elected President of NARST, a worldwide organization for improving science teaching and learning through research. She is an Elected Fellow of the American Association for the Advancement of Science (AAAS). She earned three degrees from The University of Michigan, including a bachelor's degree in Microbiology, a master's degree in Biology, and a PhD in Educational Studies/Science Education. Her research interests include inquiry-based science teaching and learning, teacher professional development, knowledge of inquiry and nature of science, models and modeling, authentic science, and student-scientist collaborations. The ultimate goal of her research is to facilitate the majority of students in classrooms in developing images of science consistent with current practice, and in understanding what science is and what science is not, and the relevancy of science to society.

M. Ali Çorlu PhD, is a retired Professor of Physics Education. In the early years of his career, Dr. Corlu worked as a high-school teacher in Turkey. Later, he decided to move on to academia as a physicist and worked in various institutions in Turkey and Europe, including Karl Ruprecht Üniversitaet Heidelberg, Max Planck Institut für Kernphysik, and Yıldız University Faculty of Engineering and Architecture. During his years as the Founding Chair of Physics Education Department at Marmara University and the first person who was awarded with a full physics education professorship in Turkey, he helped science education grow as an independent area of inquiry in the country. Although he retired from Istanbul Commerce University and active teaching in 2017, he continues to be a productive member of the science education community in Turkey.

M. Sencer Corlu PhD (Texas A&M University), is an Associate Professor of Mathematics Education and the Director of BAUSTEM Center at Bahçeşehir University, İstanbul-Turkey. Dr. Corlu was a faculty member at the Graduate School of Education, Bilkent University between 2012 and 2016. Prior to that, he had worked as a school teacher at International Baccalaureate schools in Africa and Europe, teaching middle and high-school mathematics, science and technology courses. Dr. Corlu has published over 30 academic articles and books on mathematics teacher education, STEM education, and international schools. He is the principal investigator of the externally funded Integrated Teaching Project and serves as the editor of a book series on STEM education.

Kent J. Crippen is a Professor of STEM Education in the School of Teaching and Learning at the University of Florida and serves as Editor-in-Chief of the *Journal of Science Education and Technology*. His research involves the design, development, and evaluation of STEM cyberlearning environments as well as teacher professional development. Operating from a design-based research perspective, this work focuses on using innovative, iterative, and theoretically grounded design for the dual purpose of addressing contemporary, complex, in situ learning problems while generating new theoretical insight related to the process of learning and the relationships among the people, tools, and context of the social space. The complex and situated nature of contemporary STEM learning problems dictates a collaborative team approach to design that leverages expertise along multiple dimensions. The practice of deep, on-going collaboration is a core tenant of the philosophy of Dr. Crippen's research group. This philosophy, as well as the design-based approach, is best illustrated by the current projects, which include improving the retention of underrepresented students as undergraduate engineering majors through a reform of the curriculum for general chemistry (ChANgE Chem; NSF-DUE-1245068), building participation in the science of paleontology and informal STEM learning with a community of practice that includes amateur and professional paleontologists (FOSSIL; NSF-DRL-1322725), and supporting teacher development and adoption of the Common Core through scientist-teacher partnership as professional development (Bench to Bedside; NIH-SEPA-1R25OD016551–01). Among his honors and distinctions, Dr. Crippen has been recognized as a Fellow of the American Association for the Advancement of Science (AAAS) and has served as the Educational Technology Strand Co-coordinator for the National Association for Research in Science Teaching.

Dov Dori is a Harry Lebensfeld Chair in Industrial Engineering and Head of the Enterprise System Modeling Laboratory at the Faculty of Industrial Engineering and Management, Technion – Israel Institute of Technology. He also lectures at the Systems Design and Management Program at MIT. Prof. Dori received his PhD in Computer Science in 1988 from Weizmann Institute of Science, MSc in Operations Research from Tel Aviv University in 1981, and BSc in Industrial Engineering and

Management from the Technion in 1975. His research interests include model-based systems engineering, conceptual modeling of complex systems, systems architecture and design, software and systems engineering, and systems biology. Prof. Dori invented and developed Object-Process Methodology (OPM), recently adopted as ISO 19450. He has authored over 300 publications, including journal and conference papers, books, and book chapters. He chaired or was cochair of nine international conferences and workshops. Among his editorial duties, Prof. Dori was Associate Editor of IEEE Transaction on Pattern Analysis and Machine Intelligence, and currently he is Associate Editor of Systems Engineering. He is a Fellow of INCOSE (International Council on Systems Engineering), Fellow of IAPR (International Association for Pattern Recognition), Member of Omega Alpha Association – International Honor Society for Systems Engineering, and Senior Member of IEEE and of ACM.

Emily Dringenberg PhD, is an Assistant Professor in the Department of Engineering Education at Ohio State University. She is a former public school teacher who specialized in mathematics and engineering. Her research focuses on using qualitative methods to explore the experiences of students engaging in engineering design problems. More specifically, she focuses on exploring the beliefs that undergraduate students hold about intelligence and engineering decision-making. Her work aims to leverage engineering education research to shift the culture of engineering to be more inclusive. Additionally, her research interests include neuroscience, growth mindset, epistemology, and design learning. She holds a BS in Mechanical Engineering (Kansas State '08), an MS in Industrial Engineering (Purdue '14), and a PhD in Engineering Education (Purdue '15).

Ingo Eilks studied chemistry, mathematics, philosophy and education. He is a full trained grammar school teacher, having a PhD and habilitation in chemistry education. Since 2004 he is a Full Professor for chemistry education at the Institute of Science Education at the University of Bremen, Germany. His research interests encompass societal-oriented science education, action research in science education, teaching methodology, ICT in education, teacher education, and innovations in higher chemistry teaching. He is the organizer of the biannual international symposia on science education at the universities of Dortmund and Bremen and member of several boards of national and international journals. Currently he is the Editor-in-Chief of *Chemistry Education Research and Practice*, the highest ranked chemistry education journal in the world. Ingo Eilks received several awards for his scientific work and teaching, among them the Johann-Friedrich-Gmelin Award of the division of chemistry education within the German Chemical Society (GDCh), the Berninghausen Award for excellence in teaching, the STEM for Tomorrow School Award, and the Senior Fellowship in the Kolleg Didaktik: digital. He has received three awards for projects being recognized as official projects of the United Nations World Decade of Education for Sustainable Development (DESD).

Lianghuo Fan is a Professor in Education and Head of Mathematics, Science, and Health Education Research Centre at Southampton Education School, University of

Southampton. He holds an MSc from East China Normal University and a PhD from the University of Chicago, USA, and has extensive experience in education and research in China, the USA, Singapore, and now the UK. His research interests include mathematics classroom instruction and assessment, teacher education and knowledge development, curriculum studies and textbook research, and comparative education. His publications include acclaimed books, such as *How Chinese Learn Mathematics: Perspectives from Insiders* and *How Chinese Teach Mathematics* and *Investigating the Pedagogy of Mathematics: How Do Teachers Develop Their Knowledge* by Imperial College Press. He was Chair of the International Programme Committee of the inaugural International Conference on Mathematics Textbook Research and Development (ICMT-1) held in Southampton, 2014, and Cochair of the IPC of ICMT-2 to be held in Brazil, 2017. He currently leads the Shanghai Maths Project, a UK-China joint project for developing a series of students' practice books from Year 1 to Year 11, based on Shanghai's award-winning learning resource series, *One Lesson One Exercise*, by East China Normal University Press.

Dorina Gnaur is Associate Professor at the Dpt. of Learning and Philosophy, The Faculty of Humanities, Aalborg University. She holds an Industrial PhD in Workplace Learning, and has had a special interest in continuous competence development in knowledge-intensive contexts, such as universities. She has conducted a number of internal as well as external pedagogical professional development programs within higher and further education. In this connection, she had a close collaboration with the Department of Civil Engineering at The Faculty of Engineering and Science, Aalborg University, in the period 2012–2015, in which she had the opportunity to investigate in depth the field of engineering education. Together with colleagues from the Department of Engineering, she conducted various field studies and published several papers on student learning related to various engineering education programs, including the topics of interdisciplinarity, professional skills development and digitally enhanced teaching and learning.

Kelsey Greene is a Graduate Research Assistant at the University at Buffalo, State University of New York. Kelsey is a producer, educator, and scholar who combines her production talents and academic knowledge to advance instructional practices, learning experiences, and organizations' media outreach. Having received her degrees in both media production and education, Kelsey is currently serving as the Manager of Learning Resources for Convergence Academies.

Zehavit Kohen is Lecturer at the Faculty of Education in Science and Technology, Technion – Israel Institute of Technology. Her research interests encompass professional development of mathematics teachers, integrating educational technology into teaching and learning of mathematics, promoting meta-cognition in mathematics education, and assessing STEM (science, technology, engineering, and mathematics) students' learning outcomes with focus on quantitative analysis methods. Recently, she had been a Visiting Scholar at the Center to Support Excellence in Teaching (CSET) at Stanford University. Her research involved the investigation of the Hollyhock professional development (PD) program for promoting the quality of teaching among mathematics teachers. Her studies are published in referred journals and in international conferences in the field of STEM education, such as NARST, EARLI, and ICME. Her last article was accepted for publication in the prestigious journal of Metacognition and Learning.

Bracha Kramarski is an Associate Professor in School of Education in Bar-Ilan University in Israel. Her work encompasses a number of interrelated academic fields, which combines her expertise as an educational researcher with extensive professional experience in mathematical education and teacher professional education. She developed the innovative IMPROVE method for learning mathematics, based on metacognitive and SRL principles, a cooperative learning approach, and feedback-corrective theories that initially were developed for school students and modified for teachers' professional development in many programs. Her major fields of research are mathematical education, metacognition and SRL, teachers' professional development, and students' growth with advanced technology learning environments. She was the Chief Researcher of the International Students Assessment of Reading, Mathematics, and Science Literacy (OECD/PISA) in Israel. Currently she is involved in TPCK (Technology Pedagogy Content Knowledge) of STEM school teachers along with video analysis and simulations funded by the Chief Scientist of the Ministry of Education. She is an active member in many International Professional Organizations, Professional Committees and Review Board of Research Proposals for ISF and GIF, focusing on Learning, Teaching and Thinking. For the past five years, she has been a member of the Editorial Board of the Metacognitive and Learning Journal.

Rea Lavi is a PhD student at the Faculty of Education in Science and Engineering at the Technion – Israel Institute of Technology, under the supervision of Prof. Yehudit Judy Dori. His research concerns systems thinking, problem-solving, and conceptual modeling in science education and engineering education. Rea holds a MEd from Bar-Ilan University (2013) and two BSc degrees in Biology and in Psychology (2009) from Tel-Aviv University. He was awarded the ZEF grant for first-year PhD students at the Technion for 2015–2016 as well as the Miriam and Aaron Gutwirth grant for PhD students for 2016–2017. Rea has co-published two chapters and several papers and has presented papers and posters in STEM education international conferences.

Tamara J. Moore PhD, is an Associate Professor of Engineering Education and Director of STEM Integration in the INSPIRE Research Institute for Pre-College Engineering at Purdue University. Dr. Moore's research agenda focuses on defining STEM integration and investigating its power for student learning. She is creating and testing innovative, interdisciplinary curricular approaches that engage students in developing models of real-world problems and their solutions. Her research also involves working with educators to shift their expectations and instructional practice to facilitate effective STEM integration. For her work with underrepresented, urban youth in STEM learning, Dr. Moore was awarded a Presidential Early Career Award for Scientists and Engineers in 2012.

Senay Purzer PhD, is an Associate Professor in the School of Engineering Education at Purdue University and the Director of Assessment Research at the INSPIRE Research Institute for Pre-college Engineering. Her research focuses on design learning and assessment in complex design learning environments with specific focus on student decision making and design innovation. She is a governing board member of the global Research in Engineering Education Network (REEN) representing the North American zone and an editorial board member of *Science Education* and the *Journal of Pre-College Engineering Education* (JPEER). She received an MA in Science Education (2002) and a BSE in Engineering (2009) with distinction, both from Arizona State University. She also has a BS degree in Physics Education in 1999 from Hacettepe University. She received a PhD in Science Education with an engineering focus from Arizona State University.

Michael Radosta is a Doctoral Candidate in the Curriculum and Instruction Science of Learning Program at the University at Buffalo, State University of New York. Michael has leveraged his experiences and knowledge with media production, simulated gaming environments, mobile devices, live data collection science tools, and his insight with quantitative and qualitative educational research methods to impact classroom instruction across the Western New York region. Michael is the recipient of numerous awards including the ML & SD Farr Scholarship as well as a Presidential Fellowship from the University at Buffalo.

Jesper Sjöström studied chemistry, biology, mathematics, pharmacy, environmental sciences, and education. He is a full trained upper-secondary school teacher oriented toward chemistry and general science teaching. After his PhD in chemistry in 2002 at Lund University, Sweden, he was post-doc for 3 years in the area of Science and Technology Studies (STS). Since 2015 he is an Associate Professor in science and chemistry education at the Department of Science-Environment-Society at Malmö University, Sweden. His current research interests encompass sustainability and sociopolitically oriented science education, with a particular focus on the molecular sciences and their links to philosophy of education, media, and environmental issues. Examples of journals which have published his research are *Studies in Science Education, Science and Education* and *Chemistry Education Research and Practice*.

Susan Staats is an Associate Professor in the Department of Curriculum and Instruction, at the University of Minnesota. Her research is grounded in the anthropological field known as the ethnography of speaking, which takes verbal communication as the primary perspective on social life. One focus of this research is to understand the linguistic basis of collaborative learning, particularly through the analysis of indexicality and poetic structures in mathematics classroom discourse. A second focus is designing interdisciplinary contexts for college algebra. She teaches college algebra and interdisciplinary writing seminars to first-year undergraduate students, with emphasis on themes such as internationalizing the curriculum, incorporating service learning into algebra coursework, and modeling as an equity pedagogy. She is a member of the University of Minnesota Academy of Distinguished

Teachers. She works with high school teachers to deliver the university's modeling-based college algebra class through the concurrent enrollment program known as College in the Schools. This college access and equity program encourages ethnically diverse students, English language learners, and middle-achieving high school students to earn college credit during their high school careers. The program provides over 30 Minnesota high school teachers with professional development in modeling pedagogy and serves over 700 high school students annually.

Kjeld Svidt is an Associate Professor and head of the Division of Architectural Engineering in the Department of Civil Engineering, Aalborg University, Denmark. For more than 25 years, Kjeld Svidt has worked with research and education regarding information technology and computer simulations related to the built environment. This includes CAD and Building Information Modeling (BIM) as well as building simulation including Computational Fluid Dynamics (CFD), and Virtual Reality. Kjeld Svidt is involved in teaching and course development and he is coordinator of the MSc program in Building Informatics, which is based largely on project and problem-based learning (PBL). In the first few years of this new education, he has been supervisor of 40 MSc projects in Building Informatics and Construction Management.

Noemi Waight is an Associate Professor of Science Education in the Department of Learning and Instruction in the Graduate School of Education at the University at Buffalo, SUNY. She earned a PhD and a Masters of Science in Science Education from the University of Illinois at Urbana-Champaign. Her research examines the design, development, implementation, adoption, and enactment of technological tools (e.g., computer-based models, bioinformatics tools, databases) in the context of reform-based, K–12 science teaching approaches. Two complimentary perspectives guide this research: First, she examines the enactment of technological tools by documenting the full cycle from design and development to actual implementation in science classrooms. Second, to fully understand the implications of the above cycle, her research seeks to elucidate the theoretical underpinnings of the Nature of Technology (NoT) as it pertains to K–12 science education and empirically examine the factors, conditions, and agencies that impact and mediate enactment of technology in science education. Dr. Waight currently serves as an Associate Editor for the *Journal of Research in Science Teaching*. She has also served as Co-PI of a National Science Foundation grant: *Connected Chemistry as Formative Assessment*. In addition to her research projects in the US context, Dr. Waight is also engaged in international collaborative research in Belize and Japan.

Niva Wengrowicz holds a PhD from Bar Ilan University and was a postdoctoral fellow in the Department of Education in Science and Technology at the Technion – Israel Institute of Technology and served as a Visiting Scholar in the System Design and Management Program at the Massachusetts Institute of Technology. She also holds a bachelor's degree in Computer Science and Education from Bar Ilan University, an EMBA degree from Tel Aviv University, and she graduated summa cum laude from Bar Ilan University with another master degree in Education.

Currently she is an adjunct lecturer and a researcher in the Faculty of Industrial Engineering and Management at the Technion. In addition, she serves as a researcher and lecturer at the Research & Development Authority at Levinky College, Israel. Her research interests include system thinking, model-based systems engineering, alternative assessment, collaborative thinking and technology-enhanced learning environments.

Randy Yerrick is a Professor of Science Education at the State University of New York at Buffalo where he also has served as the Associate Dean for the Graduate School of Education. Professor Yerrick's work as a researcher and teacher educator focuses on two central questions: (1) how do scientific discourse practices get enacted in classrooms and (2) how can historical barriers to science learning be traversed for STEM students through expert teaching practices? Professor Yerrick's career in higher education has been devoted to improving STEM education in K–20 contexts and coordinates his university and school collaborations for maximum impact and relevance. Dr. Yerrick has received awards from science education organizations such as the *Excellence in College Science Teaching Award* from STANYS and the *Teaching Innovation Award* from the State University of New York as well as receiving recognition as an *Apple Distinguished Educator*. He has also received the *Journal of Research in Science Teaching Outstanding Research Paper Award.*

Anat Zohar holds the Besen Family Chair in Integrated Studies in Education at the Seymour Fox School of Education, the Hebrew University of Jerusalem, Israel. Professor Zohar's main academic interests are the development of students' higher-order thinking (HOT, including argumentation and inquiry learning), metacognition, teachers' knowledge about fostering students' HOT and metacognition, science education, and gender issues in science learning. She is also interested in the challenges involved in scaling up pedagogicalinnovation across whole school systems. Between 2006 and 2009, she served as the Director of Pedagogical Affairs in the Israeli Ministry of Education. In this role, she led a pedagogical reform in Israeli schools aimed at teaching for higher-order thinking and deep understanding throughout the school curricula (K–12). In 2009–2010, she was a member in the Institute for Advanced Study, Princeton, NJ. Since 2010, she is also a faculty member in the Mandel Leadership Institute.

Author Index

A
Abd-El-Khalick, F., 3, 5, 139–163, 233, 332, 349
Abell, S.K., 12
Ackerman, R., 36
Adams, R.S., 171, 173
Aguirre, J.M., 311
Alexander, P., 37
Anderson, D., 41, 44, 46
Anderson, R.D., 12
Anhalt, C.O., 3, 7, 307–327, 332, 351
Antonenko, P.D., 3, 4, 89–110, 185, 332, 348, 358
Arthur, W.B., 142
Avargil, S., 3, 33–60, 81, 332, 348

B
Bader, B., 74
Baker, D.R., 1–8, 108, 135, 161, 185, 217–234, 331–352
Bandura, A., 35
Bauer, W., 76
Bell, D., 117
Bencze, L., 75
Ben-David, A., 41, 47
Biesta, G., 71
Bishop, A., 309
Blank, L.M., 41, 47
Bohlin, H., 72, 73
Boiarsky, C., 220
Borat, O., 3, 6, 241–257, 332, 350
Bransford, J.D., 182
Brint, S., 220
Brodeur, D.R., 194
Brown, S., 221
Bucciarelli, L., 220
Burman, A., 68
Bybee, R.W., 147

C
Cantwell, A.M., 220
Capps, D.K., 2, 3, 9–29, 185, 332
Carberry, A.R., 108, 135, 161, 185, 217–234, 332, 350
Carless, D., 210
Carter, L., 75
Ceci, S., 229
Cheng, W., 210
Chiu, M.H., 41, 47, 51
Christensen, C., 75
Christensen, H.P., 191, 194
Civil, M., 3, 7, 307–327, 332, 351
Clark, A., 144
Cobb, C.E., 310
Cochran-Smith, M., 11
Cohen, E., 125
Çorlu, M.A., 3, 6, 241–257, 332, 350
Corlu, M.S., 3, 6, 185, 233, 241–257, 332, 350
Cortez, R., 3, 7, 307, 332
Crawford, B.A., 2, 3, 9–29, 185, 332
Crippen, K.J., 29, 81, 89–110, 185, 332, 348, 360
Crismond, D.P., 171, 173
Cross, D.R., 192
Cuban, L., 118

Y.J. Dori et al. (eds.), *Cognition, Metacognition, and Culture in STEM Education*, Innovations in Science Education and Technology 24,
https://doi.org/10.1007/978-3-319-66659-4

D
Dagher, Z.R., 171
de Figueiredo, A.D., 169
De Graaff, E., 191, 194
Dennison, R.S., 35, 45
Dewey, J., 69, 72, 195
Doerr, H.M., 314
Dori, D., 5, 6, 191–212, 332, 333
Dori, Y.J., 1–8, 33–60, 191–212, 331–352, 348
Dori, Z., 148, 151
Dorst, K., 174
Dringenberg, E., 3, 5, 167–185, 332
Dubberly, H., 173
Du, X.-Y., 230, 231

E
Eilks, I., 3, 4, 65–82, 332, 333
Einstein, H.H., 220
Ellström, P.E., 248
Ellul, J., 142
Else-Quest, M., 229
English, L.D., 313, 314
Erduran, S., 171

F
Fairweather, J., 221
Fan, L., 3, 7, 261–276, 332, 333, 350
Farley, J., 40–42, 57
Fisher, P., 223
Flavell, J.H., 17, 35, 192, 341
Fleck, L., 75, 76
Font Strawhun, B., 310, 322
Ford, C.L., 193
Frank, K.A., 149, 150, 154, 162
Frank, M., 141, 154
Frezza, Z., 172
Fuller, S., 79

G
Gadamer, H.-G., 67
Garfunkel, S., 313
Garner, R., 37
Georghiades, P., 39, 41, 48, 55
Gnaur, D., 3, 6, 241–257, 332, 350
Godfrey, E., 221, 223
Goldsmith, M., 36
González, N., 308, 309, 311, 324
Graham, R., 221–223
Greene, J.A., 3, 5
Greene, K., 117–135, 332, 348
Greer, B., 310
Grotzer, T., 41, 48, 52
Gunstone, R.F., 41, 48, 77, 152
Gustavsson, B., 68, 72
Gutstein, E., 310

H
Hameiri, M., 148, 151
Hand, B., 41, 48
Hanneman, R.A., 220
Hare, R., 72
Hart, P., 74
Hassan, I.E., 228
Hasslöf, H., 72
Herrenkohl, L.R., 100
Herscovitz, O., 36, 41, 48, 51, 57, 205
Hewitt, N., 222
Hilton, M.L., 262, 271
Hodson, D., 65, 66
Hopmann, S., 70
Houwer, R., 73
Hsu, Y., 152

J
Jaakkola, T., 153
Jacobs, J.E., 35
Johnson, A.C., 222, 224
Johnson, D.W., 71
Johri, A., 191
Jonassen, D.H., 101, 174, 179, 180
Jones, B.D., 230

K
Kaberman, Z., 148
Kansanen, P., 70
Kant, I., 71
Kemp, P., 71
King, C.J., 222
Kivelä, A., 69
Klafki, W., 67, 69–71, 73
Koch, A., 41, 49, 57
Kohen, Z., 3, 7, 275, 279–302, 332, 350
Kramarski, B., 3, 7, 265, 266, 268, 269, 275, 279–302, 332, 350
Kuh, G., 220
Kuhn, D., 194
Kunda, G., 233

L
Laberge, Y., 74
Laird, N., 220

Laugksch, R.C., 65
Lave, J., 119
Lavi, R., 3, 4, 6, 33–60, 241–257, 332, 348, 350
Lederman, N.G., 19
Lesh, R., 313
Linn, M.C., 41, 47, 51
Lin, X., 194
Lipka, J., 310
Littledyke, M., 75
Liu, N.-F., 210
Lu, C.-C., 173
Lundqvist, E., 78
Lytle, S.L., 11

M
Malmberg, C., 72
Marbach-Ad, G., 153
Marks, R., 70, 75
Marra, R., 222
Marshall, H., 218
Martin, B.L., 36, 41, 45, 59
McDermott, J., 142
McRobbie, C.J., 41, 49, 53
McVee, M.B., 128
Mevarech, Z.R., 1–8, 261–276, 331–352
Michalsky, T., 41, 49
Miller, S.M., 128
Mittlefehldt, S., 41, 48, 52
Mogensen, F., 70
Montfort, D., 221
Montgomery, M., 313
Moore, T.J., 3, 5, 167–185, 332
Morin, O., 75
Morris, P., 71
Moses, R.P., 310
Moshman, D., 35

N
Nashon, S., 41, 44, 46, 55, 59
Nazir, J., 80
Newell, J., 194, 195
Nielsen, W.S., 41, 44
Norris, S.P., 41, 43
Nussbaum, M., 68

O
Olds, B.M., 191

P
Pacey, A., 143, 144, 146, 147, 154, 157
Pajares, M.F., 10
Palincsar, A.S., 100
Paris, S.G., 35, 192
Parker, L., 221
Pedretti, E., 80
Pellegrino, J.W., 262, 271
Perry, N., 285
Pintrich, P.R., 193, 205
Polya, G., 263, 268
Prensky, M., 117, 118, 128
Purzer, S., 3, 5, 167–185, 332

R
Radosta, M., 108, 117–135, 348
Radosta, R., 3, 5
Rapp, F., 142
Reis, P., 72
Ricoeur, P., 69
Roberts, D.A., 65, 77, 78, 82
Rutherford, F.J., 19

S
Sadler, T.D., 66
Sandi-Urena, S., 41, 46, 49, 53, 56
Sandoval, W.A., 101
Santos, W.L.P., 66, 73
Schaffar, B., 71
Schnack, K., 70
Schoenfeld, A.H., 268
Schraw, G., 35, 41, 45, 58, 205, 280, 289
Schwartz, D.L., 182
Schwartz, R.Z., 19
Schwarz, M.J., 220
Seymour, E., 222
Shinew, D., 221
Shin, N., 41, 44, 46
Shor, I., 72
Shoup, R., 220
Shulman, L.S., 11, 12, 70
Simonneaux, L., 66, 67, 77, 79, 80
Sjöström, J., 3, 4, 65–82, 333
Smith, D.V., 77
Sriraman, B., 313
Staats, S., 3, 7, 307
Stains, M., 151
Straub, D.W., 228
Strobel, J., 94

Stuckey, M., 74, 76
Sullivan, F.R., 148
Svidt, K., 3, 6, 241–257, 332, 350
Szelenyi, K., 231

T
Taasoobshirazi, G., 40–42, 57
Talanquer, V., 80, 151
Tao, P., 152
Tate, W., 310
Tenner, E., 142
Terenzini, P.T., 220
Thomas, G.P., 41, 42, 46, 49–51, 53
Tonso, K.L., 222
Turner, E., 310
Tylor, E., 310

U
Uljens, M., 71
Urhahne, D., 151

V
van Driel, J.H., 12
van Heuvelen, A., 99
Veenman, M.V.J, 35, 39, 54, 194
Verma, G., 148
Vieira, R.M., 72
Volti, R., 142, 145, 147, 148
von Humboldt, W., 67, 68, 343
Vos, H., 194
Vygotsky, L.S., 93

W
Waight, N., 3, 5, 135, 139–163, 233, 332, 349
Walser, A.D., 220
Wandersee, J.H., 41, 50, 51, 57
Wang, C.Y., 41, 43, 50, 55, 58, 60
Wang, S.K., 118, 120
Ward, R.E., 41, 50, 51
Warren, M., 210
Wellman, H.M., 192
Wenger, E., 119
Wengrowicz, N., 5, 6, 191–212, 332, 333
Weniger, E., 67
Wickman, P.O., 66, 77–80, 332
Williams, W., 229
Wimmer, M., 66
Wu, H., 151

Y
Yerrick, R., 3, 5, 108, 117–135, 160, 161, 332, 348
Yore, L.D., 193

Z
Zavala, M., 311
Zawojewski, J.S., 313
Zeidler, D.L., 66
Zhao, Y., 141, 149, 150, 154, 162
Zimmerman, B.J., 280, 281, 289, 298
Zion, M., 41, 50, 52
Zohar, A., 41, 47, 58
Zoller, U., 77

Subject Index

A

Abductive reasoning, 174–175
Academic class, 2–3
Accreditation Board for Engineering and Technology (ABET), 168–169, 253–254
American Indian Science and Engineering Society (AISES), 232
Authentic collaboration, 94–95
Authentic practices, 338

B

Bildung
  Anglo-American/liberal education, 68
  chemistry education, 80
  classical, 68
  cognitive and affective domains, 75
  connections, 77–79
  critical-constructive *Didaktik*, 70–71
  critical-reflexive, 69
  critical thinking, 71–73
  Dewey's democratic education, 69
  EfS, 74–75
  ethical perspectives, 73
  international educational literature, 67
  modes, 79
  post-critical pedagogy, 74
  postmodern risk-oriented education, 75
  praxis-oriented vision, 77
  Scandinavian folk-Bildung-tradition, 68–69
  science-to-society link, 75–76
  socio-political perspectives, 73
  SSIs, 73–74, 79–80
  techno-scientific rationality, 80
  traditional *vs.* alternative approach, 77
  Vision I, 65–66
  Vision II, 65–66
  Vision IIa, 78
  Vision III, 66–67
Biology Guided Inquiry Learning Environment (BGuILE), 97
Biology Student Workbench (BSW), 140–141
Biotic and abiotic factors, 149–150
Blogs, 123
Bologna cycle, 249

C

Coexisting communities of practice, 243
Complex, unfamiliar, and non-routine (CUN) tasks, 265–266
Computer-Supported Collaborative Learning (CSCL), 99–101
Conceive-design-implement-operate (CDIO) methodology, 249
Course reflection tool, 205
Critical approach, 72
Critical praxis, 72–73
Critical theory, 71
Critical thinking (CT), 71–73
Culturally relevant pedagogy (CRP)
  Funds of Knowledge approach, 311–312
  homogeneous classrooms, 310–311
  Ladson-Billings' model, 311
  lesson analysis tool, 311
  mathematical modeling
    community contexts, 317–319
    connection to academic success, 321

Y.J. Dori et al. (eds.), *Cognition, Metacognition, and Culture in STEM Education*, Innovations in Science Education and Technology 24,
https://doi.org/10.1007/978-3-319-66659-4

Culturally relevant pedagogy (CRP) *(cont.)*
connection to critical consciousness, 320
connection to culture and community, 320
content motivation, 315–316
Fences task, 319–320
implications, 321–322
integrate contexts relevant to students, 316
problem situation, 317
promoting discussion between students, 316
pedagogical approaches, 310
Culturally responsive mathematics teaching (CRMT), 311
Culture, 2, 351
academic mathematics, 309
communities, 308–309
country-level influences, 333
CRP (*see* Culturally relevant pedagogy (CRP))
culture of technology, 344
definition, 308
engineering culture, 345–346
engineering education, 345
ethnomathematics, 309–310
fairness and equity, 310
flipped classroom, 344–345
incorporate social perspectives, 309
international mathematical ideas, 309
local mathematical concepts, 309
NoT, 344
PBL, 346
scaffolding, 345
technology, 344
Tylor's definition, 308
and values, 343
Cyberlearning
blended learning, 97–98
cognitive apprenticeship, 91
designers, 97
features, 91
problem solving (*see* Problem solving)

D
Danish Society of Engineers (IDA), 245–246
Declarative knowledge, 35
Design-based research (DBR), 101–102
Design strategies
behaviors, 173
classroom studies, 172–173
expert and novice behaviors, 172
features, 173, 174
longitudinal studies, 172
Didactical analysis, 70–71
Digital tools, 121–122
Discussion board, 121, 127
Dublin Descriptors, 249

E
Educational focus, 2–3
Education for Sustainability (EfS), 74–75
Education for sustainable development (ESD), 74–75
Electricity, 153
Engaging children
recommendations, 29
teacher cognition
beliefs and knowledge, 10–11
Calderhead's definition, 10
classroom problems, 9–10
knowledge base, 11–12
scientific practices (*see* Scientific practices (SPs))
Engineering cognition
clarification, 170
contemporary STEM, 169
definition, 168–169
design
ill-defined and ill-structured problem solving, 171, 172
reasoning, 174–175
strategies, 172–174
fragmented approaches, 179, 180
K-12 education, 168
knowledge acquisition, 175–176
knowledge application, 176
knowledge production, 177
normative and nonnormative views
background knowledge, 181–182
egg-drop challenge, 181, 182
features, 180
problem context, 177–178
recommendations, 183–185
traditionally defined STEM, 169
Engineering culture
changes, 222–223
document analysis, 231
factors, 218
NGSES, 232
perception, 217–220
portfolio, 231
production, 218–220
professional societies, 232
recommendations, 233–234

societal culture, 226–227
model selection, 228
teaching methods, 221–222
TIMSS data, 229
traditional engineering programs, 220–221
women, 229, 230
workplace experience, 231
Engineering design cognition, 340–341
Engineering education, 5–6, 337, 350
cognition (*see* Engineering cognition)
culture (*see* Engineering culture)
in Denmark
accreditation systems, 245
CDIO methodology, 249
challenges, 244
curriculum design, 248
democratization movement, 246
disciplinary congestion, 247–248
doctoral programs, 246
globalization, 247
massive mergers, 247
present-day education, 255–256
problem-solving approach, 247
professional elements, 245
project management skills, 247
quality assurance, 245, 249
science-focused program, 246–247
theoretical and technical qualifications, 246
3.5-year undergraduate program, 244–245
youth protest, 246
historical development
Anglo-American traditions, 243–244
Continental Europe, 243–244
European Higher Education Area, 244
innovation, 242–243
knowledge society, 242–243
metacognition (*see* Metacognition)
in Turkey
applications, 251–252
European standards, 253
foundations, 250–251
global perspective, 252
interdisciplinary learning environment, 252–253
present-day education, 255–256
public service, 251
Quality Board, 254
scientific research projects, 251–252
women engineers, 253
UK
history, 254
initial professional development, 254–255
present-day education, 255–256
undergraduate students, 333
Environment for Collaborative Learning Integrating Problem Solving Experiences (ECLIPSE)
general-purpose technologies, 103
scaffolds for instructors, 103–104
scaffolds for students
collaborative process, 104–105
discursive practices, 102, 107–108
high school Biology and Physics, 108
learner module, 106–107
materials and tools, 107–108
participant structures, 107–108
task structures, 107–108
technology-supported, 105–106
undergraduate Entomology, 108
Epistemic scripts, 100
EUR-ACE Framework Standards, 253
Exotic species, 150

F
Face-to-face instruction, 128
Flash and NetLogo modeling tools, 140
Flipped-style classroom format, 127
Funds of Knowledge, 311–312, 338–339

G
Global citizenship education, 339

H
Hierarchical Analysis Tool (HAT), 99
Higher mental processes (HMP), 272–273
Higher-order thinking, 339–340
Holistic approach, 160

I
iMovie, 131
IMPROVE method
eight-item posttest, 272
HMP, 272–273
intact classrooms, 271–272
LMP, 272–273
regular math problems, 271
ten-item examination, 272
time and group, 272–273

Information age, 45
Information, communication, and technology (ICT), 117–119, 125, 132, 268–269
Inquiry teaching, *see* Scientific practices (SP)
Intellectual technology, 117
Invading species, 150

J
Joint problem space, 100
Junior Metacognitive Awareness Inventory (Jr. MAI), 45

K
Knowledge acquisition, 175–176, 179, 337
Knowledge application, 176, 337
Knowledge-based life, 243
Knowledge of cognition (KC), 281, 341
Knowledge of Nature of Science (NOSK), 14–16, 19–20
Knowledge of Scientific Practices (SPK), 14–16
Knowledge production, 177, 338

L
Literature search
  classification, 38
  empirical papers
    assessment tools, 40, 42–43
    discipline-based features, 40
    investigated population, 40–41
    metacognition learning processes, 41, 43–46
    pedagogical intervention, 41, 46–53
  journals, 37
  reviews, 38–39
  self-regulated learning, 38
  theoretical and non-empirical, 39
Lower mental processes (LMP), 272–273

M
Mastery Learning, 29
Mathematical knowledge for teaching (MKT) model, 12
Mathematical modeling
  components, 314–315
  CRP
    community contexts, 317–319
    connection to academic success, 321
    connection to critical consciousness, 320
    connection to culture and community, 320
    content motivation, 315–316
    Fences task, 319–320
    implications, 321–322
    integrate contexts relevant to students, 316
    problem situation, 317
    promoting discussion between students, 316
  definition, 313–314
  nature's principles, 312
  perspectives, 313–314
  preservice teachers
    culture and Funds of Knowledge, 324
    haircuts and styles, 324
    limitations, 325
    mathematics course, 323–324
    modeling problems, 325
    open-ended problems, 324
    person's shoe, 324–325
  teacher education, 323
Mathematics education, 7–8, 337
  mathematical modeling (*see* Mathematical modeling)
  mathematics literacy (*see* Mathematics literacy (ML))
  pedagogical metacognition (*see* Metacognitive learners)
  word clouds, 334–335
Mathematics literacy (ML), 340
  changes and relationships, 263
  classroom observations, 274
  contexts, 265
  CUN tasks, 265–267
  cyber-learning environments, 274
  data analysis, 264
  deeper learning, 262
  definition, 262–263
  ICT and non-ICT environments, 274
  IMPROVE, 267–275
  meaningful learning, 262
  metacognition
    higher-order thinking program, 267
    ICT, 268–269
    MOE of Singapore, 269–270
    object level, 267
    positive correlations, 267
    reading comprehension, 270–271
  problem-solving, 262–263, 270
  processes, 264
  quantity, 264
  space and shape, 264
  uncertainty, 264

Meta-assessment, 197–198, 204–208, 210
Metacognition
assessment tools, 342
course reflection tool
peer assessment task, 207–208
teamwork, 205, 207
definition, 341
disciplines, 193
effective and efficient assessment, 196–197
formative assessment, 196
goal of, 194
IMPROVE, 342
knowledge of, 192, 193, 341
mathematics education, 331–333
MBSE
graduate course, 203
OPM, 199–200, 208
peer assessment and meta-assessment, 200
semantics-oriented criteria, 198–199
settings, 200
size, level, and culture, 200
UML, 199, 208
undergraduate course, 201–202
PBL, 195–196
pedagogical intervention, 341
peer assessment
definition, 197
distinctive cognitive, 209
"hands-on" training, 211
instructors, 211
metacognitive elements, 209
student's peer assessment, 209
tool, 203–204
pentagon framework, 269–270
principles, 194, 195
regulation of, 193, 341–342
research and development, 342
science learning, 342
self-regulation, 341
SRL, 192
students' reflections, 212
summative assessment, 196
Metacognition Baseline Questionnaire (MBQ), 44
Metacognitive Activities Inventory (MCAI), 46
Metacognitive-guided Inquiry within Networked Technology (MINT), 52–53
Metacognitive learners
data analysis, 289–290
educators and researchers, 280
multidimensional Cog/Meta_T model
backward and forward thinking, 288
dual pedagogical dimensions, 299
engagement activation, 295–297
explicit strategies, 293–295
implementation, 287–288
incidence, quality, and sequential pattern, 290–293
metacognitive support, 298
pedagogical metacognition, 281–282
phases, 283
potential contributions, 300
pre-action, in-action, and post-action, 288–289
self-regulation model, 281, 300
teaching episodes, 289
teaching instruction, 283–285, 293–294
teaching practice, 299–300
training and follow-up, 301
web-based learning environments, 285–286
Schraw's and Zimmerman's models, 280–281
Metacognitive Orientation Learning Environment Scale-Science, 46
Metacognitive pedagogy, 267–270
Model-based systems engineering (MBSE)
graduate course, 203
OPM, 199–200, 208
peer assessment and meta-assessment, 200
semantics-oriented criteria, 198–199
settings, 200
size, level, and culture, 200
UML, 199, 208
undergraduate course, 201–202
Model clarity and understandability (MCU), 199

N

National Science Foundation (NSF), 191
National Society of Black Engineers (NSBE), 232
Nature of science (NOS), 13, 19–20, 339
Nature of technology (NoT)
adoption and enactment, 158–159
artifact, process, and activity, 142
cognitive processes, 151
computer-based technology, 152–153
diffusion, 147
dimensions, 141
expertise, 148
financial, professional, and social, 161–162

Nature of technology (NoT) (*cont.*)
implementation, adoption, and enactment, 149–151
implication, 162
intent
classroom conditions, 157
content knowledge, 156–157
cultural activity, 157
PD, 158
teacher preparation, 158
technical knowledge, 156
machine's construction, 143
outcomes, 159–161
PCK, 152
preplanning, 155
progression, 147
rationality, 145
3D representations, 151–152
Next Generation Science and on Engineering Standards (NGSES), 232
Next generation science standards (NGSS), 13
North Atlantic Treaty Organization (NATO), 251

O
Object-Process Methodology (OPM), 199–200
Overscripting, 100–101

P
Pedagogical content knowledge (PCK), 11, 152
Pedagogical intervention
automated instrument, 46
classroom learning environment, 46
follow-up study, 52
IMMEX problem, 53
metacognitive moves, 52
multi-method assessment, 46
scientific skills/knowledge, 41, 46–50
small groups collaboration, 46
Pedagogical knowledge (PK), 11, 20
Peer assessment
definition, 197
distinctive cognitive, 209
"hands-on" training, 211
instructors, 211
metacognitive elements, 209
student's peer assessment, 209
tool, 203–204
Performance knowledge, 128
Permeable disciplinarity, 243
Personal authenticity, 94
Physics Education Technology Project (PhET), 97
Physics Metacognition Inventory (PMCI), 40, 42–43
PISA, 261–265
Population region, 2–3
Portfolio, 231
Preservice elementary teachers (PETs)
class activities, 120
contextualize technological tasks, 133–134
cultural contexts, 132–133
digital natives and immigrants, 117, 118
inquiry approach, 121
intellectual technology, 117
interests, goals, and research based approaches, 134
online electronic resources, 121
online tasks, 127–128
participants' assessment, 121, 122
pedagogical technology, 120
prior experiences and attitudes
blogs, 123
disfavor technology, 124
inquiry-based learning, 123–124
inquiry task, 125–126
Internet, 123
prevalence and frequency technology, 122
professional learning, 124–125
research-based assessment, 120
safe contexts, 134–135
scaffold public collaboration, 133
social networking tools, 121
vodcasting, 121
Preservice teachers (PTs)
culture and Funds of Knowledge, 324
haircuts and styles, 324
modeling problems, 325
open-ended problems, 324
Problem solving
authenticity, 94–95, 97
case-based reasoning, 93
DBR, 101–103
difficult problems, 96
dimensions, 95
domain-specific practices, 92–93, 96
dynamicity, 96
ECLIPSE (*see* Environment for Collaborative Learning Integrating Problem Solving Experiences (ECLIPSE))
goal-based scenarios, 93
ill-structured problems, 95
implications, 108–109

inquiry learning, 93
inter-related activity system, 92
learning as conceptual change, 93
levels of difficulty, 91
open-ended learning, 93
problem-based learning, 93
scaffolding, 98–101
Professional development (PD) program, 140
research scientist, 15
SMK, 17–18
Project-based learning (PBL), 191, 195–196, 248, 346
Propositional knowledge, 128
Public Financial Management and Controlling Law, 253
Public perception, 218–220

R
Reading comprehension, mathematics literacy and metacognition, 270–271
Reading Comprehension of Science Test (RCST), 43
Regulation of cognition (RC), 281, 341–342
Renewable energy sources, 226–227
Research Experiences for Teachers (RETs), 19
Research Experiences for Undergraduates (REUs), 19
Restrained teaching, 70
Robert College engineering school (1912), 250
Roskilde University Center, 247

S
Scaffolding
collaboration, 99–101
discipline-based education researchers, 98
ECLIPSE (*see* Environment for Collaborative Learning Integrating Problem Solving Experiences (ECLIPSE))
evidence, 98
examples, 99
mathematical representation, 99
Science education, 2, 4, 337
engaging children (*see* Engaging children)
metacognition, 331–333
definition, 34–35
evaluation, 54–57
general education, 36–37
literature search (*see* Literature search)
recommendations, 59–60
research gaps, 54, 58
scientific literacy (*see* Scientific literacy (SL))
word clouds, 334–335
Scientific inquiry, 338, 340
Scientific literacy (SL)
classical, 68
cognitive and affective domains, 75
connections, 77–79
critical-constructive *Didaktik,* 70–71
critical-reflexive, 69
critical thinking, 71–73
ethical perspectives, 73
science-to-society link, 75–76
socio-political perspectives, 73
SSIs, 73–74, 79–80
techno-scientific rationality, 80
traditional *vs.* alternative approach, 77
elements, 65
Scientific practices (SPs)
CoP, 20–21
deep and integrated knowledge, 13
fifth to ninth grade children, 14–15
Fossil Finders
background lessons, 21
environmental changes, 21
pedagogical practices, 21
pre and post-tests of knowledge, 21–22
metacognition, 17
NGSS, 13
pedagogical knowledge, 20
reflection, 16–17
research scientist, 15
teacher's knowledge
aspects, 15–16
NOS, 19–20
reform-based documents, 18–19
traditional teacher preparation programs, 15
United States, 338
variations of inquiry, 12–13
Self-Efficacy Metacognition Learning Inventory-Science (SEMLI-S), 42–43
Self-regulated learning (SRL), 192
Singapore Mathematics Assessment Project, 266
Social bookmarking tool, 127
Society of Women Engineers (SWE), 232
Sociocultural theory, 119
Socio-scientific issues (SSI), 73–74, 79–80
Story Problem-Solving Environment (SPSE), 99

Subject matter knowledge (SMK)
  assumption, 13–14
  NOSK, 14
  PD program, 17–18
  reform-based teaching practice, 14
  SPK, 14
Systems Modeling Language (SysML), 203

T
Teacher cognition, 338, 347–348
  beliefs and knowledge, 10–11
  Calderhead's definition, 10
  classroom problems, 9–10
  knowledge base, 11–12
  scientific practices (*see* Scientific practices (SPs))
Technical knowledge, 156
Technology education, 4–5, 337
  cyberlearning (*see* Cyberlearning)
  metacognition, 331–332
  NoT (*see* Nature of technology (NoT))
  PETs (*see* Preservice elementary teachers (PETs))
Technology enhanced learning (TEL), 152
Theoretical-practical model, 281
Transformative learning, 73, 340
Transforming Undergraduate Education in Engineering (TUEE), 168
Trends in International Mathematics and Science Study (TIMMS), 229

U
Undergraduate Research Experience (URE), 23
Unified Modeling Language (UML), 199
University Ranking by Academic Performence (URAS), 254

V
Value authenticity, 94
Vision I, 65–66
Vision II, 65–66
Vision IIa, 78
Vision III, 66–67
Vodcasting, 121, 130

W
Washington Accord Graduate Attributes, 253
Web-Based Inquiry Science Environment (WISE), 97
Women in Science and Engineering (WISE), 232
Word clouds, 334–335

The manufacturer's authorised representative in the EU is Springer Nature Customer Service Centre GmbH, Europaplatz 3, 69115 Heidelberg, Germany. If you have any concerns regarding our products, please contact ProductSafety@springernature.com

Printed and bound by CPI Group (UK) Ltd, Croydon, CR0 4YY
30/09/2026
02225237-0012